Forests and Forestry: Diversity and Management

Forests and Forestry: Diversity and Management

Editor: Lee Zieger

www.callistoreference.com

Callisto Reference,
118-35 Queens Blvd., Suite 400,
Forest Hills, NY 11375, USA

Visit us on the World Wide Web at:
www.callistoreference.com

ISBN: 978-1-63239-820-8 (Hardback)

Cataloging-in-publication Data

Forests and forestry : diversity and management / edited by Lee Zieger.
    p. cm.
Includes bibliographical references and index.
ISBN 978-1-63239-820-8
1. Forests and forestry. 2. Forest management. 3. Forest management--Environmental aspects.
4. Forest biodiversity. 5. Forest biomass. 6. Forest ecology. I. Zieger, Lee.
SD131 .F67 2017
634.9--dc23

Table of Contents

Preface

Forestry and forest management are fields that are concerned with the preservation and expansion of forest cover on the earth's surface. Forests and Forestry are important subjects of study that aid in the analysis and historiographical accounting of various aspects of forests including area and spread, species of flora and fauna and impact of policy on pasturage and forage practices. This book is compiled in such a manner that it will provide in-depth knowledge about the theory and practice of forestry. Different approaches, evaluations, methodologies and advanced study have been included herein. This book on forests and forestry seeks to cover topics that stress on management, biodiversity and preservation of ecosystems. This book will help new researchers by foregrounding their knowledge in this branch.

Over the recent decade, advancements and applications have progressed exponentially. This has led to the increased interest in this field and projects are being conducted to enhance knowledge. The main objective of this book is to present some of the critical challenges and provide insights into possible solutions. This book will answer the varied questions that arise in the field and also provide an increased scope for furthering studies.

I hope that this book, with its visionary approach, will be a valuable addition and will promote interest among readers. Each of the authors has provided their extraordinary competence in their specific fields by providing different perspectives as they come from diverse nations and regions. I thank them for their contributions.

Editor

Supply and Demand Determine the Market Value of Access to Infants in the Golden Snub-Nosed Monkey (*Rhinopithecus roxellana*)

Wei Wei[1][9]**, XiaoGuang Qi**[1][9]**, Paul A. Garber**[3]**, SongTao Guo**[1]**, Pei Zhang**[1]**, BaoGuo Li**[1,2]*

1 Key Laboratory of Resource Biology and Biotechnology in Western China, College of Life Sciences, Northwest University, Xi'an, China, 2 Institute of Zoology, Shaanxi Academy of Sciences, Xi'an, China, 3 Anthropology Department, University of Illinois, Urbana, Illinois, United States of America

Abstract

According to a biological market paradigm, trading decisions between partners will be influenced by the current 'exchange rate' of commodities (good and services), which is affected by supply and demand, and the trader's ability to outbid competitors. In several species of nonhuman primates, newborn infants are attractive to female group members and may become a desired commodity that can be traded for grooming within a biological market place. We investigated whether grooming was interchanged for infant handling in female golden snub-nosed monkeys (*Rhinopithecus roxellana*) inhabiting the Qinling Mountains of central China. *R. roxellana* exhibit a multilevel social organization characterized by over 100 troop members organized into 6–11 one-male units each composed one adult male and several adult females and their offspring. Behavioral data were collected over the course of 28 months on grooming patterns between mothers with infants less than 6 months old (N = 36) and other adult female troop members. Our results provide strong evidence for the interchange of grooming for access to infants. Grooming for infant access was more likely to be initiated by potential handlers (nonmothers) and less likely reciprocated by mothers. Moreover, grooming bout duration was inversely related to the number of infants per female present in each one-male unit indicating the possibility of a supply and demand market effect. The rank difference between mothers and handlers was negatively correlated with grooming duration. With increasing infant age, the duration of grooming provided by handlers was shorter suggesting that the 'value' of older infants had decreased. Finally, frequent grooming partners were allowed to handle and maintain access to infants longer than infrequent groomers. These results support the contention that grooming and infant handling may be traded in *R. roxellana* and that the price individuals paid for access to infants fluctuated with supply and demand.

Editor: Ronald Noë, Université de Strasbourg, France

Funding: This research was supported by the grants from the Key Program of National Nature Science Foundation of China (31130061); the National Nature Science Foundation of China (31270438, 31270441, 31201721); the Research Fund for the Doctoral Program of Higher Education of China (20106101110005); Specialized Foundation of Department of Education of Shaanxi Province, China (12JK0825); Key Program of Academy of Sciences of Shaanxi Province (2012k-01). The funding organizations had no role in study design, data collection and analysis, decision to publish, or preparation of the manuscript.

Competing Interests: The authors have declared that no competing interests exist.

* E-mail: baoguoli@nwu.edu.cn

[9] These authors contributed equally to this work.

Introduction

Biological market theory [1,2] offers a set of predictions regarding dynamic changes in the value and exchange of goods and services among members of a social group. This explanatory framework offers a new perspective on the costs and benefits of cooperative behavior under conditions in which individuals have the opportunity to interact with a diverse set of partners, and the different individuals within a given trader class compete to outbid each other for access to valued commodities [3]. The principles of a biological market reflect the natural interactions of group-living animals, in which potential cooperators differ in the quality or type of services they can offer in exchange for the quality and type of services they require [1,2]. In this regard, two primary trader classes have been described: 'sellers', provide a specific commodities, and 'buyers', choosing the most valuable commodity. [1,2]. Potential cooperators also may differ in their access to or preference for particular trading partners based on age, domi-

nance status, kinship, and previous social histories. Within the 'biological market' paradigm, traders are expected to seek partners who offer the optimal commodity value based on current supply and demand as well as their ability to effectively outbid competitors for these goods and services [2]. Scarcity in terms of the availability of goods, services, or particular traders (i.e. access to a high ranking coalitionary partner) and increased individual need, should lead to an increase in the cost of obtaining that commodity, whereas under conditions of surplus there is an expected reduction in cost [1,2].

Grooming, is the most common affiliative behavior engaged in by nonhuman primates [4,5]. And, although grooming is likely to serve several social functions, it plays a primary role in establishing and reinforcing social bonds [6,7,8] and in reducing social tension and promoting reconciliation between group mates [9,10,11].Therefore, grooming represents a beneficial service that primates may exchange or interchange for other commodities [7,12,13].

Previous research has demonstrated that primate 'reciprocal traders' commonly exchange or reciprocate grooming for grooming [14,15,16,17,18,19,20]. In addition, grooming may be swapped for other commodities or services among 'interchange traders'. Across primate species, grooming has been interchanged for coalitionary support [21], food [22,23], tolerance [24], mating opportunities [25,26,27], food sharing [28] and access to information concerning female reproductive status [19,29].

In several species of primates, newborn infants are very attractive to group members, especially females, and mothers with young infants receive high levels of social attention [13,30,31]. Females often groom mothers in an attempt to touch, handle, nuzzle and inspect their infants [32,33]. Maternal restrictiveness and the attraction that females have for infants make them a valuable and restricted commodity [34,35]. Moreover, given that the postnatal costs of infant care (nursing, transporting, food sharing, and protecting infants over a relatively extended developmental period) for female primates can be high [30], mothers in many primate species are tolerant of aunts or helpers who contribute to infant care [36,37,38,39,40]. Using a biological market framework, females without infants may be expected to interchange grooming or other service for access to infants (hereafter handlers who may gain critical infant care experience that increases the survivorship of their future offspring or develop strong social bonds with the infant's mother). This creates a demand for access to infants [41,42,43,44]. Henzi & Barrett (2002) [41] provided the first test of the biological market model on grooming and infant handling in the context of interchange trading. In the case of chacma baboons (*Papio hamadryas ursinus*) these authors found that mothers were groomed for longer periods of time when there were fewer newborns in the group, a finding confirmed in several primate studies [42,43,45], but not in others [44,46]. In the present study, we use a biological market framework to examine social interactions, grooming relationships, and access to infants in female golden snub-nosed monkeys (*Rhinopithecus roxellana*). Observations of this species indicate that females without infants are attracted to females with young infants and groom them frequently [13].

The golden snub-nosed monkey (*Rhinopithecus roxellana*) is a rare and endangered colobine primate endemic to China. Colobines are foregut fermenters and possess adaptations of their stomach (multichambered, enlarged, low PH, highly diverse microbacterial biome) that enable them to exploit a diet composed primarily of leaves, lichen, seeds, and other difficult to digest plant material [39,47]. The social organization of *R. roxellana* is characterized by a multi-level society [48] composed of several one-male units (OMUs) and an associated all-male band. Troop size commonly exceeds 100 individuals [49,50]. The OMU is the basic social and reproductive unit and consists of a single resident male, 3–7 adult females, subadults, juveniles, and infants [13,48,51]. Troop size is reported to fluctuate across seasons, with a decrease in the number of one-male units that foraging together during periods of low food availability (November-nest year January). Males leave their natal OMU before sexual maturity and join an all-male band. Although some females remain in their natal OMU and breed, other females disperse into different OMUs within the troop or on occasion into a new troop [52]. *R. roxellana* is a strictly seasonal breeder [53]. The mating season is from September to December, and the birth season is from March to May [54]. Gestation is approximately 6–7 months, and females with surviving offspring give birth once every two years. Given that the average number of adult females per OMU is 4.76±1.57 (mean ± SD) and females breed once every two years, the number of infants (<6 months old) in each OMU is approximately 1.8±1.3 (mean ± SD).

Female golden snub-nosed monkeys without infants are attracted to newborn infants. Access to young infants requires than that nonmothers cautiously approach mothers and either groom them or sit near them [13]. In the biological market place that is the golden snub-nosed monkey OMU, access to infants is a commodity desired by females without infants, and therefore, differences in rank between mother and handler, as well as infant age, are expected to influence the 'value' of goods and services interchanged during social interactions. In order to examine these relationships more clearly, we test the following predictions:

(1) Grooming sessions between nonmothers and mothers are primarily unidirectional and nonreciprocal. Nonmothers are expected to groom mothers for a longer duration than they groom other females.

(2) Grooming as a means of 'purchasing' access to infants is less likely to occur between mother–mother dyads than between mother-nonmother dyads.

(3) The ratio of infants per adult female in an OMU should be inversely related to duration of grooming bouts in which nonmothers groom mothers. Specifically, nonmothers are expected to groom mothers for a longer duration when fewer infants are available.

(4) Assuming that nonmothers are more attracted to younger and more dependent infants than to older infants, we expect infant age to be negatively correlated with the amount of grooming received by a mother to obtain access to her infant.

(5) The length of grooming bouts provided by nonmothers is expected to be negatively correlated with the rank difference between mothers and handlers. In addition, female handlers that outrank mothers are expected to maintain access to infants for a longer period of time than handlers who are of lower rank than mothers.

(6) The amount of time that nonmothers have access to infants is expected to be positively correlated with the amount of grooming they provide that infant's mother.

Methods

All research protocols reported in this manuscript were reviewed and approved by the Chinese Academy of Science. This research adhered to the regulations and applicable national laws of the Zhouzhi National Nature Reserve (ZNNR), China, where the study took place, and to the American Society of Primatologists principles for the ethical treatment of primates. This research received clearance from and complied with the protocols approved by animal care committees of the Wildlife Protection Society of Shaanxi Province, China (permit number: SX43537ACC). Legal permission to conduct this research also was provided by ZNNR.

Study Site

The study site is located in the Yuhuangmiao region of Zhouzhi National Nature Reserve on the northern slopes of the Qinling Mountains, in Shaanxi province, China (108°14′–108°18′E,33°45′–33°50′N). The ZNNR was established in 1985 to protect 52,931 km^2 of temperate forest and is characterized by a semi-humid montane climate. Elevation ranges from 1,400 to 2,890 m above sea level. The composition of the forest varies with altitude, from deciduous broad-leaf forest at low elevations to mixed coniferous broad-leaf forest above 2,200 m, to coniferous forest above 2,600 m [55]. The terrain is extremely mountainous. During the study period, the average annual temperature was 10.7°C with a maximum of 31.5°C, in July and a minimum of −

Table 1. Definitions of Infant handling behaviors.

Behavior	Definition
Nuzzle	Handler's nose contacted the infant's body.
Cradle	Handler supported the infant using one or both arms and brought the infant to her chest.
Carry	Handler maintained ventral-ventral contact with an infant while walking >1 m.
Groom	Handler manipulated an infant's hair with its hand and/or mouth (except for momentary touching), sometimes removing and eating small items found in the infant's fur
Touch	Handler gently used its hand to made contact with the infant
Hold	Handler grabbed an infant using one or both hands.

14.3°C in January. The diet of the golden snub-nosed monkey is characterized by considerable seasonal variation, but consists principally of 29.4% fruit/seeds, 29.0% lichen, 24.0% leaves, 11.1% bark, 4.2% buds, 1.3% twigs and 1.0% unidentified items (based on time spent feeding, [56]).

Study Troop

Two polygynous groups of golden snub-nosed monkeys are present ZNNR, the east ridge troop (ERT) and the west ridge troop (WRT). The WRT was chosen as the study troop and is characterized by a multi-level social structure consisting of one to two all-male bands and 6–11 one-male units (OMU). Details of the study troop have been reported previously [50,57]. All members were individually recognized by prominent physical characteristics, such as differences in body size, facial features, pelage color, hair patterns, physical disabilities, or the shape of granulomatous flanges on both sides of the upper lip [54].

Data Collection

Data presented in this study are based on 1,956 hr of behavioral observations collected from March 2009 to 2011 July. Given that individuals of the same one-male unit, especially females, usually remain in close spatial proximity, we were able observe all resident adult OMU females simultaneously. Continuous focal animal behavioral sampling [58] during three hour intervals was used to collect data on all grooming sessions that occurred between females. Data also were recorded on the timing, duration, initiator, and recipient of other social interactions such as approaches, infant handling, threats, fighting, and avoidance. The beginning and end of each grooming bout was recorded to the nearest second, along with the identity of the groomer and the groomee. One-male units were observed from 10:00 to 16:00 in an alternating sampling schedule across days to ensure that a relatively equal amount of data were collected on each OMU. Depending on the local conditions of the terrain, we were able to observe the focal OMU from a distance of between 0.5 and 50 m. If members of the target OMU were lost from view for a period of greater than 20 minutes, the next OMU on our sampling schedule was selected for study and observed for a period of 3 hrs. Each observation day we targeted two one-male units for study. Overall, we collected an average of 196 hr of behavioral data on each of the 6 OMUs in our study troop.

During the study period, 26 females in 6 OMUs gave birth. One of these females lost her infant a few days after birth, probably due to natural causes. Ten of these females gave birth both in 2009 and 2011. Five of the 26 females were primiparous and the remaining 21 females were multiparous. Infants under the age of 6 months were selected for study. Infants of this age class possess a distinctive

natal coat and are the most attractive to other animals. In total we observed 36 infants during the study period.

Given that we have been observing this troop for over a decade, maternal kinship and non-kin relationships are known for all group members. Approaches were scored when potential handlers or mothers moved to within one arm's length of the infant. We also recorded the time and social context during periods when handlers or mothers approached or moved away from the infant. A grooming bout was defined as a continuous period of allogrooming lasting at least 10 seconds. A groom bout was scored as ending when either the direction of grooming changed (A groomed B became B groomed A) or when there was a cessation in grooming of more than 60 s. A grooming session is a series of bouts of grooming in either direction between two individuals that was not interrupted for more than 10 min. Infant-handling behaviors recorded included visual inspection, nuzzle, hold, carry, nurse, cradle, groom, and touch (Table 1).

Data Analysis

To evaluate the effects that the presence of an infant had on female-female social interactions, we calculated hourly rates of approaches and the proportion of time a female received or gave grooming 5 months before and 6 months after giving birth. Two-way repeated measures ANOVAs were used to examine whether there were differences in the rate of approaches or frequency of grooming related to maternal status (mother vs. nonmother) and direction of behavior (given and received). Individual data were compared using the Paired sample t-test to analyze whether females' were approached more frequently and groomed for longer periods after giving birth than before parturition. The Paired sample t-test also was used to analyze whether there was a difference between the proportion of grooming mothers' received and the proportion of grooming mothers gave on each day. We used a binomial test to determine whether grooming sessions were more likely to be initiated by nonmothers. A G test was used to compare the proportion of reciprocal grooming sessions in mother-nonmother dyads and nonmother –nonmother dyads. Grooming sessions directly linked to sexual interactions were excluded. A Wilcoxon signed-ranks test was conducted to analyze whether potential handlers preferentially selected mothers as grooming partners. The observed proportion of grooming episodes received by mothers from potential handlers for each day was compared with the proportion of grooming episodes mothers were expected to receive in relation to their availability ($N = 197$ data collection day).

We used multilevel mixed-effects logistic regressions to determine whether grooming by handlers affect maternal responses to attempts by handlers to interact with infants. Maternal responses were recorded as 'positive' when mother allowed a handler to interact with her infant or 'negative' when mother left the area or moved her infant away from an approaching handler. Maternal responses were entered as dependent variables and grooming given by potential handlers, the age of the infant (months) at the time the grooming bout was observed, rank difference between potential handler and mother, the current number of newborn infants (≤ 6 months) per female in each one-male unit were included as independent fixed effect variables. Grooming given by, potential handlers was tested either as a categorical (yes/no occurrence) variable. Data points here were all dyadic interactions between mothers and potential handlers in which the individuals from the initial approach to the final leaving of either the mother or the potential handler in observation ($N = 897$). A G test was used to investigate whether handlers in mother–mother dyads were significantly more likely to obtain access to infants without first grooming the mother compared to handlers in mother– nonmother dyads.

To investigate whether the predictions 3–5 (factors affecting the grooming time nonmothers invested in mothers) were supported, multilevel mixed-effects linear regressions was used which examines each of the fixed effects. The duration of grooming (s) given by handlers was used as the dependent variable. For each grooming point, fixed effects consisted of the current number of newborn infants (≤ 6 months) per female in each one-male unit, the age of the infant (months) when the grooming bout was observed, and the rank difference between the mother and the handler. The position of females in their OMU dominance hierarchy was determined based on the outcome of decided agonistic events between females [59]. Dominance rank was assessed using the dominance index method and used to analyze the direction and the amount of aggressive and submissive behaviors observed [47,60]. Rank difference was defined as the recipient's dominance rank subtracted from the initiator's dominance rank. The rank of the alpha female's was recorded as '1'. Thus, rank difference could range from negative to positive values [13]. To prevent pseudoreplication, the identity of handlers and mothers were inserted as random effects, crossed with each other and nested in the OMU identity. Data points entered in this analysis were dyadic interactions between mothers and potential handlers that the handlers groomed mothers and exchanged for accessing to infants ($N = 756$). Grooming data were standardized by subtracting the mean and dividing by the standard deviation to show whether the duration of grooming in each OMU fluctuated with the number of infants present during each of the three different birth seasons. The mean grooming time for each OMU was transformed by subtracting mean grooming bout length from the length of each grooming bout and dividing by the standard deviation in each of the 3 birth seasons over which this study was conducted. A multilevel mixed-effects linear regression was also used to investigate whether high-ranking potential handlers were better able to get access to younger or relatively less infants. The rank difference between the mother and the handler were entered as the dependent variable, and age of infants and the number of infants per female were included as independent fixed-effect variables. The identity of handlers and mothers were inserted as random effects, crossed with each other and nested in the OMU identity. Data points were the number of that the handlers got accessing to infants from mothers ($N = 781$). To test whether predictions 5 and 6 were supported (impact of grooming time and rank difference on infant handling time), we also used multilevel mixed-effects linear regressions. The duration of infant handling (s) were defined as the dependent variable. Fixed effects included the duration of the handler's grooming bout, the current number of infants (≤ 6 months) per female in each OMU, the age of the

infant when the grooming bout occurred, and the rank difference between the mother and the handler. The results showed that rank difference did not have a significant effect on handling time. However, we noted that certain females in each OMU were permitted to handle infants for longer periods of time when rank difference interacted with infant's age and the number of infants presented in an OMU. In addition, previous analyses performed on grooming sessions outside birth and mating seasons revealed that even if females had as many as five partners to select from, most grooming occurred between two closely ranked females who interacted as frequent groomers [13]. Using criteria outlined in Barrett et al. (2000) [61], if individuals in the same dyad spent more than 5% of their total active and passive grooming time grooming, they were defined as frequent partners. Considering this, the effect 'frequent groomer' was used instead of 'rank difference' to test the prediction that female handlers who were frequent grooming partners maintained access to infants for a longer period of time than handlers who were infrequent partners. A dichotomous variable was used to represent mother and handler dyads for all handling times: 'positive' when the females were frequent grooming partners outside the birth and mating seasons, and 'negative' when females were not frequent grooming partners outside of the birth and mating seasons. In order to prevent distortion of the results due to pseudo-replication, random effects included in this analysis were the identity of the handlers and mothers, crossed with each other and nested in the OMU identity. Dyadic interactions as data points entered here were between mothers and potential handlers in which the potential handler had access to the infant ($N = 756$). Data were analyzed using the SAS 9.2 statistical package.

Results

The presence of infants significantly influenced the rate at which handlers approached mothers. We found that the rate at which mothers or nonmothers were approached differed significantly ($F_{1,70} = 23.17$, $p < 0.05$; Fig. 1). Based on our sample (N = 897) potential handlers were responsible for 81.7% of total approaches in nonmother-mother dyads. Females were approached at a significantly higher rate by nonmothers in the 6 months after they gave birth (paired t test: $t_{35} = -10.01$, $p < 0.05$) compared to the 5 months prior to giving birth (paired t test: $t_{35} = 12.05$, $p < 0.05$). In contrast, nonmothers were approached less frequently than mothers (paired t test: $t_{47} = -25.31$, $p < 0.001$), and the rate of approaches among nonmother-nonmother dyads decreased during the birth season compared to the period prior to infant births (paired t test: $t_{47} = 15.11$, $p < 0.05$). During the 6 month period after mothers' gave birth, the rate nonmothers approached mothers was significantly higher than the rate at which they approached nonmothers (paired t test: $t_{47} = -14.17$, $p < 0.001$).

Grooming also was significantly affected by the presence of infants ($F_{1,70} = 37.12$, $p < 0.01$). A total of 756 grooming-infant handling interchanges were recorded between mothers and nonmothers. Of these, 89.3% were initiated by potential handlers and only 10.7% initiated by mothers. We found that grooming sessions were more likely to be initiated by nonmothers (binomial test: $N = 756$, observed $P = 0.89$, theoretical $P = 0.5$, $p < 0.001$) and less likely to be reciprocated by mothers ($G_1 = 901.73$ $p < 0.001$). Mothers were chosen as grooming partners significantly more than expected based on their relative presence within their OMU ($Z = -8.41$, $N = 197$, $P < 0.05$). An analysis of grooming sessions indicated that 83% were unidirectional and mothers reciprocated grooming on only 17% of occasions. In contrast, 91.7% of the total 673 grooming session between nonmother dyads were bidirec-

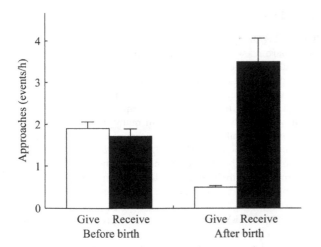

Figure 1. The ratio of approaches per hour (mean ± SE) mothers gave (■) and received (□) before parturition and after giving birth. Mother received significantly more approaches from nonmothers in the months following infant births than in the 5 months prior to giving birth (paired *t* test: $t_{35} = -10.01$, $p < 0.05$). New mothers approached other females less after giving birth than before parturition (paired *t* test: $t_{35} = 12.05$, $p < 0.05$).

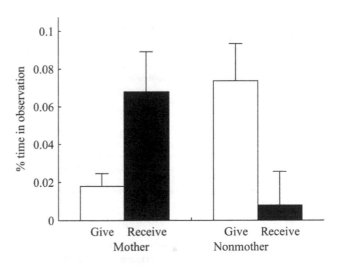

Figure 2. Percentage of time grooming (mean ± SE) mothers gave (■) and received (□) compared to the grooming nonmothers gave and received. The proportion of grooming mothers received from nonmothers was significantly higher than the proportion of grooming mothers gave to nonmothers on each day after parturition (paired *t* test: $t_{35} = -9.43$, $p < 0.01$). Nonmothers gave significantly more grooming to mothers than they received from mothers (paired *t* test: $t_{47} = -13.72$, $p < 0.01$).

tional and nonmothers commonly reciprocated by grooming their partners. The average duration of grooming bouts in which nonmothers groomed mothers was 241.4±292.8 s (mean ± SD). Mothers were groomed by nonmothers for a longer duration after they gave birth (241.4±292.8 s) than before parturition (197.1±121.7 s) (paired *t* test: $t_{35} = -8.521$, $p < 0.001$). However the proportion of time mothers spent grooming others decreased across these two periods (paired *t* test: $t_{35} = 11.41$, $p < 0.05$) while the proportion of time nonmothers spent grooming mothers increased (paired *t* test: $t_{47} = -19.21$, $p < 0.05$). Similarly, the proportion of grooming mothers with young infants received from nonmothers was significantly higher than the proportion of grooming these mothers directed at nonmothers (paired *t* test: $t_{35} = -9.43$, $p < 0.001$). Similarly, nonmothers groomed mothers for a significantly longer period of time than they were groomed by mothers (paired *t* test: $t_{47} = -13.72$, $p < 0.05$; Fig. 2), and the proportion of total grooming time nonmothers invested in mothers was significantly higher than the proportion of total grooming time they invested in nonmothers (paired *t* test: $t_{47} = -12.63$, $p < 0.05$). Finally, nonmothers groomed mothers for significantly longer periods of time than they grooming other females (paired *t* test: $t_{47} = -3.01$, $p < 0.05$).

Whether nonmothers groomed mothers prior to handling infants has significant effect on maternal responses to attempts by nonmothers to interact with infants. Grooming by nonmothers prior to handling infants increased the probability of nonmothers accessing to the infants (Table 2) and decreased the probability of nonmothers either being threatened or chased away by mothers (Table 3). When grooming preceded handling, mothers were more likely to sit passively and allow nonmothers to handle their infant (93%, 756/813). In cases in which nonmothers attempted to handle infants without first grooming the mother, mothers commonly threatened or moved away from the potential handler (90%, 223/248). The longer nonmothers groomed, the more likely they were allowed to remain in close proximity to mothers and handle their infants ($\mathcal{Z} = 1.49$, $N = 756$, $P = 0.012$).

We recorded only 29 cases infant handling that occurred across mother-mother dyads. The duration of mother-mother grooming

Table 2. Probability of positive maternal responses in relation to grooming given by potential handler.

Independent variables	Coefficient	z	p
Grooming given	0.72	3.91	0.001
Number of infants	−0.13	−0.49	0.126
Age of infants	0.38	0.74	0.301
Rank difference	0.25	0.66	0.249

Positive maternal responses – mother allowed a handler to interact with her infant. Other independent variables are the number of infants per OMU female, the age of the infant and the rank difference between the mother and the handler. $N = 897$.

was shorter (112.2±79.3 s vs. 241.4±292.8 s) than nonmother-mother grooming ($t_{257} = -18.94$, $p < 0.001$). Handlers in mother-mother dyads were significantly more likely to obtain access to infants without first grooming the mother than were handlers in mother-nonmother dyads (mother-mother: with grooming: 20.7%, $N = 6$; without grooming: 79.3%, $N = 23$; mother-nonmother: with grooming: 92.4%, $N = 699$; without grooming: 7.6%, $N = 57$; $G_1 = 83.09$, $p < 0.001$).

The number of infants per female, age of infants, and rank difference had a significant impact on the amount of time nonmothers invested in grooming mothers. Grooming duration increased significantly when the number of available infants per female in each OMU decreased (Coefficient = −0.45, $\mathcal{Z} = -2.12$, $N = 756$, $P < 0.001$, Fig. 3a). The grooming nonmothers gave to mothers before accessing to infants was significantly affected by the current supply of infants in each OMU. Moreover, within a single OMU the duration of grooming varied consistently with the number of infants per

Table 3. Probability of in relation to grooming given by potential handlers.

Independent variables	Coefficient	z	p
Grooming given	−0.93	−3.79	0.017
Number of infants	−0.32	−1.61	0.138
Age of infants	0.28	0.67	0.274
Rank difference	0.58	1.34	0.103

Negative maternal responses – mother left the area or moved her infant away from an approaching handler. Other independent variables are the number of infants per OMU female, the age of the infant and the rank difference between the mother and the handler. $N = 897$.

female across three breeding periods (Fig. 3b). In each case, as the number of infants per female decreased, the longer handlers were found to groom mothers. When the number of infants per OMU increased, grooming durations were shorter. The infants' age was associated with a decreased duration of grooming nonmothers gave to mothers (Coefficient $= -0.11$, $\mathcal{Z} = -1.20$, $\mathcal{N} = 756$, $P = 0.021$, Fig. 4). The duration of grooming decreased significantly as infants grew older. Handling older infants required less grooming over shorter durations. Similarly, there was evidence that increasing rank difference negatively influenced grooming duration of nonmother-to-mother grooming bouts that was required before handlers were able to touch or groom infants (Coefficient $= -0.57$, $\mathcal{Z} = -0.94$, $\mathcal{N} = 756$, $P = 0.017$). The higher ranking the handler was compared to the mother, the less time was invested in grooming the mother. The greater the rank difference between a lower ranked female and a mother, the greater the duration of time that female groomed the mother in order to obtain access to her infant. In addition, the rank difference between nonmothers and mothers was significantly related to the age of the infant (Coefficient $= 0.67$, $\mathcal{Z} = 3.18$, $\mathcal{N} = 781$, $P = 0.015$, controlling for the number of infants per female). When infants were young, mothers and handlers were closely ranked or handlers were of higher rank than mothers, supporting the notion that access to young infants was a valued commodity sought after by nonmothers. However, nonmothers of lower rank than mothers were not excluded from getting access to infants when the number of infants per female was relatively low in the OMUs (Coefficient $= -0.98$, $\mathcal{Z} = 2.12$, $\mathcal{N} = 781$, $P = 0.157$, controlling for the age of infant).

The number of infants per female significantly predicted the duration of infant handling. The duration of infant handling decreased significantly when the number of available infants increased (Coefficient $= -0.27$, $\mathcal{Z} = -1.61$, $\mathcal{N} = 756$, $P = 0.011$). Grooming times only had a significant influence on infant handling time when controlling for the number of infants per female in each OMU (Coefficient $= 0.66$, $\mathcal{Z} = 2.39$, $\mathcal{N} = 756$, $P = 0.031$). Nonmothers who groomed mothers longer increased their access to infants only when infant availability also increased. Frequent female grooming partners significantly affected the infant handling duration (Coefficient $= 0.76$, $\mathcal{Z} = 2.19$, $\mathcal{N} = 756$, $P < 0.001$). Frequent groomers were allowed to handle infants significantly longer than infrequent grooming partners and the duration of infant handling decreased significantly when females were not frequent groomers. The duration of infant handling was not significantly affected by the infant age (Coefficient $= 0.58$, $\mathcal{Z} = 1.95$, $\mathcal{N} = 756$, $P = 0.812$).

Discussion

The biological market model predicts that trading decisions between partners will be influenced by the current 'exchange rate' of particular commodities which is a function of each partners' need for the commodity and the ability to outbid or outcompete other group members who also can provide a needed good or service [1,2]. Thus, the 'price' of a commodity is expected to fluctuate with supply and demand. In many species of primates, mothers and their infants represent a central focus of interest for females in the group [30,31,62,63,64]. Mothers are generally tolerant of approaches that result in nonmothers being in close spatial proximity to their offspring. However, direct physical access to offspring requires that nonmothers invest time and effort in first grooming the infant's mother.

The results of this study indicate that the presence of newborn infants has a significant effect on the dynamics of adult female social interactions in wild golden snub-nosed monkeys (*Rhinopithecus roxellana*). Females received more approaches and were groomed for longer periods of time in the months following the birth of their infant than in the months prior to parturition. Females without infants (nonmothers) were strongly attracted to newborn infants and preferentially groomed mothers as a social tool to obtain access to their infants. Nonmothers were found to groom mothers more frequently and for a greater duration than mothers groomed them in return. Previous research on *R. roxellana* [13] indicated asymmetries in female grooming relationships with factors such as rank and the number of available grooming partners having a significant effect on the degree to which grooming was reciprocated. In the present study, grooming an infant's mother increased the potential handlers' chances of establishing contact with the infant. Given that grooming initiated by nonmothers was rarely reciprocated by the infant's mother, our results support the contention that in golden snub-nosed monkeys females without infants interchange grooming for access to infants (Prediction 1 supported). In contrast, we found that grooming between mother-mother dyads was reciprocal, and occurred significantly less frequently than grooming between mother-nonmother dyads (Prediction 2 supported). Thus, it appears that individual females adjust their social strategies and grooming relationships depending on several factors including their reproductive status, availability of young infants, and position in the dominance hierarchy. For example, the duration of grooming bouts that nonmothers directed toward mothers increased significantly when fewer infants were present in an OMU, and with an increase in rank difference between the mother and infant handlers (Prediction 3 supported and Prediction 5 partly supported). In contrast, the duration of grooming traded for access to infants decreased significantly with infant age (Prediction 4 supported). The fewer infants present, the greater amount of time handlers groomed mothers for access to their infant. Such evidence implies that similar amounts of grooming had a decreasing effect on the probability of getting access to infants when the number of infants per female in the OMUs decreased. Our findings suggest that in golden snub-nosed monkeys the grooming of mothers by nonmothers is a service that is actively interchanged for infant handling privileges, and that the 'value' of infants changed depending on supply and demand in the biological market place of the *R. roxellana* OMU. These results support the biological market prediction that the scarcity of a resource or a service increases its value, while the increased availability of that resource or service decreases its value [3]. Studies of several other primate species also have demonstrated that the relative number of infants present in a group can influence

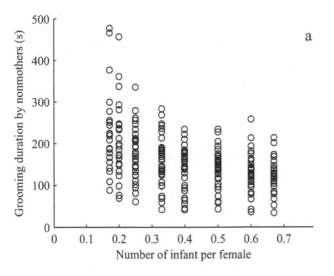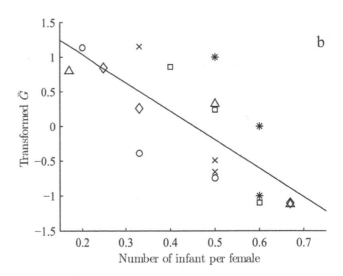

Figure 3. (a) The duration of grooming nonmothers invested in mothers before getting access to the infants in relation to the number of infants per female in each one-male unit. When the number of infants per female in each one-male unit increased the amount of grooming given to mothers also decreased. (b) The duration of grooming varied with the number of infants per female during each of the 3 infant birth seasons in a given OMU (total 6 OMUs). Similarly the amount of grooming that nonmothers gave to mothers was negatively correlated with the number of infants in the OMU. The 6 OMUs are identified as follows: JB, FP, PK, JZT, BB, RX. Each sign represents one OMU in one birth season–JB (○) FP (*) PK (□) JZT (◇) BB (△) RX (×).

the duration of grooming received by mothers who interchanged grooming for access to infants [26,41,43,45].

Allowing others access to her offspring is likely to be a source of anxiety for the mother especially during the infants first few months of life [11], which is a period of infant vulnerability [40]. Moreover, inexperienced females may mishandle infants, carry infants away from the mother, or continue to retain access to infants during periods when they need to nurse. It has been argued that given the important role that grooming plays in reducing stress, affiliative social contact with the mother prior to attempting to handle her infant may serve to reduce maternal anxiety and facilitate the groomer's access to the infant [28,41,65]. This appears to be an important factor affecting the interchange of grooming for infants contact in chacma baboons [41]. However,

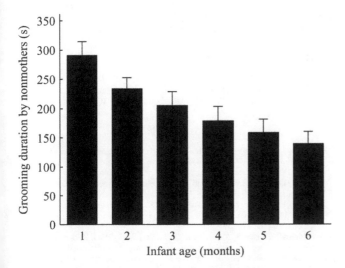

Figure 4. The amount of grooming that nonmothers (mean ± SE) gave to mothers was negatively correlated with infant age.

factors in addition to supply and demand ratios, may affect the 'value' of handling infants. For example, we found an effect of infant age on the interchange of grooming for infant access. Nonmothers groomed mothers longer for access to infants less than 6 months of age than for access to infants older than 6 months of age. What remains less clear is whether this is the result of a decrease in the attractiveness of older infants to handlers (supply of infants remains the same but handler demand/motivation decreased) or whether mothers of older infants are less anxious or more permissive and allow others to handle their offspring (supply of infants remains the same but cost required by mothers for access to infants has decreased) or whether the main goal of handling is to get to know the new group member and form a (positive) relationship with this individual and this becomes less necessary when the infant grows older.

We also found an effect of female rank on patterns of grooming associated with access to infants. In golden snub-nosed monkeys, rank appears to shift the balance for many individuals from dyadic interactions based on reciprocity to dyadic interactions that are characterized by the unequal exchange of the same service or the interchange of differences services. For example, lower-ranked females groomed higher-ranked mothers for longer duration than vice versa. On average higher-ranking females grooming lower ranking mothers for 213.7±171.3 seconds (mean ± SD) whereas lower-ranking females groom higher ranking mothers an average of 261.2±234.1 seconds in order to obtain access to their infants (Prediction 5 partly supported). This suggests that lower-ranked and higher-ranked females are characterized by different social strategies and these change with infant attractiveness. Higher-ranked nonmothers were able to 'pay a lower price' than the one set by the current supply/demand ratio or paid back in a different 'currency', e.g. tolerance at food sources or support in conflicts. When infants were very young, handlers were more likely to be closely ranked or of higher rank than the infant's mother. However, this pattern did not continue once infants reached 6 months of age. This suggests that when infants were most highly valued, higher-ranked nonmothers successfully outbid lower-ranked competitors for infant access. As these infants aged, however, their 'value' appeared to decrease and lower-ranked handlers had greater access to infants by increasing the time they invested in grooming higher-ranking mothers. If access to higher-ranked mothers and their infants are of greater value in the biological market place than access to lower-ranked females and their infants, then the primary goal of lower ranking females may have been to use access to infants as a means of developing a social relationship with a female of higher rank. These same low-ranking females could gain equal infant caregiving experience at a lower cost (reduced time grooming) by preferentially grooming mothers of low social rank. In contrast, it is likely that higher ranked females who groom lower ranking mothers do so for the primary purpose of obtaining infant caregiving experience, as social bonds or alliances with these low ranking mothers offer limited benefits. This may be analogous to data obtained in studies of captive female tamarins (*Saguinus*) and marmosets (*Callithrix*) [17]. In these primates experience that young females obtain in caring for another's offspring is argued to be a critical factor increasing the survivor ship of that female's future offspring [66]. Taken together, our results indicate that the market value of infants depends on multiple factors, some of which are generally stable over the birth season (OMU size and composition, female social rank, experience caring for offspring) and others that may fluctuate based on changes in outbidding competition, infant age, and supply and demand ratios.

Finally, several previous studies [41,42,43,67] have focused on infants as a tradable commodity with their value represented by the costs incurred (grooming time and energy) by nonmothers. Less attention has been paid to the quality of the infant handling experience obtained by nonmothers and its potential benefits to nonmothers in alliance formation, change in dominance status, and future reproductive success [45]. In order to best understand the social dynamics of the biological market place greater emphasis needs to be laid on understanding and better estimating variation in the cost and value to individual traders (mothers and handlers) of access to an infant and access to that infant's mother. In the present study, the duration of grooming did not directly correlate with infant handling times, and thus Prediction 6 was not supported. Only when the number of available infants increased, did an increase in the duration of a grooming bout result in increased access to infants. When the number of infants in an OMU was limited, longer grooming bouts did not necessarily result in increased handling time. The results also revealed that when the number of available infants increased, handling time decreased. When females and mothers were infrequent grooming partners, the duration of infant handling also decreased. However, nonmother frequent grooming partners spent a significantly greater amount of time handling infants than did nonmother infrequent groomers regardless of the number of infants present in

each OMU. This suggests that persistent or long-term social partners are characterized by a preferred or special relationship that may enable them to receive a discount in the biological market place. Female *R. roxellana* commonly form alliances in the context of grooming frequently, moving, feeding, stay and even fighting together, and thus long-term social partnerships (whether kin or nonkin based) offer advantages in fitness (Wei Wei, unpublished data).

Our results suggest that in *R. roxellana* access to infants represents a commodity that can be traded for grooming. The 'value' of infant handling varied based on supply and demand, the social status of the individual traders, and whether potential handlers had a long-term relationship with the infant's mother. Given that golden snub-nosed monkeys are strictly seasonal breeders, more than one similarly aged infant is commonly present in an OMU. Nevertheless nonmothers from one OMU do not groom mothers from neighboring OMUs or interact with their infants. Therefore, the availability of infants in a female's resident OMU has a strong impact on female social interactions. The presence of young infants appears to have a direct influence on female partner choice, grooming patterns, and the formation of long-term alliances. Mother's with newly born infants were groomed for a longer period of time than nonmothers or mothers with older offspring. Differences in the amount of grooming females traded for access to another female's infant appeared to be based on market 'rules' of supply and demand as well as female rank. Although in the case of higher ranking females, reciprocity or long-term alliances may help explain the value or costs of exchanging goods and serves, for many other females directly responding to the current value of goods and services may reduce the need for complex calculations, cost-benefit analyses or bookkeeping. We plan to continue our research on social interactions and interchange trading in golden snub-nosed monkeys in order to better identify how individuals adjust their social behavior in the face of short-term changes in the value of goods and services.

Acknowledgments

We thank Zhouzhi National Nature Reserve for permission to carry out this study. We greatly appreciate Wang XW, Wang CL, Zhao HT, Huang K, Fu WW, Zhang D, Wu LL, Zhang HY (College of Life Science, Northwest University) for their indispensable and sincere support during this study. WW expresses her special thank to Hui Cao for his sincere help. PAG wishes to thank Chrissie, Sara, and Jenni for their love and support.

Author Contributions

Conceived and designed the experiments: WW XGQ BGL. Performed the experiments: WW. Analyzed the data: WW. Contributed reagents/materials/analysis tools: STG PZ. Wrote the paper: WW PAG.

References

1. Noë R, Hammerstein P (1994) Biological markets: supply and demand determine the effect of partner choice in cooperation, mutualism and mating. Behav Ecol Sociobiol 35: 1–11.
2. Noë R, Hammerstein P (1995) Biological markets. Trends Ecol Evol 10: 336–339.
3. Noë R, Hooff JARAM, Hammerstein P (2001) Economics in nature: social dilemmas, mate choice and biological markets: Cambridge University Press.
4. Dunbar RIM (1988) Primate social systems. London: Chapman & Hall.
5. Cheney DL, Seyfarth RM (1990) How monkeys see the world: Inside the mind of another species. Chicago: University of Chicago Press.
6. Manson JH, Rose LM, Perry S, Gros-Louis J (1999) Dynamics of female-female relationships in wild *Cebus capucinus*: data from two costa rican sites. Int J Primatol 20: 679–706.
7. Barrett L, Henzi SP (2006) Monkeys, markets, and minds: Biological markets and primate sociality. In: Kappeler PM, van Schaik C, editors. Cooperation in primates and humans: Mechanisms and evolution. Berlin: Springer 209–232.
8. Schino G (2007) Grooming and agonistic support: a meta-analysis of primate reciprocal altruism. Behav Ecol 18: 115–120.
9. Aureli F, Preston SD, de Waal F (1999) Heart rate responses to social interactions in free-moving rhesus macaques (*Macaca mulatta*): A pilot study. J Comp Psychol 113: 59.
10. Schino G, Scucchi S, Maestripieri D, Turillazzi PG (1988) Allogrooming as a tension-reduction mechanism: a behavioral approach. Am J Primatol 16: 43–50.
11. Arnold A, Fraser O, Aureli F (2011) Postconflict reconciliation. In: Campbell CJ, Fuentes A, MacKinno KC, Bearder S, Stumpf R, editors. Primates in Perspective Vol. 2. New York: Oxford University Press. 608–625.
12. Schino G, Aureli F (2008) Grooming reciprocation among female primates: a meta-analysis. Biol Lett 4: 9–11.

13. Wei W, Qi X-G, Guo S-T, Zhao D-P, Zhang P, et al. (2012) Market powers predict reciprocal grooming in Golden Snub-nosed Monkeys (*Rhinopithecus roxellana*). PLoS ONE 7: e36802.

14. Henzi SP, Barrett L (1999) The value of grooming to female primates. Primates 40: 47–59.

15. Stopka P, Graciasová R (2001) Conditional allogrooming in the herb-field mouse. Behav Ecol 12: 584–589.

16. Payne HFP, Lawes MJ, Henzi SP (2003) Competition and the exchange of grooming among female samango monkeys (*Cercopithecus mitis erythrarchus*). Behaviour 140: 453–471.

17. Lazaro-Perea C, Arruda MF, Snowdon CT (2004) Grooming as a reward? Social function of grooming between females in cooperatively breeding marmosets. Anim Behav 67: 627–636.

18. Manson JH, David Navarrete C, Silk JB, Perry S (2004) Time-matched grooming in female primates? New analyses from two species. Anim Behav 67: 493–500.

19. Port M, Clough D, Kappeler PM (2009) Market effects offset the reciprocation of grooming in free-ranging redfronted lemurs, *Eulemur fulvus rufus*. Anim Behav 77: 29–36.

20. Chancellor RL, Isbell LA (2009) Female grooming markets in a population of gray-cheeked mangabeys (*Lophocebus albigena*). Behav Ecol 20: 79–86.

21. Seyfarth RM, Cheney DL (1984) Grooming, alliances and reciprocal altruism in vervet monkeys. Nature 308: 541–543.

22. de Waal FBM (1997) The Chimpanzee's service economy: Food for grooming. Evol Hum Behav 18: 375–386.

23. Fruteau C, Voelkl B, van Damme E, Noë R (2009) Supply and demand determine the market value of food providers in wild vervet monkeys. Proc Natl Acad Sci 106: 12007–12012.

24. Kutsukake N, Clutton-Brock TH (2006) Social functions of allogrooming in cooperatively breeding meerkats. Anim Behav 72: 1059–1068.

25. Barrett L, Henzi SP (2001) The utility of grooming in baboon troops. In: Noë R, Hooff JARAM, Hammerstein P, editors. Economics in nature: social dilemmas, mate choice and biological markets. Cambridge: Cambridge University Press. 119–145.

26. Gumert MD (2007) Payment for sex in a macaque mating market. Anim Behav 74: 1655–1667.

27. Norscia I, Antonacci D, Palagi E (2009) Mating first, mating more: biological market fluctuation in a wild prosimian. PLoS ONE 4: 46–79.

28. Watts DP (2002) Reciprocity and interchange in the social relationships of wild male chimpanzees. Behaviour: 343–370.

29. Stopka P, Macdonald DW (1999) The market effect in the wood mouse, *Apodemus sylvaticus*: selling information on reproductive status. Ethology 105: 969–982.

30. Altmann J (1980) Baboon Mothers and Infants. Cambridge: Harvard University Press.

31. Nicolson NA (1987) Infants, mothers, and other females. In: Smuts BB, Cheney DL, Seyfarth RM, Wrangham RW, Struhsaker TT, editors. Primate Societies. Chicago: University of Chicago Press.

32. Lancaster JB (1971) Play-mothering: the relations between juvenile females and young infants among free-ranging vervet monkeys (*Cercopithecus aethiops*) Folia Primatol 15: 161–182.

33. Maestripieri D (1994) Social structure, infant handling, and mothering styles in group-living Old World monkeys. Int J Primatol 15: 531–553.

34. Manson JH (1999) Infant handling in wild *Cebus capucinus*: testing bonds between females? Anim Behav 57: 911–921.

35. Barrett L, Gaynor D, Henzi SP (2002) A dynamic interaction between aggression and grooming reciprocity among female chacma baboons. Anim Behav 63: 1047–1053.

36. Goldizen AW (1987) Tamarins and marmosets: communal care of offspring. In: Primate Societies. In: Smuts BB, Cheney DL, Seyfarth RM, W WR, T ST, editors. Chicago: University of Chicago Press. 34–43.

37. Garber PA (1997) One for all and breeding for one: cooperation and competition as a tamarin reproductive strategy. Evolutionary Anthropology 5: 187–199.

38. Silk JB, Alberts SC, Altmann J (2003) Social Bonds of Female Baboons Enhance Infant Survival. Science 302: 1231–1234.

39. Kirkpatrick R (2011) The Asian colobines: diversity among leaf-eating monkeys. In: Campbell CJ, Fuentes A, MacKinnon KC, Bearder SK, Stumpf RM, editors Primates in perspective, 2nd ed New York: Oxford university Press p: 189–202.

40. Xi W, Li B, Zhao D, Ji W, Zhang P (2008) Benefits to female helpers in wild Rhinopithecus roxellana. Int J Primatol 29: 593–600.

41. Henzi SP, Barrett L (2002) Infants as a commodity in a baboon market. Anim Behav 63: 915–921.

42. Gumert MD (2007) Grooming and infant handling interchange in *Macaca fascicularis*: The relationship between infant supply and grooming payment. Int J Primatol 28: 1059–1074.

43. Slater KY, Schaffner CM, Aureli F (2007) Embraces for infant handling in spider monkeys: evidence for a biological market? Anim Behav 74: 455–461.

44. Frank RE, Silk JB (2009) Grooming exchange between mothers and non-mothers: the price of natal attraction in wild baboons (*Papio anubis*). Behaviour 146: 889–906.

45. Fruteau C, van de Waal E, van Damme E, Noë R (2011) Infant access and handling in sooty mangabeys and vervet monkeys. Anim Behav 81: 153–161.

46. Tiddi B, Aureli F, Schino G (2010) Grooming for infant handling in tufted capuchin monkeys: a reappraisal of the primate infant market. Anim Behav 79: 1115–1123.

47. Li BG, Li HQ, Zhao DP, Zhang YH, Qi XG (2006) Study on dominance hierarchy of Sichuan snub-nosed monkeys (*Rhinopithecus roxellana*) in Qinling Mountains. Acta Theriol Sin 26: 18–25.

48. Ren RM, Yan KH, Su YJ, Zhou Y, Li JJ, et al. (2000) A field study of the society of *Rhinopithecus roxellana*. Beijing: Beijing University Press.

49. Qi XG, Li BG, Tan CL, Gao YF (2004) Spatial structure in a Sichuan golden snub-nosed monkey (*Rhinopithecus roxellana*) group in Qinling Mountains while being no-locomotion. Acta Zool Sin 50: 697–705.

50. Zhang P, Watanabe K, Li BG, Tan CL (2006) Social organization of Sichuan snub-nosed monkeys (*Rhinopithecus roxellana*) in the Qinling Mountains, Central China. Primates 47: 374–382.

51. Chen FG, Min ZL, Lou SY, Xie WZ (1983) An observation on the behavior and some ecological habits of the Golden snub-nosed monkeys (*Rhinopithecus roxellana*) in Qinling Mountains. Acta Theriol Sin 3: 141–146.

52. Qi XG, Li BG, Garber PA, Ji WH, Watanabe K (2009) Social dynamics of the golden snub-nosed monkey (*Rhinopithecus roxellana*): female transfer and one-male unit succession. Am J Primatol 71: 670–679.

53. Zhang SY, Liang B, Wang LX (2000) Seasonality of matings and births in captive Sichuan golden monkeys (Rhinopithecus roxellana). Am J Primatol 51: 265–269.

54. Li BG, Zhao DP (2007) Copulation behavior within one-male groups of wild *Rhinopithecus roxellana* in the Qinling Mountains of China. Primates 48: 190–196.

55. Li BG, Chen C, Ji WH, Ren BP (2000) Seasonal home range changes of the Sichuan snub-nosed monkey (*Rhinopithecus roxellana*) in the Qinling Mountains of China. Folia Primatol 71: 375–386.

56. Guo ST, Li BG, Watanabe K (2007) Diet and activity budget of *Rhinopithecus roxellana* in the Qinling Mountains, China. Primates 48: 268–276.

57. Qi XG, Li BG, Ji WH (2008) Reproductive parameters of wild female *Rhinopithecus roxellana*. Am J Primatol 70: 311–319.

58. Altmann J (1974) Observational study of behavior: sampling methods. Behaviour 49: 227–267.

59. Zumpe D, Michael RP (1986) Dominance index: a simple measure of relative dominance status in primates. Am J Primatol 10: 291–300.

60. Zhao H (2011) Study on the correlationship between female dominance hierarchy and mating behavior of Sichuan snub-nosed monkeys (Rhinopithecus roxellana). Xi'an: Northwest University.

61. Barrett L, Henzi SP, Weingrill T, Lycett JE, Hill RA (2000) Female baboons do not raise the stakes but they give as good as they get. Anim Behav 59: 763–770.

62. Hrdy SB (1976) Care and exploitation of nonhuman primate infants by conspecifics other than the mother. In: Rosenblatt JS, Hinde RA, E S, Bier C, editors. Advances in the study of behavior vol 6. New York: Academic Press. 101–158.

63. Seyfarth RM (1976) Social relationships among adult female baboons. Anim Behav 24: 917–938.

64. Maestripieri D (1994) Influence of infants on female social relationships in monkeys. Folia Primatol 63: 192–202.

65. Clutton-Brock T, Parker G (1995) Sexual coercion in animal societies. Anim Behav.

66. Bales K, Dietz J, Baker A, Miller K, Tardif S (2000) Effects of allo-care givers on fitness of infants and parents in callitrichid primates. Folia Primatol 71: 27–38.

67. Schaffner C, Aureli F (2005) Embraces and grooming in captive spider monkeys. Int J Primatol 26: 1093–1106.

2

Variation in Carbon Storage and Its Distribution by Stand Age and Forest Type in Boreal and Temperate Forests in Northeastern China

Yawei Wei[1,4], Maihe Li[1,2], Hua Chen[1,3], Bernard J. Lewis[1], Dapao Yu[1], Li Zhou[1], Wangming Zhou[1], Xiangmin Fang[1,4], Wei Zhao[1,4], Limin Dai[1]*

1 State Key Laboratory of Forest and Soil Ecology, Institute of Applied Ecology, Chinese Academy of Sciences, Shenyang, China, 2 Tree Physiology Group, Swiss Federal Research Institute WSL, Birmensdorf, Switzerland, 3 University of Illinois at Springfield, Springfield, Illinois, United States of America, 4 University of Chinese Academy of Sciences, Beijing, China

Abstract

The northeastern forest region of China is an important component of total temperate and boreal forests in the northern hemisphere. But how carbon (C) pool size and distribution varies among tree, understory, forest floor and soil components, and across stand ages remains unclear. To address this knowledge gap, we selected three major temperate and two major boreal forest types in northeastern (NE) China. Within both forest zones, we focused on four stand age classes (young, mid-aged, mature and over-mature). Results showed that total C storage was greater in temperate than in boreal forests, and greater in older than in younger stands. Tree biomass C was the main C component, and its contribution to the total forest C storage increased with increasing stand age. It ranged from 27.7% in young to 62.8% in over-mature stands in boreal forests and from 26.5% in young to 72.8% in over-mature stands in temperate forests. Results from both forest zones thus confirm the large biomass C storage capacity of old-growth forests. Tree biomass C was influenced by forest zone, stand age, and forest type. Soil C contribution to total forest C storage ranged from 62.5% in young to 30.1% in over-mature stands in boreal and from 70.1% in young to 26.0% in over-mature in temperate forests. Thus soil C storage is a major C pool in forests of NE China. On the other hand, understory and forest floor C jointly contained less than 13% and <5%, in boreal and temperate forests respectively, and thus play a minor role in total forest C storage in NE China.

Editor: Ben Bond-Lamberty, DOE Pacific Northwest National Laboratory, United States of America

Funding: This work was supported by the "Strategic Priority Research Program" of the Chinese Academy of Sciences, Grant No. XDA05060200, National Key Technologies R&D Program of China (2012BAD22B04) and the Chinese Academy of Sciences visiting professorship for senior international scientists (2012T1Z0006). The funders had no role in study design, data collection and analysis, decision to publish, or preparation of the manuscript.

Competing Interests: The authors have declared that no competing interests exist.

* E-mail: lmdai@iae.ac.cn

Introduction

Temperate and boreal forests cover 1.9 billion hectares worldwide and account for approximately 46% of global forest carbon (C) storage [1]. Field and modeling studies suggest that these forests function as significant carbon sinks [2,3], although the magnitude, location, and mechanisms of C sequestration remain uncertain [1,4]. It is widely recognized that temperate and boreal forests are much more susceptible to global warming than tropical forests [5,6], and that high northern hemisphere latitudes are experiencing a relatively rapid and significant change in climate [5]. The northeastern forest region of China (NE China) encompasses a forest area of more than 50×10^4 km^2, ranging from temperate forests in the south to boreal forests in the far north. These forests play an important role in the global carbon budget [7]. Thus a more thorough assessment of forest ecosystem C stocks and their dynamics in the country's temperate and boreal forests is clearly worthwhile.

In northeastern China, numerous studies have been conducted to analyze spatial and temporal patterns of C storage on regional scales [8,9,10], to examine the effects of wildfire and human logging activities on changes in C storage [11,12,13], and to investigate C storage and its distribution across forest types via plot analyses [14,15]. Such studies have advanced our knowledge of forest C storage and its variation at different scales. Stand age also has been shown to be a key factor in regulating C storage and its partitioning in different forest components (vegetation, debris and soil) [16,17,18]. To our knowledge, however, stand age has seldom been considered with respect to carbon dynamics and the pattern of carbon distribution in different forest components (tree, understory, forest floor and soil) in northeastern China.

The focus of this study therefore was to quantify the partitioning pattern of C storage in different forest components – tree, understory, forest floor and soil – across different aged forests (from young to over-mature) for the major natural temperate and boreal forests in NE China. The overall goal was to better understand C sequestration potential in boreal and temperate forests, and to provide information on carbon balances that might be used to improve forest management practices intended to increase carbon storage.

Materials and Methods

Ethics Statement

All necessary permits for the described field investigation were obtained at the start of the study from the provincial and locally state-owned forestry bureaus. The study forests refer neither to privately-owned field and biosphere nature reserves, nor to endangered or protected species.

Study Area

The study was conducted in state-owned forests in the northeastern forest region of China, which includes Heilongjiang and Jilin provinces and the eastern-most part of the Inner Mongolia Autonomous Region ($41°42'\sim53°34'$N, $115°37'\sim135°5'$E, 109.04×10^4 km^2). Three major mountain ranges (Daxing'an, Xiaoxing'an and Zhangguangcai-Changbai) occur in the study region (Fig. 1). The climate is controlled by high latitude East Asian monsoons. Mean annual temperatures range from $-2.5°C$ (north) to $4.8°C$ (south) and mean precipitation ranges from 250 mm (west) to 1100 mm (east). From south to north, the forest region is divided into a temperate coniferous and broadleaved mixed forest zone (Changbai and Xiaoxing'an mountains), and a boreal coniferous zone (Daxing'an mountain range). Dark-brown soils are predominant in the temperate zone and brown coniferous forest soils in the boreal zone.

Field Design

Four representative sites in NE China forests were selected for study – the Lushuihe site in the Changbai mountain area; the Yichun site in the Xiaoxing'an mountains; and the Genhe and Huzhong sites in the Daxing'an mountains (Fig. 1). The Lushuihe and Yichun sites include three major temperate natural forest types – coniferous mixed forest (CMF), coniferous and broad-leaved mixed forest (CBF), and broadleaved mixed forest (BMF); While the Genhe and Huzhong sites include two major boreal natural forest types – larch forest (LF) and birch forest (BF) (Table 1). In each type, four stand age classes were delineated according to the ages of dominant trees. For CMF, CBF, and LF, age classes were defined as young (<40 years), mid-aged (41–80 years), mature (81–140 years), and over-mature (>141 years). For BMF and BF forests, stand ages were defined as young (<30 years), mid-aged (31–50 years), mature (51–80 years), and over-mature (>81 years). During the field investigation, stand ages were based on the predominant tree species and other information (i.e. forest maps, forest management or logging history, and so on) provided by local forestry bureaus.

Each of the 179 study plots was 20×20 m (Table 1). Tree biomass, understory biomass, forest floor biomass and soil C were measured within each plot. A small subset of plots within forest types of the same or similar age class, species composition, and geographical conditions served as replicates within each age class and forest type replicated three times.

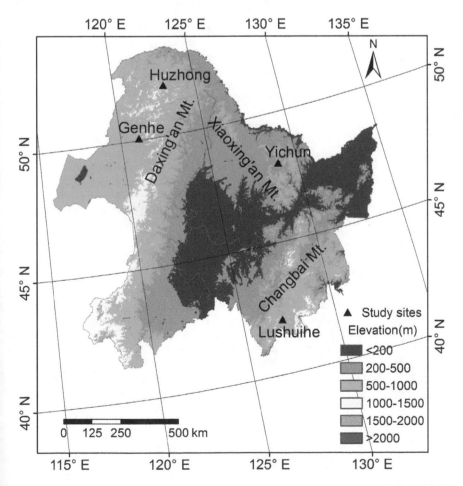

Figure 1. Geographic location of the study sites in the northeast forest region of China.

Table 1. Characteristics of the study sites and stands in northeastern China.

	Geographic factors			Climatic factors		Community characteristics		Stand density (trees·ha⁻¹)	No. of plots
Sites	Latitude(N)	Longitude (E)	Elevation (m,asl)	MAT (°C)[a]	MAP (mm)	Forest types[b]	Dominant tree species	Stand density (trees·ha⁻¹)	No. of plots
Boreal zone									
Huzhong	51°4'~52°2'	122°2'~124°0'	446~990	−0.8~1.1	359~636	LF, BF	Larix gmelinii, Betula platyphylla	1000~2567	38
Genhe	50°3'~50°6'	120°6'~121°3'	446~1011	−2.5~−0.7	208~381	LF, BF	Larix gmelinii, Betula platyphylla	1353~3460	41
Temperate zone									
Yichun	47°1'~48°2'	128°1'~129°2'	259~599	1.2~2.8	421~823	CMF,CBF, BMF	Pinus koraiensis, Picea jezoensis, Abies nephrolepis, Populus davidiana, Tilia amurensis, Quercus mongolica	613~2644	57
Lushuihe	42°3'~42°4'	127°5'~128°0'	582~1039	2.6~4.8	509~810	CMF,CBF, BMF	Pinus koraiensis, Picea jezoensis, Abies nephrolepis, Populus davidiana, Tilia amurensis, Quercus mongolica	650~1663	43

[a]MAT = mean annual temperature; MAP = mean annual precipitation.
[b]Forest types: LF: larch forest; BF: birch forest; CMF: coniferous mixed forest; CBF: coniferous and broadleaved mixed forest; BMF: broadleaved mixed forest.

Field Sampling and Forest Carbon Storage Estimation

Trees. Within each 400 m² plot, all trees (standing and fallen) with a diameter at breast height (DBH, 1.37 m above ground) of ≥5 cm were identified in terms of species, height, DBH, and living or dead status. Individual tree biomass (above and below-ground) was estimated using species-specific allometric equations, developed by Chen & Zhu [19] for the Changbai area, Wang [20] for the Xiaoxing'an area, and Han & Zhou [21] for the Daxing'an area. For a few tree species where no species-specific allometric equations were available, we used equations for similar species. Given that such trees were seldom encountered, the influence of surrogate equations was considered insignificant.

Understory. Understory vegetation included shrubs, herbs, and small trees with a DBH <5 cm. This component was measured in three randomly established 2×2 m subplots within each plot. All vegetation in subplots was harvested and weighed with an electronic balance (accuracy: ±1 g). The fresh weight was recorded for small trees (foliage, branches and stems), shrubs (leaves and branches) and herbs. Understory biomass was estimated via fresh weight multiplied by previously established dry-wet ratios for different understory vegetation in NE China [22].

Forest floor. For this study we defined forest floor biomass as woody debris, surface litter, organic matter above the mineral soil, and undifferentiated organic matter. Only fine woody debris, snags and fallen wood were measured, since coarse woody debris (CWD) with a mid-length diameter >2.5 cm had been almost completely removed by local farmers. All of these components were collected in three randomly selected 1×1 m quadrats within each plot, and their fresh weights obtained with an electronic balance (accuracy: ±1 g). Subsamples of fresh weights were taken to the laboratory and oven-dried at 65°C to constant weight (0.1 g) to obtain fresh mass/dry mass ratios for calculating the dry mass of all samples. The oven-dried samples were also utilized to measure organic carbon content using the $K_2Cr_2O_7$-Oxidation method [23].

Soil C storage. Soil samples were obtained from two randomly selected vertical profiles within each 20×20 m plot. Soils were sampled to depths that either reached the parent material or did not exceed 1 m. Each soil profile was divided into the following vertical layers of 0–10, 10–20, 20–30, 30–50 and 50–100 cm. In this study, the average soil profile depth for the temperate zone forests was approximately 100 cm and that of the boreal zone forests was 40 cm (Table 2). For each plot, soil samples were extracted and mixed in order to obtain a 0.5-kg sample for each layer. Soil cores (100 cm³, 5.0 cm in diameter and 5.0 cm in depth) were collected for bulk density (BD) estimation. Rocks and gravel (>2 mm in diameter) were sieved and their content (%) estimated for each soil layer. When BD could not be measured directly due to a large amount of stones, it was estimated from adjacent additional profiles within the plot.

Soil organic C content was determined using the $K_2Cr_2O_7$-Oxidation method [23]. Soil organic C was estimated from the following equation [24]:

where SOC is the total soil organic C storage (Mg·C·ha⁻¹) of a given profile; SOC_i is the SOC content (g·kg⁻¹) in soil layer i, BD_i is the bulk density (g·cm⁻³) in soil layer i, H_i is the thickness (cm) in the soil layer i, and R_i is the volumetric fraction (%) of stones >2 mm in the soil layer i (Table 2).

Data Analysis

Tree biomass C storage was calculated as the product of biomass multiplied by carbon conversion coefficients, which for NE China range from 0.49 for broadleaved mixed forest to 0.52

Table 2. Soil bulk density, soil C content and soil C density in boreal and temperate forest soils (mean ±1SD) in northeastern China.

| | Boreal forests | | | | Temperate forests | | | |
| | Soil property | | | | | Soil property | | | |
Soil depth (cm)	No. of samples	Bulk density (g·cm^{-1})	C content (g·kg^{-1})	C density (MgC·ha^{-1})	No. of samples	Bulk density (g·cm^{-1})	C content (g·kg^{-1})	C density (MgC·ha^{-1})
0–10	147	0.85±0.26	64.5±3.50	50.6±16.33	197	0.66±0.17	73.1±2.43	46.3±13.51
10–20	118	1.16±0.28	33.9±1.89	31.9±18.45	197	1.06±0.19	32.7±1.58	33.4±13.62
20–30	56	1.32±0.30	21.5±1.39	23.4±13.66	188	1.27±0.21	19.6±1.01	23.9±11.51
30–50	10	1.41±0.13	18.1±1.01	33.1±27.19	179	1.41±0.18	12.5±0.75	29.6±15.72
50–100	5	1.35±0.10	14.9±0.42	71.6±20.35	107	1.51±0.17	7.9±0.51	45.9±30.92

for larch forest [25]. Understory biomass C storage was obtained by utilizing the standard biomass-C storage transformation coefficient of 0.5 [26], and forest floor biomass C was estimated through its organic C content. Total C storage was the sum of tree biomass, understory biomass, forest floor biomass, and soil C storage.

Statistical analyses were conducted using SPSS 16.0 software. Two-way ANOVA was used to test for effects of forest zone (boreal vs. temperate) and stand age on C storage by total C, tree C, understory C, forest floor C, and soil C. Within each forest zone, effects of forest type and stand age on C storage and its proportion in different components were tested using two-way ANOVA, followed by one-way ANOVA (LF vs. BF) or Tukey's HSD test, to compare the means across the four stand age classes. Significance levels were set at P<0.05 for all analyses.

Results

Carbon Storage across Forest Zones

Carbon storage was strongly affected by forest zone and stand age (Table 3). Total C, tree C, forest floor C and soil C storage all differed with forest zone (all P-values were <0.001; Table 3), with the exception of understory C (F = 0.30, P = 0.59; Table 3). Stand age significantly affected total C, tree C, and forest floor C but not understory C and soil C storage (Table 3). There were significant interactions between forest zone and stand age associated with tree C and forest floor C storage (both interactions with P<0.001, Table 3).

Carbon Storage in Boreal Forests

In the boreal forests, forest type was significantly correlated only with tree C, whereas stand age was significantly correlated with tree C, forest floor C, and total C storage (Table 4). The latter increased with increasing stand age, with highest storage levels (295.8–434.0 Mg·C·ha^{-1}) occurring in over-mature forests (Fig. 2u, v). Tree biomass C was significantly greater in LF (49.3–266.7 Mg·C·ha^{-1}) than in BF (42.4–137.0 Mg·C·ha^{-1}) (Fig. 2a, b; Table 4). The contribution of tree biomass C to total C storage varied significantly by forest types and stand age (Table 4). Percentage C in trees ranged from 27.7% (young) to 46.6% (over-mature) for BF, and from 39.3% (young) to 62.8% (over-mature) for LF (Fig. 3).

Understory biomass C ranged from 1.3 to 2.9 Mg·C·ha^{-1} across age classes and forest types (Fig. 2f, g, Table 4). It accounted for only a small proportion of total C storage (0.6%–1.3%) (Fig. 3).

Forest floor biomass C in over-mature forests (up to 26.6–33.7 Mg·C·ha^{-1}; Fig. 2k, l) was significantly greater than that in any other age classes (P<0.01, Table 4). The proportion of forest floor C to total C storage was significantly correlated with stand age and ranged from 6.5%–12.0% (Table 4; Fig. 3).

Both soil C content and density (per 10 cm soil layer) decreased with soil profile depth, whereas soil total C storage increased with depth (Table 2). Soil C storage did not vary significantly with either forest type or stand age (Fig. 2p, q; Table 4). However, its contribution to total C storage significantly differed among forest types and significantly decreased from 62.5%–48.4% in young forests to 40.4%–30.1% in over-mature stands (Fig. 3; Table 4).

Table 3. Effects of forest zone and stand age on forest C storage in northeastern China.

		Total		Tree		Understory		Forest floor		Soil	
Factors	df1/df2	F-value	P	F-value	P	F-value	P	F-value	P	F-value	P
Forest zone (Z)	1/177	44.06	0.00	48.51	0.00	0.30	0.59	178.25	0.00	16.80	0.00
Age class (A)	3/175	43.47	0.00	93.54	0.00	1.39	0.25	15.39	0.00	0.51	0.67
Z×A	3/171	1.49	0.22	9.09	0.00	0.73	0.54	7.38	0.00	0.45	0.72

Note: df1 and df2 are the numerator and denominator degrees of freedom, respectively. Statistical significances were tested using two-way ANOVA based on F-values; a P value of <0.05 indicates significance of differences at the 0.05 level. The two forest zones are boreal zone and temperate zone; the four stand age classes are young, mid-aged, mature, and over-mature.

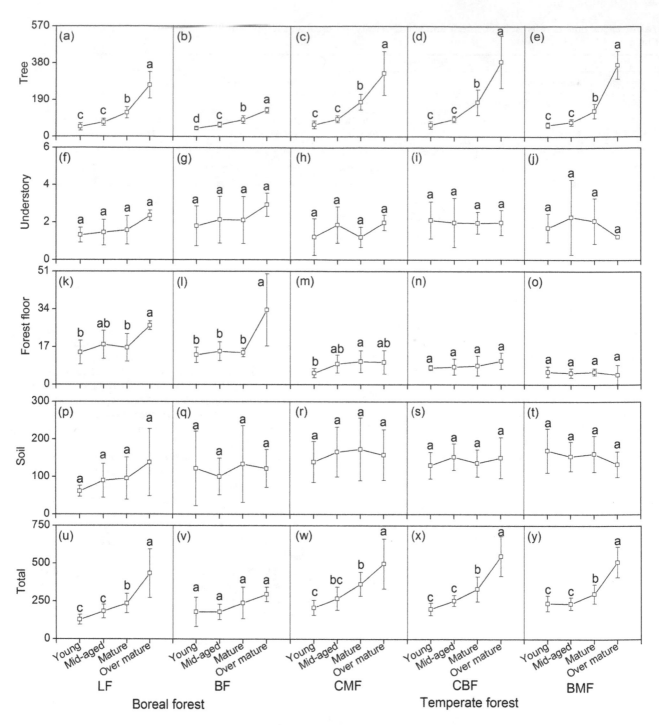

Figure 2. C storage (mean values ±1SE, Mg·C·ha⁻¹) in young, mid-aged, mature and over-mature stands in the boreal and temperate forest zones in northeastern China. Notes: (1) Forest types in boreal forest zone: larch forest (LF); birch forest (BF); Forest types in temperate forest zone: coniferous mixed forest (CMF); coniferous and broadleaved mixed forest (CBF); broadleaved mixed forest (BMF). (2) Different letters within each cell indicate significant differences among the four age classes.

Carbon Storage in Temperate Forests

In temperate forests, forest type was significantly related only to forest floor C whereas stand age significantly affected both total C and tree C storage (both P<0.001, Table 4). Total C storage was greatest in the over-mature forests (498.5–549.8 Mg·C·ha⁻¹) (Fig. 2w–y). Tree C storage increased with increasing stand age (P<0.05, Fig. 2c–e), resulting in a significant positive correlation

between percent of tree C contributions to total C storage and stand age (Table 4); percentages ranged from 26.5%–31.3% in young stand to 65.7%–72.8% in over-mature stands (Fig. 3).

Understory C, which ranged from 1.2 to 2.3 Mg·C·ha⁻¹, did not vary significantly with either stand age or forest type (Fig. 2h–j, Table 4). In contrast, its contribution to total C storage (0.3%–

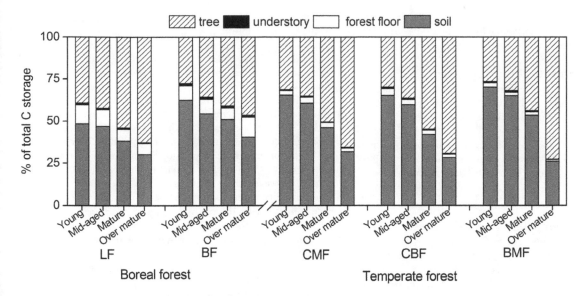

Figure 3. Distribution pattern of C storage among forest components in young, mid-aged, mature, and over-mature stands in boreal and temperate forests of northeastern China. Forest types in boreal forest zone: larch forest (LF); birch forest (BF). Forest types in temperate forest zone: coniferous mixed forest (CMF); broadleaved mixed forest (BMF); coniferous and broadleaved mixed forest (CBF).

1.0%) decreased significantly with increasing stand age (Fig. 3; Table 4).

Forest floor C was significantly affected by forest type (Table 4), with lowest values occurring in BMF (Fig. 2o). The contribution of forest floor C to total C storage (ranging from 1.0% to 4.1%)

varied significantly with both forest type and stand age (Fig. 3, Table 4).

As in the boreal forest, both soil C content and density decreased with soil profile depth (Table 2), but soil C did not vary significantly with forest type or stand age in the temperate forests (Table 4; Fig. 2r–t). At the same time, soil C contribution to total

Table 4. Effects of forest type and stand age on C storage by forest components (tree, understory, forest floor and soil) and their percent of total C storage in boreal and temperate forests in northeastern China.

| | | Boreal forest[a] | | | | | Temperate forest | | | | |
| | | | C storage | | C percent | | | C storage | | C percent | |
Components	Factors	df1/df2[b]	F-value	P[c]	F-value	P	df1/df2	F-value	P	F-value	P
Tree	Forest type (F)	1/78	48.80	**0.00**	10.84	**0.00**	2/97	0.88	0.42	1.21	0.30
	Age class (A)[d]	3/75	73.90	**0.00**	7.06	**0.00**	3/96	86.03	**0.00**	65.80	**0.00**
	F×A	3/71	10.29	**0.00**	0.44	0.72	6/88	1.01	0.43	0.84	0.54
Understory	Forest type (F)	1/78	3.69	0.06	3.09	0.08	2/97	1.06	0.35	0.59	0.56
	Age class (A)	3/75	1.41	0.25	0.68	0.57	3/96	0.55	0.65	3.12	**0.03**
	F×A	3/71	0.04	0.99	0.16	0.92	6/88	0.58	0.75	0.46	0.83
Forest floor	Forest type (F)	1/78	0.01	0.92	0.21	0.65	2/97	7.26	**0.00**	5.46	**0.00**
	Age class (A)	3/75	10.48	**0.00**	5.28	**0.00**	3/96	1.62	0.19	4.10	**0.00**
	F×A	3/71	1.09	0.36	2.59	0.06	6/88	1.20	0.32	0.99	0.44
Soil	Forest type (F)	1/78	1.26	0.27	7.43	**0.00**	2/97	0.80	0.45	1.70	0.19
	Age class (A)	3/75	0.77	0.51	3.82	**0.01**	3/96	0.28	0.84	53.29	**0.00**
	F×A	3/71	0.65	0.58	0.28	0.84	6/88	0.49	0.81	0.63	0.71
Total	Forest type (F)	1/78	1.01	0.32			2/97	0.20	0.82		
	Age class (A)	3/75	12.42	**0.00**			3/96	41.98	**0.00**		
	F×A	3/71	1.95	0.13			6/88	0.87	0.52		

[a]Boreal forests are composed of larch and birch; temperate forests include coniferous mixed forest, broadleaved mixed forest, coniferous and broadleaved mixed forest.
[b]Refers to numerator (df1) and denominator (df2) degrees of freedom, respectively.
[c]Statistical significances was tested using two-way ANOVA; a P value of <0.05 indicates significance of difference at the 0.05 level.
[d]Stand age classes are young, mid-aged, mature, and over-mature.

C storage decreased significantly with stand age (Table 4), and ranged from 65.2%–70.1% in young forests to 26.0%–31.7% in over-mature forests (Fig. 3).

Carbon Storage in the 0–20 cm Soil Layer

C storage in the top 20 cm of the soil did not vary significantly among forest types or stand ages. It ranged from 53.6 to 101.1 Mg·C·ha^{-1} in boreal forests and from 66.8 to 85.8 Mg·C·ha^{-1} in temperate forests (Fig. 4). Its contribution to total soil C storage also did not vary significantly between forest types or among stand ages (Fig. 4), and accounted for 67.8%–90.6% in boreal forests and 45.0%–60.8% in temperate forests.

Discussion

Carbon Storage in Temperate vs. Boreal Forests

With the exception of understory C, carbon storage of other forest components (tree, forest floor, and soil), as well as total C varied with forest zone (temperate vs. boreal) which is associated with climate conditions (Table 3). Thus, for example, tree C in temperate forests exceeded that in boreal forests (Fig. 2a–e), possibly due to greater forest ecosystem net primary productivity (NPP) associated with higher temperature and a longer growing season in the former [27,28], given the general absence of water deficit in NE China [9,29]. Previous studies conducted in other regions also found that biomass C storage in temperate forests exceeded that of tropical and boreal forests, leading to the conclusion that cool temperatures in combination with moderate precipitation favors biomass carbon accumulation [30,31]. Moreover, the relatively high plant species diversity or tree species composition in the temperate mixed forests compared to the boreal forests may also lead to greater biomass productivity and accumulation in the former in our study region [32].

Forest floor biomass is determined by the net balance between litter fall input and decomposition output. Climate can influence forest floor biomass by controlling the rates of these two processes [33,34]. Although variables such as forest type, stand age and disturbance regime are important in controlling forest floor biomass [35], the greater level of forest floor C in boreal as opposed to temperate forests (Figs. 2k–o; Table 3) suggests that low temperature plays a more important role than these other factors in determining forest floor biomass in NE China [34].

Soil C storage decreased from temperate forests to the more northern boreal forests (Fig. 4d), which is inconsistent with the findings of previous studies [36,37]. However, our results may have been influenced by specific edaphic factors that differed across the two forest zones. Others have found that soil profile total C storage increases with increasing soil thickness [24,38]. The relatively shallow soils of the boreal forest (approximately 40 cm) compared to the temperate forests (approximately 100 cm) (Table 2) is the likely cause of the lower soil C storage level in boreal forests. Correspondingly, the ratio of total forest C to total soil C in the upper soil layers (0–20 cm) was higher in boreal forests (67.8%–90.6%) than in temperate forests (45.0%–60.8%) (Fig. 4). However, these fractions are still much lower than those found in other boreal forests [16,39], while our results may have been influenced by the relatively large proportion of stones and shallower soils in the boreal forests we observed.

Total C storage was much higher in temperate forests (198.9–549.8 Mg·C·ha^{-1}) than in boreal forests (128.6–434.0 Mg·C·ha^{-1}), which agrees closely with the regional-scale study results of Pregitzer et al. [37] (239 vs. 143 Mg·C·ha^{-1}). The greater C storage in temperate forests was related to the high tree biomass and soil carbon storage as mentioned above. Similarly, large C storage occurred in the temperate forests of the Pacific Northwest Region of North America [30,31]. But on a global scale, Pan et al. [1] showed that the average total C storage in boreal forests (239 Mg·C·ha^{-1}) exceeded that of temperate forest (155 Mg·C·ha^{-1}). Such observed differences may be the result of study region size and heterogeneity and related issues involving scaling that can, by themselves, produce uncertainty and high spatial variation in forest ecosystem C storage. Such differences also may be related to climatic and edaphic factors, human disturbance, and stand age structure [4,31,37].

Effects of Forest Type on Carbon Storage

Neither total C storage nor that of forest components (tree, understory, forest floor and soil) varied significantly with forest type, with the exception of tree C in boreal forests and forest floor C in temperate forests (Table 4). Given that tree C significantly differed by forest type in boreal but not temperate forests may be associated with differences in tree species composition or species diversity [32]. In our study, boreal forests formed nearly pure stands with relatively low species diversity, unlike the temperate mixed forests we observed with their higher species diversity

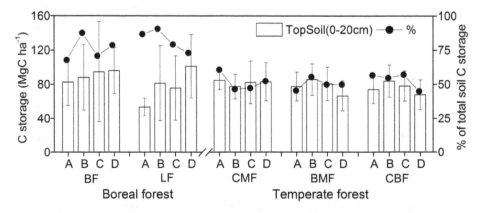

Figure 4. C storage in 0–20 cm soil layers (mean values ±1SE, Mg·C·ha^{-1}) and its contribution to total soil C storage across four stand age classes in boreal and temperate forests in northeastern China. Forest types in boreal forest zone: larch forest (LF) and birch forest (BF). Forest types in temperate forest zone: coniferous mixed forest (CMF), broadleaved mixed forest (BMF), coniferous and broadleaved mixed forest (CBF). Age classes: young (A), mid-aged (B), mature (C), over-mature (D).

(Table 2). Although forest type and tree species composition have been found to influence both forest floor C and soil C storage through litter production, litter quality, and the surrounding decomposition environment [40,41], human disturbance also can play a role in regulating C storage in the forest floor and soil [42]. Bunker et al. [32] and Saha et al. [43] nevertheless reported that forest type and tree species composition are important in regulating potential C storage even when anthropogenic disturbances are excluded.

In addition, we compared our estimates of biomass C storage with previous studies for similar temperate forest types. Zhu et al. [14] and Zhou et al. [15] reported that in northeastern China biomass C storage of coniferous and broadleaved mixed forest ranged from 182.7 and 191 $Mg \cdot C \cdot ha^{-1}$; Smithwick et al. [30] found that biomass C storage reached 506–627 $Mg \cdot C \cdot ha^{-1}$ in the temperate forests of the American Pacific Northwest; Pregitzer & Euskirchen [37] reported that average global temperate forest biomass C was 114.7 $Mg \cdot C \cdot ha^{-1}$; while this study estimated biomass C storage of temperate forest ranged from 58.9 to 386.5 $Mg \cdot C \cdot ha^{-1}$ across stand ages in NE China. These discrepancies in biomass C storage among those studies are probably related to stand age structure and human disturbance [4,31,37].

In our study, patterns of C storage in tree, understory, forest floor and soils differed within boreal forests but not with temperate forests (Table 4). Tree C contribution to the total C storage in the temperate mature and over-mature mixed forests ranged from 43.8% to 72.8% (Fig. 3), respectively, which was higher than that observed in the boreal mature and over mature forests (41.1% and 62.8%, respectively). Similar tree C proportions have been reported by others [14,30]. The higher tree C proportion in temperate forests may reflect the effects of the relatively high growth temperatures in the temperate zone, leading to higher net primary production than that in the boreal zone [28]. As mentioned above, forest floor C is determined by the balance between litter input and decomposition output, where the latter is positively correlated with low temperature [34]. Thus, forest floor C accounted for a relatively greater proportion of total forest C storage (6.5%–12.0%) in the boreal forests (Fig. 3). This emphasizes the importance of the forest floor in maintaining the C pool in that forest zone [1,37].

Effect of Stand Age on Carbon Storage

Our study provides comprehensive estimates of forest C storage for tree, understory, forest floor and soil components and their distributional patterns on a regional scale. We found that stand age significantly influences tree C and total ecosystem C (Tables 3 and 4). As stand age increases, tree biomass C storage also increases (Fig. 2a–e). This result highlights the longer temporal accumulation of net primary production and the important role of tree biomass C in determining a forest ecosystem's potential for carbon storage [32]. Thus the high tree biomass of over-mature forests (Fig. 2a–e) also portends a high potential for young, mid-aged and mature forests to sequester C.

Compared to increases tree C storage with stand age, C storage for the other observed forest components remained relatively stable with increasing stand age. The exception was forest floor C in boreal forests (Fig. 2k, l; Table 4). There, soil C, apparently remains in place over years to centuries [17,44]. Similarly, previous studies also have failed to find significant differences in soil C storage among forests of different ages [18,45]. In upland

ecosystems, soil C storage is primarily determined by the balance between carbon input from litter production and output through decomposition [46]. But that balance can be affected by logging and other human disturbances, which in turn depend on the amount of biomass removed, quality of debris remaining, and soil fertility [16]. Martin et al. [17] reported that harvesting may not necessarily affect total soil C content, which suggests that soil C tends to remain relatively stable with increasing stand age provided that forests are not severely disturbed [15,47].

The various forest components (tree, understory, forest floor and soil) have different C turnover times, and thus play different roles in C sequestration [1,37]. For example, tree C contributions to total C storage increases with increasing stand age (Fig. 3). A greater proportion of tree C with respect to total C storage in mature and over-mature forests emphasizes the importance of maintaining mature and over mature forests [30]. In contrast, the soil C pool accounted for a greater proportion of total C storage in young and mid-aged forests than in old growth forests, reflecting a decreasing contribution of soil C to the total C storage with increasing stand age (Fig. 3) [48]. Thus there is an apparent shift in forest C partitioning as stands age. Over-mature forests maintained substantial C pools in both forest zones. Similarly, previous studies have shown that undisturbed old-growth forests continue to function as C sinks [47,49], even though their rates of C incorporation into soil layers are low [45]. Hence, these observations have implications with respect to protecting mature and over-mature forests from human disturbance, limiting CO_2 emissions, and ultimately global warming.

Conclusion

Our study represents an early step in understanding the carbon pool size and its distribution varies among different ecosystem components, and across stand ages in the forests of NE China. We found that total forest C storage was greater in the temperate than in the boreal forests, as well as in older than in younger forests. Tree biomass C was the main component of total forest C storage, and its fraction increased with increasing stand age. This supports the great C sequestration potential of old-growth forests as observed by Zhou et al. (2006) and Luyssaert et al. (2008). Tree biomass C was significantly affected by forest zone (temperate vs. boreal), stand age, and forest type, which in turn are associated with climate, biomass accumulation rate, and stand composition and diversity. The large fraction of soil C within the total forest C storage indicated that soil C storage is also an important C pool in the forests of NE China. On the other hand, understory and forest floor C storages jointly contributed to <13% in boreal and <5% in temperate forests play a minor role to total forest C storage in NE China.

Acknowledgments

We would like to thank Qing-jun Wang (Yichun's Academy of Forestry), and Fei Wang (Inner Mongolia Agricultural University) for their help with fieldwork conducted from July through September of 2011.

Author Contributions

Conceived and designed the experiments: YW HC BL DY LZ W. Zhou XF W. Zhao LD. Performed the experiments: YW W. Zhou XF W. Zhao. Analyzed the data: YW ML HC LD. Contributed reagents/materials/analysis tools: DY LZ W. Zhou LD. Wrote the paper: YW ML HC BL LD. Language proofread: ML BL LD.

References

1. Pan Y, Birdsey RA, Fang J, Houghton R, Kauppi PE, et al. (2011) A large and persistent carbon sink in the world's forests. Science 333: 988–993.

2. Khatiwala S, Primeau F, Hall T (2009) Reconstruction of the history of anthropogenic CO$_2$ concentrations in the ocean. Nature Letter 462: 346–349.

3. Canadell JG, Le Quéré C, Raupach MR, Field CB, Buitenhuis ET, et al. (2007) Contributions to accelerating atmospheric CO2 growth from economic activity, carbon intensity, and efficiency of natural sinks. Proceedings of the National Academy of Sciences 104: 18866–18870.

4. Goodale CL, Apps MJ, Birdsey RA, Field CB, Heath LS, et al. (2002) Forest carbon sinks in the Northern Hemisphere. Ecological Applications 12: 891–899.

5. Interngovernmental Panel on Climate Chang (IPCC) (2007) Climate change 2007: The Physical Science Basis. Solomon, S., Qin, D., Manning, M, et al. eds. Contribution of Working Group I to the Fourth Assessment Report of the Intergovernment Panel on Climate Change. Cambridge, UK and New York, USA: Cambridge University Press. 996 p.

6. Piao S, Friedlingstein P, Ciais P, Peylin P, Zhu B, et al. (2009) Footprint of temperature changes in the temperate and boreal forest carbon balance. GEOPHYSICAL RESEARCH LETTERS 36: L0704, doi:10.1029/2009GL037381.

7. Bousquet P, Ciais P, Peylin P, Ramonet M, Monfray P (1999) Inverse modeling of annual atmospheric CO$_2$ sources and sinks 1. J Geophys Res 104: 26,161–126,178.

8. Piao S, Fang J, Zhu B, Tan K (2005) Forest biomass carbon stocks in China over the past 2 decades: Estimation based on integrated inventory and satellite data. J Geophys Res 110: G01006, doi:10.1029/2005JG000014.

9. Tan K, Piao S, Peng C, Fang J (2007) Satellite-based estimation of biomass carbon stocks for northeast China's forests between 1982 and 1999. Forest Ecology and Management 240: 114–121.

10. Yang Y, Mohammat A, Feng J, Zhou R, Fang J (2007) Storage, patterns and environmental controls of soil organic carbon in China. Biogeochemistry 84: 131–141.

11. Wang CK, Gower ST, Wang YH, Zhao HX, Yan P, et al. (2001) The influence of fire on carbon distribution and net primary production of boreal Larix gmelinii forests in north-eastern China. Global Change Biology 7: 719–730.

12. Wang H, Shao G-f, Dai L-m, Xu D, Gu H-y, et al. (2005) An impacts of logging operations on understory plants for the broad-leaved/Korean pine mixed forest on Changbai Mountain, China. Journal of Forestry Research (Harbin) 16: 27–30.

13. Bu R, He HS, Hu Y, Change Y, Larsen DR (2008) Using the LANDIS model to evaluate forest harvesting and planting strategies under possible warming climates in Northeastern China. Forest Ecology and Management 254: 407–419.

14. Zhu B, Wang X, Fang J, Piao S, Shen H, et al. (2010) Altitudinal changes in carbon storage of temperate forests on Mt Changbai, Northeast China. Journal of Plant Research 123: 439–452.

15. Zhou L, Dai L, Wang S, Huang X, Wang X, et al. (2011) Changes in carbon density for three old-growth forests on Changbai Mountain, Northeast China: 1981–2010. Annals of Forest Science 68: 953–958.

16. Law BE, Sun OJ, J.Campbell, Vantuyl S, Thornton PE (2003) Changes in carbon storage and fluxes in a chronosequence of ponderosa pine Global Change Biology 9: 510–524.

17. Martin JL, Gower ST, Plaut J, Holmes B (2005) Carbon pools in a boreal mixedwood logging chronosequence. Global Change Biology 11: 1883–1894.

18. Fonseca W, Rey Benayas JM, Alice FE (2011) Carbon accumulation in the biomass and soil of different aged secondary forests in the humid tropics of Costa Rica. Forest Ecology and Management 262: 1400–1408.

19. Chen CG, Zhu JF (1989) A handbook for main tree species biomass in Northeast China. Beijing: China Forestry Publishing House. (in Chinese).

20. Wang C (2006) Biomass allometric equations for 10 co-occurring tree species in Chinese temperate forests. Forest Ecology and Management 222: 9–16.

21. Han MZ, Zhou XF (1994) A study on biomass and net primary production in a Dahurian larch birch forest ecosystem. Long-term research on China's forest ecosystems, Northeast Forestry University Press, Harbin,: 451–458.(in Chinese).

22. Xing Y (2005) Regional Forest Biomass and Carbon Storage Estimation Study for Northeast Natural Forest Based On RS and GIS: Ph. D. dissertation. Harbin: Northeast Forestry University. (in Chinese).

23. Lu R (1999) The methods of soil and agricultural chemistry analysis. Beijing: China Agriculture Science and Technology Publishing House. (in Chinese).

24. Wiesmeier M, Spörlein P, Geuß U, Hangen E, Haug S, et al. (2012) Soil organic carbon stocks in southeast Germany (Bavaria) as affected by land use, soil type and sampling depth. Global Change Biology 18: 2233–2245.

25. Wang X, Huang G, Sun Y, Fu X, Han A (2008) Forest carbon storage and dynamics in Liaoning province from 1984 to 2000. Acta Ecologica Sinica 28: 4757–4764. (in Chinese).

26. Brown S, Lugo AE (1984) Biomass of tropical forests: a new estimate based on forest volumes. Science(Washington) 223: 1290–1293.

27. Wang X, Fang J, Tang Z, Zhu B (2006) Climatic control of primary forest structure and DBH–height allometry in Northeast China. Forest Ecology and Management 234: 264–274.

28. Zhang N, Yu G, Yu Z, Zhao S (2003) Analysis on factors affecting net primary productivity distribution in Changbai Mountain based on process model for landscape scale. Chinese Journal of Applied Ecology 14: 659–664. (in Chinese).

29. Wang X, Fang J, Zhu B (2008) Forest biomass and root-shoot allocation in northeast China. Forest Ecology and Management 255: 4007–4020.

30. Smithwick EAH, Harmon ME, Remillard SM, Acker S, Franklin J (2002) Potential upper bounds of carbon stores in forests of the Pacific Northwest. Ecological Applications 12: 1303–1317.

31. Keith H, Mackey BG, Lindenmayer DB (2009) Re-evaluation of forest biomass carbon stocks and lessons from the world's most carbon-dense forests. Proceedings of the National Academy of Sciences 106: 11635–11640.

32. Bunker DE, DeClerck F, Bradford JC, Colwell RK, Perfecto I, et al. (2005) Species loss and aboveground carbon storage in a tropical forest. Science 310: 1029–1031.

33. Liu C, Westman CJ, Berg B, Kutsch W, Wang GZ, et al. (2004) Variation in litterfall-climate relationships between coniferous and broadleaf forests in Eurasia. Global Ecology and Biogeography 13: 105–114.

34. Zhang X, Wang X, Zhu B, Zong Z, Peng C, et al. (2008) Litter fall production in relation to environmental factors in northeast China's forests. Journal of Plant Ecology 32: 1031–1040. (in Chinese).

35. Yanai RD, Currie WS, Goodale CL (2003) Soil carbon dynamics after forest harvest: an ecosystem paradigm reconsidered. Ecosystems 6: 197–212.

36. Melillo JM, McGuire AD, Kicklighter DW, Moore B, Vorosmarty CJ, et al. (1993) Global climate change and terrestrial net primary production. Nature 363: 234–240.

37. Pregitzer KS, Euskirchen ES (2004) Carbon cycling and storage in world forests: biome patterns related to forest age Global Change Biology 10: 2052–2077.

38. Jobbágy EG, Jackson RB (2000) The vertical distribution of soil organic carbon and its relation to climate and vegetation. Ecological Applications 10: 423–436.

39. Wang C, Bond-lamberty B, Gower ST (2003) Carbon distribution of a well-and poorly-drained black spruce fire chronosequence. Global Change Biology 9: 1066–1079.

40. Wynn JG, Bird MI, Vellen L, Grand-Clement E, Carter J, et al. (2006) Continental-scale measurement of the soil organic carbon pool with climatic, edaphic, and biotic controls. Global Biogeochemical Cycles 20: GB1007, doi:10.1029/2005GB002576.

41. Vivanco L, Austin AT (2008) Tree species identity alters forest litter decomposition through long-term plant and soil interactions in Patagonia, Argentina. Journal of Ecology 96: 727–736.

42. Jonsson M, Wardle DA (2010) Structural equation modelling reveals plant-community drivers of carbon storage in boreal forest ecosystems. Biology letters 6: 116–119.

43. Saha SK, Nair PKR, Nair VD, Kumar BM (2009) Soil carbon stock in relation to plant diversity of homegardens in Kerala, India. Agroforestry systems 76: 53–65.

44. Lal R (2004) Soil carbon sequestration impacts on global climate change and food security. Science 304: 1623–1627.

45. Ostertag R, Marín-Spiotta E, Silver WL, Schulten J (2008) Litterfall and decomposition in relation to soil carbon pools along a secondary forest chronosequence in Puerto Rico. Ecosystems 11: 701–714.

46. Smith P, Fang C (2010) Carbon cycle: A warm response by soils. Nature 464: 499–500.

47. Zhou G, Liu S, Li Z, Zhang D, Tang X, et al. (2006) Old-growth forests can accumulate carbon in soils. Science 314: 1417.

48. Jandl R, Lindner M, Vesterdal L, Bauwens B, Baritz R, et al. (2007) How strongly can forest management influence soil carbon sequestration? Geoderma 137: 253–268.

49. Luyssaert S, Schulze ED, Borner A, Knohl A, Hessenmoller D, et al. (2008) Old-growth forests as global carbon sinks. Nature 455: 213–215.

3

Assessing the Spatiotemporal Variation in Distribution, Extent and NPP of Terrestrial Ecosystems in Response to Climate Change from 1911 to 2000

Chengcheng Gang[1], Wei Zhou[1], Jianlong Li[1]*, Yizhao Chen[1], Shaojie Mu[1], Jizhou Ren[2], Jingming Chen[3], Pavel Ya. Groisman[4]

1 Global Change Research Institute, School of Life Science, Nanjing University, Nanjing,Jiangsu, P. R. China, 2 State Key Laboratory of Grassland Agro-ecosystems, College of Pastoral Agriculture Science and Technology, Lanzhou University, Lanzhou, Gansu, P. R. China, 3 Department of Geography, University of Toronto, Toronto, Ontario, Canada, 4 NOAA National Climatic Data Center, Asheville, North Carolina, United States of America

Abstract

To assess the variation in distribution, extent, and NPP of global natural vegetation in response to climate change in the period 1911–2000 and to provide a feasible method for climate change research in regions where historical data is difficult to obtain. In this research, variations in spatiotemporal distributions of global potential natural vegetation (PNV) from 1911 to 2000 were analyzed with the comprehensive sequential classification system (CSCS) and net primary production (NPP) of different ecosystems was evaluated with the synthetic model to determine the effect of climate change on the terrestrial ecosystems. The results showed that consistently rising global temperature and altered precipitation patterns had exerted strong influence on spatiotemporal distribution and productivities of terrestrial ecosystems, especially in the mid/high latitudes. Ecosystems in temperate zones expanded and desert area decreased as a consequence of climate variations. The vegetation that decreased the most was cold desert (18.79%), while the maximum increase (10.31%) was recorded in savanna. Additionally, the area of tundra and alpine steppe reduced significantly (5.43%) and were forced northward due to significant ascending temperature in the northern hemisphere. The global terrestrial ecosystems productivities increased by 2.09%, most of which was attributed to savanna (6.04%), tropical forest (0.99%), and temperate forest (5.49%). Most NPP losses were found in cold desert (27.33%). NPP increases displayed a latitudinal distribution. The NPP of tropical zones amounted to more than a half of total NPP, with an estimated increase of 1.32%. The increase in northern temperate zone was the second highest with 3.55%. Global NPP showed a significant positive correlation with mean annual precipitation in comparison with mean annual temperature and biological temperature. In general, effects of climate change on terrestrial ecosystems were deep and profound in 1911–2000, especially in the latter half of the period.

Editor: Gil Bohrer, The Ohio State University, United States of America

Funding: This work was supported by the "The Key Project of Chinese National Programs for Fundamental Research and Development (973 Program, 2010CB950702)", "APN (Asia Pacific Network) Global Change Fund Project (APCR2013-16NMY-Li)", "The National High Technology Project (2007AA10Z231)", the National Natural Science Foundation of China (40871012, J1103512, J1210026), and the Public Sector Linkages Program supported by Australian Agency for International Development (64828). The funders had no role in study design, data collection and analysis, decision to publish, or preparation of the manuscript.

Competing Interests: The authors have declared that no competing interests exist.

* E-mail: jlli2008@nju.edu.cn

Introduction

Studies regarding the interactions between global change and terrestrial ecosystems are becoming widespread in the current body of global change research. Increasing atmospheric CO_2 concentration in the past decades has been accompanied by other global changes. Rising air temperatures and altered precipitation patterns are among the most prominent of the predicted changes that, along with elevated CO_2, have affected ecosystem structure and function deeply and in a profound way [1,2,3]. Improving understanding of the interactions and feedback mechanisms of physical climate systems and environmental systems, predicting longer term trends, and preparing strategies for future events are grand challenges [4,5]. Climate is the main driving force in the distribution of ecosystems, and vegetation is the most distinct indicator of this distribution [6]. Climate change affects vegetation mainly through changes in precipitation and temperature which

affect the effective accumulated temperature and the content of soil organic matter [7,8]. Net primary productivity (NPP), which measures the energy fixed by the plant community through photosynthesis and indicates the growth ability in a specific natural environment, provides a link between biomes and the climate system through the global carbon and water cycles. The dynamics of NPP can reflect the variations of ecosystems in response to climate change, which is of great significance to assess disturbance of terrestrial ecosystems and evaluate terrestrial carbon sink [9,10].

Since the concept of PNV was introduced, many endeavors have been devoted to evaluating the impacts of simulated past and future climate change on ecosystems at regional-to-global scales [1,11,12], which greatly improves our ability to assess the interactions between terrestrial ecosystems and climate change. However, these models always require complicated parameters and input data to reflect the ecological processes and simulate the process of vegetation dynamics [13]. Additionally, the application of biogeographic models (e.g. Holdridge Life Zone) and equilibrium vegetation models (e.g. BIOME4) is mainly focused on forests. Based on the relationships of climate, soil and vegetation, CSCS was mainly driven by mean temperature and precipitation data which overcomes the deficiency of complicated and insufficient parameters, especially in regions that lack of collected data. After years of developed and optimized, the model has been widely used in terrestrial ecosystems classification and global change research [14,15,16,17].

NPP refers to the organic matter that is fixed by plants mainly through the process of photosynthesis, and thus can reflect the growing status of vegetation and measure the amount of trophic energy flows in food webs and chains [18]. Vast research has been conducted to evaluate terrestrial NPP at multiple levels. NPP estimation models, such as climate-based models (i.e. MIAMI model [19], Thornthwaite Memorial model [19]), process-based models (e.g. CENTURY [1], TEM [20], BIOME-BGC [21]), and light use efficiency models (e.g. CASA [22], GLO-PEM [23]), have been widely reported, and their accuracy increased resulted from the significant development of remote sensing technology. However, parameters used in process-based models are complicated which leads to difficulties in data acquisition in some regions. This makes process-based models more suited for regional NPP estimation. In contrast, light use efficiency models are much more widely used in regional and global NPP estimation due to the readily available parameters that are derived from remote sensing data either directly or indirectly. However, satellite-based parameters employed in models, e.g. normalized difference vegetation index (NDVI), have only been accessible for the past 30 years which prevents their application in evaluating NPP over the length of a century. Although simple, climate-based models are valuable and quite capable of simulating global vegetation NPP and its variation in response to climate change over the length of a century when meteorological data is available [24]. In this paper, a synthetic model [25] was used to evaluate global vegetation NPP and its variations under climate change in the period of 1911–2000.

The interactions between terrestrial ecosystems and climate change ranges in timescale from seconds to millions of years and from local to worldwide in spatial scale. The structure and functions of vegetation are strongly determined by climate change primary in terms of temperature and precipitation [26,27], yet the bidirectional influences of climate change and terrestrial vegetation are still obscure [28]. To better clarify this problem, in this paper, the dynamics of spatiotemporal distribution, extent, and NPP of global terrestrial ecosystems from 1911 to 2000 were quantitatively assessed using the CSCS and synthetic model. The correlation between NPP dynamics and climate factors in the same period was also studied to investigate its response to climate change. The results of this work provided a general outlook of the effects of climate change on terrestrial ecosystems in the past century, and the outcomes may complement the IPCC report. Furthermore, methods used in this paper can serve as a guide for studies in past and future global change in regions lacking collected data.

Materials and Methods

Global climatic data

In this paper, the global climate dataset CRU_TS 2.1 from the climate research unit (CRU) (http://www.cru.uea.ac.uk/timm/grid/CRU_TS_2_1.html) was empolyed in the CSCS to generate global PNV maps and in the synthetic model to simulate NPP. The dataset of grids extends from 1901 to 2002, covers the global

land surface (excluding Antarctica) at a 0.5° resolution, and provides best estimates of month-by-month variations in climate variables. The well-established dataset has already been widely applied [29]. In this study, the mean annual temperature (MAT) and mean annual precipitation (MAP) data in the period 1911–2000 were incorporated from monthly grid data using ArcGIS v9.3 software (ESRI, Redlands, CA, USA). Additionally, the Mollweide projection with the WGS_1984 spheroid was applied to all of the related databases for the calculation of the area of vegetation types.

The CSCS model

Based on hydrothermal conditions, the CSCS model is composed of three levels: class, subclass, and type. The class level, the basic unit, is determined by bioclimatic conditions, the subclass level is classified by edaphic conditions, and the type level is based on vegetation characteristics [15]. Subclasses are integrated into classes according to an index of moisture and temperature which captures the natural occurrence of vegetation ecosystems. The classes are mainly established by annual cumulative temperature above 0°C ($\Sigma\theta$) (Growing Degree-Days on 0°C base, GDD0) and humidity index (K), as calculated by:

$$K = MAP/\left(0.1 \times \sum \theta\right) = MAP/(0.1 \times GDD0) \quad (1)$$

where MAP is the mean annual precipitation (mm) and 0.1 is an empirical parameter. To more explicitly reflect the spatial distribution of PNV at a global scale, classes were regrouped into 10 vegetation types, i.e. tundra &alpine steppe, cold desert, semi-desert, steppe, temperate humid grassland, warm desert, savanna, temperate forest, subtropical forest, tropical forest.

Global potential natural vegetation maps

To simulate the dynamics of PNV more reasonably, the global biomes maps were produced at 30-year intervals. According to the IPCC Third Assessment Report, the periods that most obviously increased in temperature during the 20th century were from 1910 to 1945 and 1976 to 2000 [30]. We divided these 90 years into three intervals as 1911 to 1940 (T1), 1941 to 1970 (T2), and 1971 to 2000 (T3).

NPP model

NPP of natural vegetation was simulated using the synthetic model. The model was based on actual evapotranspiration which was closely related to the photosynthesis of vegetation, and was established mainly on the biomass data from 125 sets of natural mature forest in China and 23 sets of natural vegetation NPP that included forest, grassland, and desert. These data were obtained during the International Biological Program (IBP) by Efimova [31,32]. This model integrated the interaction among many variables and was developed in light of the same references as the Chikugo model [33]:

NPP

$$= RDI^2 \times \frac{MAP \times (1+RDI+RDI^2)}{(1+RDI)(1+RDI^2)}$$
$$\times \exp\left[-\sqrt{(9.87+6.25RDI)}\right] \times 100 \quad (2)$$

$$RDI = \left(0.629 + 0.237PER - 0.00313PER^2\right)^2 \quad (3)$$

$$PER = PET/MAP = BT \times 58.93/MAP \qquad (4)$$

where MAP is the mean annual precipitation (mm), RDI is the radioactive dryness index which can be calculated by PER, PER is the rate of evapotranspiration, PET is potential evapotranspiration (mm), and BT is biological temperature which is the average temperature during the vegetative growth of plants ($0\sim30°C$, temperatures below $0°C$ and above $30°C$ are excluded). NPP is calculated in units of g DW m^{-2} yr^{-1}.

Correlation between NPP and climate factors

The Pearson correlation coefficient was employed to reflect the relationship between NPP and climate factors, including MAP, MAT and BT. The spatial distribution maps of correlation coefficients between NPP and climate factors were obtained through the equation of correlative analysis, which is expressed as follows:

$$r = \frac{\sum_{i=1}^{n}(x_i - \bar{x})(y_i - \bar{y})}{\sqrt{\sum_{i=1}^{n}(x_i - \bar{x})^2 \sum_{i=1}^{n}(y_i - \bar{y})^2}} \qquad (5)$$

where y_i refers to climate factors (including MAP (mm), MAT (°C), and BT (°C)) in year i; and $\bar{y}$ represents the mean climate values over the years. When the correlation coefficient was tested for significance ($P<0.01$ or $P<0.05$), it displayed an "extremely significant" or "significant" linear correlation [34].

Results

Changes of climate factors during 1911–2000

Structure and function of ecosystems are strongly determined by climate influences, primarily through temperature ranges and precipitation available. According to our research, the global warming showed an obvious zonal distribution, especially in the mid- and high latitude on northern hemisphere (Figure.1). Based on our findings, for the past 90 years extending from 1911 to 2000, the global MAT increased by $0.23°C$. Regions that showed a decreasing trend amounted to 23.12% of total land area, and were mainly distributed in the south of Greenland, Mideast and east of America, west of Brazilian plateau, the Mediterranean Coast, Yunnan-Guizhou Plateau, part of the Qinghai-Tibet Plateau, and the north of Siberia. Accordingly, the global BT moderately increased by $0.06°C$ in the 90 years, which was closely related to the continuously increasing global temperature. 65.2% of the total land presented an increasing trend, while regions with decreased BT were mainly located in the east and southeast of the Sahara, southeast of Australia, and the Mongolian Plateau. With regards to MAP, a distinct spatial heterogeneity was observed with an overall 14.39 mm increase globally. Regions that showed increasing MAP were estimated to be 65.57% of total land area during T3 period relative to T1 period. This increase was mainly located on both sides of equator and mid- and high latitudes while regions with reduced precipitation mainly occurred 10° north of the equator and on west coast of South America. (The dataset of MAT and MAP for each of the three periods are available in the Appendix S1– S6)

Shifts of terrestrial ecosystems on the basis of the CSCS in 1911–2000

The global PNV maps of the T1, T2, and T3 periods were obtained through the CSCS. As indicated in Figure.2, the maps showed an obvious zonal distribution of terrestrial ecosystems. Tundra & alpine steppe were mainly distributed in the high latitudes and elevations of the northern hemisphere, in places such as Siberia, Greenland, the north of North America, and the Qinghai-Tibet Plateau. In the three desert types, the cold desert was restricted to the northwest of China, part of the Turanian Plateau, and scattered areas along the Andes. The semi-desert encircled the cold desert in Central Asia, Mongolia, and the Brazilian Plateau, while the warm desert was mainly localized in the Sahara Desert, the Arabian Peninsula, and central Australia. The steppe was distributed in areas neighboring the semi-desert, which mainly located in Inner Mongolia, part of West Asia, and the Great Plains of America. Temperate humid grassland was mainly localized in Canada and Eurasia adjacent to steppe. Savanna, strongly controlled by tropical savanna climate which is highly temperate and distinguished by a dry and wet season, was mainly distributed south and north of rainforest in Africa and South America and encircling warm desert in Australia. Most of the temperate forest was distributed near 60 degrees north in Asia and Europe, the east of North America, and only a little in Oceania. Subtropical forest was mainly localized in the southeast of North America and China, while tropical forest was mainly located on both sides of the equator. (The global terrestrial biomes simulated by CSCS of the three periods are available in the Appendix S7–S9)

Climate change is the main driver of the patterns and processes in global ecosystems over long time periods, and determined the succession of different vegetation types. At a global scale, the area of tundra & alpine steppe decreased significantly with 96.5×10^4 km^2 (5.43%) in the period of 1911–2000 (Table 1). Similarly, all three desert vegetation types shrunk during the same period, semi-desert, cold desert and warm desert decreased by 1.55%, 18.79%, and 6.02%, respectively. The persistently increasing area was found in savanna with a small increase from T1 to T2 (0.90%) and a marked increase from T2 to T3 (9.32%). With regards to steppe and temperate humid grassland, both decreased greatly by 9.68% and 11.36% from the T1 to T2 periods, respectively, and then increased moderately from the T2 to T3 periods (1.00% and 2.53%). Temperate forest expanded by 4.96%, tropical forest rose by just 0.17%, and subtropical forest decreased by 0.94%. Areas of different ecosystems on continental levels are shown in Table 1.

Validation of synthetic model

The simulated total NPP of the synthetic model was compared with current available data according to Ito's search [35], and the gridded simulated value was also compared with observed data. In this research, the majority of observed NPP data was gathered from the Oak Ridge National Laboratories (ORNL) Net Primary Production database (http://daac.ornl.gov/NPP/npp_home. shtml) [36], which is especially useful for model and hypothesis testing. These study sites represent a broad range of vegetation types as defined by eco-regions or climatic zones [37,38]. Data were mapped according to their associated geographic coordinates, and sites with incomplete geographic coordinates or absence of total NPP data were excluded from the study (Figure.3). Based on Table 2 and Figure.4, we can see that the simulated NPP of the synthetic model is in good agreement with available published and observed data ($R^2 = 0.8579$). Therefore, the synthetic model is

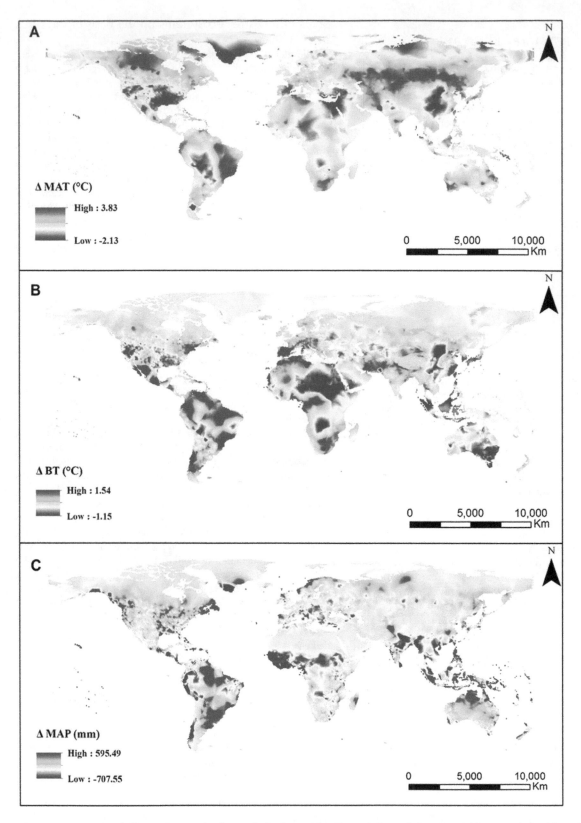

Figure 1. Dynamic of climate factors in the period 1911–2000. The variations of climatic variables were derived from CRU_TS 2.1 data. (a): mean annual temperature (MAT), (b): biological temperature (BT), (c): mean annual precipitation (MAP).

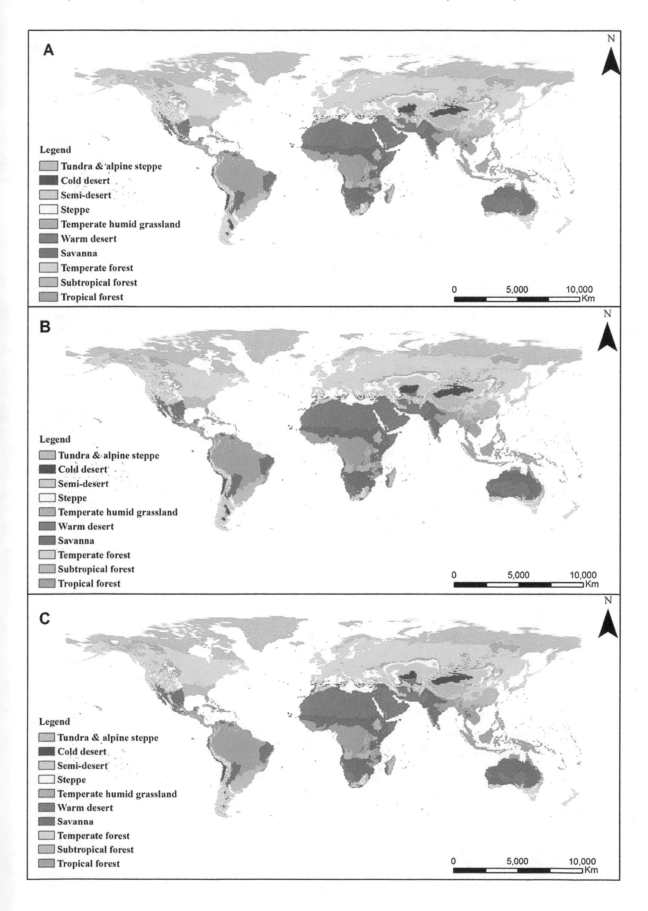

Figure 2. Spatial distribution of global natural vegetaiton biomes in the period 1911–2000. (a): T1 period (1911–1940), (b): T2 period (1941–1970), (c): T3 period (1971–2000). The different colors represent the different ecosystems, tundra & alpine steppe (12.85%), cold desert (1.56%), warm desert (13.93%), semi-desert (5.96%), savanna (17.66%), steppe (3.29%), temperate humid grassland (5.86%) and Forest (39.65%) in T3 period.

capable of evaluating global terrestrial ecosystems productivities and their variations over the length of a century.

Changes of global vegetation NPP

Climate change is expected to affect terrestrial ecosystems' NPP through disturbing structures, functions, energy flows, etc. To understand the effect of climate change on ecosystem production, the NPP was calculated using the synthetic model.

At a global scale, the total NPP of terrestrial ecosystems increased gradually by 2317.97 Tg DW yr^{-1} (2.09%) over the period of 1911–2000 (Table 3). This was mainly attributed to the increase of savanna (1740.14 Tg DW yr^{-1}, 9.42%), temperate forest (942.08 Tg DW yr^{-1}, 5.49%), and tropical forest (489.12 Tg DW yr^{-1}, 0.99%). The NPP of forest vegetation accounted for more than 70% of the global NPP. The maximum estimation was in tropical forest. In contrast to the increasing trend of temperate forest and tropical forest, subtropical forest presented a slightly descending trend of 0.70%. With regards to grasslands vegetation, the NPP of savanna, which amounted to nearly 70% of total grasslands NPP, was found to be the largest NPP increase. The NPP of tundra & alpine steppe, steppe, and temperate humid grassland all declined, by 162.78 Tg DW yr^{-1} (3.92%), 197.17 Tg DW yr^{-1} (7.65%), and 118.40 Tg DW yr^{-1} (6.22%), respectively. The productivities of desert types were relatively low, occupying less than 5% of total NPP. The NPP losses were higher in cold desert (27.33%) than in warm desert (17.78%), while NPP of semi-

desert was basically flat with a slight increase (0.15%) in the same period. Continental distributions of terrestrial NPP are shown in Table 3.

From a climatic regions perspective, vegetation NPP in the tropical region (Figure. 5(a)), which amounted to more than half of total terrestrial ecosystem NPP, increased slightly by 1.32% in the period 1911–2000. The NPP of tropical forest and savanna took up nearly 60% and 25%, respectively, of that tropical region. The NPP of savanna in this period increased remarkably by 1441.82 Tg DW yr^{-1} (8.99%), while the increasing trend of tropical forest (0.70%) was relatively low. In contrast, a declining trend of NPP was observed in other ecosystems. For instance, the decrease in warm desert was remarkable (21.40%). The NPP of ecosystems in the northern frigid zone also increased slightly overall (0.16%) over the course of the entire study period (Figure. 5(b)). Temperate humid grassland was estimated to increase by 21.21 Tg DW yr^{-1} (7.95%). The NPP of steppe and temperate forest also went up by 7.13 Tg DW yr^{-1} and 6.81 Tg DW yr^{-1}, respectively. A slight decrease (1.11%) was observed in tundra & alpine steppe, the NPP of which occupied nearly 90% of vegetation NPP in the northern frigid zone. As can be seen in Figure 5 (c) and (d), the NPP in the northern temperate zone is nearly seven times that of the southern temperate zone, and both increased gradually in this period by 3.55% and 8.27%, respectively. In the northern temperate zone, the temperate forest NPP amounted to more than 50% of the total and contributed the

Table 1. Areas of terrestrial ecosystems at continental levels in 1911–2000.

	Africa	Asia	Europe	Oceania	North America	South America	Global
Tundra & alpine steppe	×	1933.87±31.36	311.27±6.38	×	1793.97±61.70	23.73±11.05	4089.27±84.72
		(−1.00%)	(2.02%)		(−6.64%)	(−63.76%)	(−3.92%)
Cold desert	×	114.01±11.42	×	×	1.53±1.19	351.41±18.34	139.65±23.57
		(−16.93%)			(−86.82%)	(−9.95%)	(−27.33%)
Semi-desert	300.77±20.33	1495.03±53.91	149.23±22.71	265.22±13.33	502.37±32.94	226.49±18.05	3182.06±35.76
	(−12.66%)	(2.82%)	(33.03%)	(−2.02%)	(−11.98%)	(−14.38%)	(−0.15%)
Steppe	185.41±18.41	831.08±46.67	342.73±24.62	102.19±11.16	554.81±45.80	66.58±5.54	2437.90±121.60
	(−16.46%)	(−0.40%)	(−10.44%)	(−17.34%)	(−13.35%)	(−11.64%)	(−8.35%)
Temperate humid grassland	×	920.62±61.97	200.47±9.65	×	618.30±22.62	9.21±0.99	1809.96±82.44
		(−7.12%)	(−4.86%)		(−4.54%)	(−19.12%)	(−6.22%)
Warm desert	490.15±13.36	293.02±9.74	×	424.10±120.68	42.87±8.49	2471.83±90.92	1304.57±136.04
	(5.25%)	(−0.66%)		(−42.48%)	(−3.94%)	(−4.21%)	(−17.78%)
Savanna	8715.89±195.12	2400.37±247.63	42.08±6.96	3012.38±546.33	1214.97±36.09	772.24±35.37	19114.76±950.51
	(4.19%)	(18.44%)	(30.18%)	(36.71%)	(4.51%)	(−8.76%)	(9.42%)
Temperate forest	105.88±17.18	5010.41±250.77	4487.59±45.10	582.62±18.20	4895.13±358.26	2991.46±109.50	17726.47±503.00
	(−26.48%)	(5.97%)	(1.53%)	(−1.52%)	(15.77%)	(2.90%)	(5.49%)
Subtropical forest	2815.76±160.11	3429.22±48.04	82.33±10.01	283.91±15.98	1989.43±14.98	18044.70±412.52	12631.45±108.87
	(−6.14%)	(−0.79%)	(21.90%)	(1.83%)	(1.04%)	(4.50%)	(−0.70%)
Tropical forest	10375.76±451.62	10736.07±234.12	×	735.28±73.55	1757.00±28.69	24985.32±355.37	50001.87±531.72
	(−5.21%)	(−1.39%)		(20.07%)	(2.53%)	(2.73%)	(0.99%)

(Unit: × 10^4 km^2) "×" indicates that a continent did not have a type of vegetation; a negative sign within parentheses indicates a decreasing trend in the period 1911–2000.

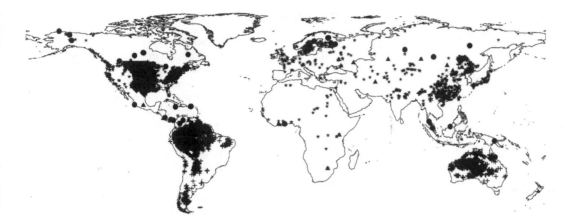

Figure 3. Map showing the geographical distribution of detailed terrestrial NPP study sites. The dataset are obtained from the Oak Ridge National Laboratory (ORNL), Distributed Active Archive Center database. ● = grassland sites included in the present analysis; ▲ = NPP study sites forests (tropical forest, temperate forest and boreal forest), ★ and + = multi biomes – EDMI (B and C) data [63]. These data, and further information about the study sites, are publicly available at www.daac.ornl.gov/NPP/.

most to the NPP increase. The increasing trend of NPP was also found in savanna (9.50%), subtropical forest (6.82%), and semi-desert (2.18%), whereas a declining trend was observed in tundra & alpine steppe (7.89%), cold desert (18.47%), steppe (7.42%), and temperate humid grassland (8.42%). In the southern temperate zone, NPP of subtropical forest and savanna went up by 193.14 Tg DW yr^{-1} (22.19%) and 183.76 Tg DW yr^{-1} (45.64%), respectively. Similarly, NPP of semi-desert and temperate humid grassland displayed an increasing tendency with 2.03% and 2.11%, respectively. In contrast, NPP losses were found in tundra & alpine steppe (2.53%), cold desert (19.04%), steppe (15.50%), warm desert (42.15%) and temperate forest (2.65%). (The dataset of global terrestrial NPP based on the synthetic model of the three periods are available in the Appendix S10–S12)

Correlations between NPP and climate factors

The maps of the spatial distribution of correlation coefficients were obtained based on the correlation coefficients between NPP and climatic factors from 1911 to 2000. In this period, the positive correlation between NPP and MAP predominated in the world (Figure 6). 73.6% of regions showed extremely significant positive correlation, and the global average correlation coefficient was 0.88. Regions displaying a negative correlation were mainly located in Alaska and some arid and semiarid areas. In contrast, the correlation between NPP and MAT showed great spatial heterogeneity, especially in northern hemisphere. Regions that displayed a positive correlation occupied 65.12% of the total area and were mainly distributed in north of North America, east of Africa, and most areas in Australia. Globally, regions with a

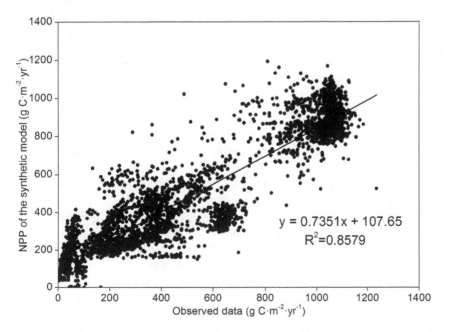

$$y = 0.7351x + 107.65$$
$$R^2 = 0.8579$$

Figure 4. Comparison of NPP value simulated by synthetic model and observed data (*r = 0.9262, P<0.001*). The observed data are collected from the Oak Ridge National Laboratory (ORNL). These data, and further information about the study sites, are publicly available at www.daac.ornl.gov/NPP/.

Table 2. Comparison of published values of present terrestrial net primary productivity (NPPT) and our research.

Reference	NPPT (Pg·C·yr^{-1})
Woodwell et al. [39]	52.8
IPCC 1st Assessment Report [40]	52
Siegenthaler & Sarmiento [41]	51.97
Sundquist [42]	60
IPCC 2nd Assessment Report [43]	61.3
Schlesinger [44]	60
Schimel et al. [45] after Cramer et al. [46]	42–68
Geider et al. [47]	56.4
IPCC 3rd Assessment Report [48]	60
Gruber et al. [49]	57
Our search	**50.78**

correlation lower than the significance level of 0.05 accounted for 34.10%. The correlation between NPP and BT was also analyzed in this paper. Based on our findings, regions that showed positive correlations made up 66.24%. Regions presenting insignificant correlation between NPP and BT/MAT were estimated to be 63.58% and 65.9% of total land covers, respectively. This implies that at global scale, vegetation productivities were more affected by precipitation than MAT/BT, but at regional scales, the significance of MAT and BT are highlighted.

Discussion

Discussion of the methodology

The meteorological dataset used in this paper was CRU TS 2.1, a well-documented dataset with over a century-long time scale and a high-resolution (0.5) latitude-longitude grid with global coverage. Such grids may be inappropriate for regional studies but are more useful for larger scales. The gridded data was constructed from the global network of meteorological observation stations, mainly obtained from seven sources [29,50]. The station network for CRU variables exhibited a gradual increase in the total number of stations from 1901 to about 1980. Distributions of precipitation stations are continentally extensive with high density in America, Australia, Europe, and eastern Asia and low density in Siberia and central Asia. There were almost half as many precipitation stations as mean temperature stations. Most of the mean temperature stations were distributed in America, Europe, and eastern Asia. Records from stations were checked and merged using the overlap function and if the records were not overlapped, a fixed reference series was constructed which is one of the main error sources of CRU data. It was indicated that side effects emerge when examining long-term average statistical properties or the frequency of extreme events using a fixed baseline period (1961–1990) [29]. The other source of error may lie in the homogenization method. Due to the need of checking for inhomogeneities, records obtained from areas and periods when density is low were added to the database without any checks. In addition, the interpolation method is also a weakness in detecting abrupt rather than gradual inhomogeneities in the data which cannot be detected unless they are widespread [51]. However, the error of the CRU dataset is substantially smaller than the temperature trends believed to have

Table 3. NPP of terrestrial ecosystems at continental levels in 1911-2000.

	Africa	Asia	Europe	Oceania	North America	South America	Global
Tundra & alpine steppe	×	1933.87±31.36	311.27±6.38	×	1793.97±61.70	23.73±11.05	4089.27±84.72
		(−1.00%)	(2.02%)		(−6.64%)	(−63.76%)	(−3.92%)
Cold desert	×	114.01±11.42	×	×	1.53±1.19	351.41±18.34	139.65±23.57
		(−16.93%)			(−86.82%)	(−9.95%)	(−27.33%)
Semi-desert	300.77±20.33	1495.03±53.91	149.23±22.71	265.22±13.33	502.37±32.94	226.49±18.05	3182.06±35.76
	(−12.66%)	(2.82%)	(33.03%)	(−2.02%)	(−11.98%)	(−14.38%)	(−0.15%)
Steppe	185.41±18.41	831.08±46.67	342.73±24.62	102.19±11.16	554.81±45.80	66.58±5.54	2437.90±121.60
	(−16.46%)	(−0.40%)	(−10.44%)	(−17.34%)	(−13.35%)	(−11.64%)	(−8.35%)
Temperate humid grassland	×	920.62±61.97	200.47±9.65	×	618.30±22.62	9.21±0.99	1809.96±82.44
		(−7.12%)	(−4.86%)		(−4.54%)	(−19.12%)	(−6.22%)
Warm desert	490.15±13.36	293.02±9.74	×	424.10±120.68	42.87±8.49	2471.83±90.92	1304.57±136.04
	(5.25%)	(−0.66%)		(−42.48%)	(−3.94%)	(−4.21%)	(−17.78%)
Savanna	8715.89±195.12	2400.37±247.63	42.08±6.96	3012.38±546.33	1214.97±36.09	772.24±35.37	19114.76±950.51
	(4.19%)	(18.44%)	(30.18%)	(36.71%)	(4.51%)	(−8.76%)	(9.42%)
Temperate forest	105.88±17.18	5010.41±250.77	4487.59±45.10	582.62±18.20	4895.13±358.26	2991.46±109.50	17726.47±503.00
	(−26.48%)	(5.97%)	(1.53%)	(−1.52%)	(15.77%)	(2.90%)	(5.49%)
Subtropical forest	2815.76±160.11	3429.22±48.04	82.33±10.01	283.91±15.98	1989.43±14.98	18044.70±412.52	12631.45±108.87
	(−6.14%)	(−0.79%)	(21.90%)	(1.83%)	(1.04%)	(4.50%)	(−0.70%)
Tropical forest	10375.76±451.62	10736.07±234.12	×	735.28±73.55	1757.00±28.69	24985.32±355.37	50001.87±531.72
	(−5.21%)	(−1.39%)		(20.07%)	(2.53%)	(2.73%)	(0.99%)

(Unit: Tg DW yr^{-1}). "×" indicates that a continent did not have a type of vegetation; a negative sign within parentheses indicates a decreasing trend in the period 1911–2000.

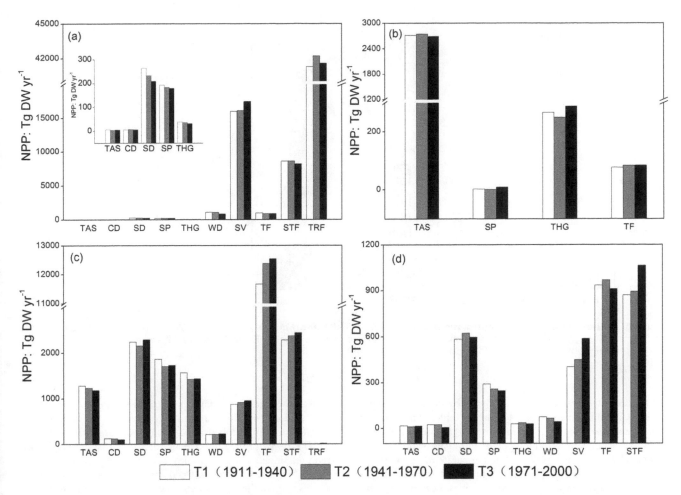

Figure 5. Variations of terrestrial ecosystems NPP in climatic zones in the period 1911–2000. (a): Tropical zone, (b): North frigid zone, (c): North temperate zone, (d): South temperate zone. South frigid zone was not involved in this research. TAS: tundra & alpine steppe; CD: Cold desert; SD: Semi-desert; SP: Steppe; THG: Temperate humid grassland; WD: warm desert; SV: savanna; TF: Tempearte forest; STF: Subtropical forest; TRF: Tropical forest.

been occurring during the twentieth century, so it is capable of correctly reflecting climate change.

The natural vegetation maps of the three intervals in 1911–2000 were obtained through the CSCS. The CSCS was established by linking vegetation with their climatic and edaphic factors [15]. Humidity index, determined by MAP and cumulative temperature, is the main parameter in the CSCS which presents promising applications in the research of simulating past and future global change, especially in regions where lack collected data. However, this system does not take the effects of underlying surfaces into account which lowers its accuracy in regions with complicated underlying surfaces such as mountain regions. In high latitude and elevation regions, it also underestimates the precipitation data due to neglecting the supply of underground water and melt water, thereby increasing the errors and bias. Although there are errors in assessment, the methodology enables a novel method of natural vegetation classification and easily demonstrates the spatial zonal distribution and dynamics of vegetation systems in various climate conditions at a global scale. Therefore, the CSCS is a feasible approach to map the global biomes and their response to climate change over the length of a century.

In this study, we employed a synthetic model to simulate terrestrial ecosystems' NPP during 1911–2000. Based on the water use efficiency of vegetation, determined by the ratio of the CO_2

flux equation to vapor flux equations, the synthetic model is an actual evapotranspiration model which provides the connection between water balance and heat balance, and reflects the effects of energy and water on the rate of evaporation. Because this model mainly focused on the effect of water and heat on plant ecophysiology, it does not take the effects of CO_2 concentration, soil nutrients, and interactions between vegetation and climate systems on NPP into consideration. Nevertheless, due to its sound mechanism and easily available data, the synthetic model is capable of detecting global NPP and variations in response to long term climate change. The synthetic model has been widely used in terrestrial ecosystems NPP estimation with techniques that are simpler, yet still useful for many cases [52,53].

Effects of climate change on distributions, extent and NPP of terrestrial ecosystems

This study demonstrated a comprehensive 90-year examination of the spatiotemporal variation in global terrestrial ecosystems and NPP based on the CSCS model and synthetic model. The simulated results showed that climate change played a crucial role in influencing the spatiotemporal distribution and NPP of terrestrial ecosystems.

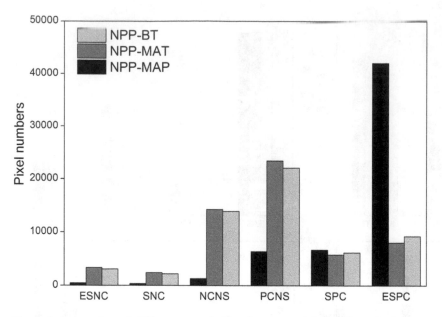

Figure 6. Proportion of different correlations between global NPP and MAP, MAT and BT. ESNC is extremely significant negative correlation; SNC is significant negative correlation; NCNS is negative correlation but none significant; PCNS is positive correlation but none significant; SPC is significant positive correlation; ESPC is extremely significant positive correlation.

The simulated results indicated that ecosystems were experiencing a gradual and irreversible change under climate change with an expansion of forest and savanna and a shrinking of tundra & alpine steppe, steppe, and temperate humid grassland [54]. As a result of climate change, vegetation in temperate zones was pushed to high elevations and latitudes which forced a readaptation to the environment and a reduction in productivities. Ecosystems in mid- and high latitudes in the northern hemisphere were more vulnerable and sensitive to climate change, and opportunities for these ecosystems to adapt to changes were limited because these systems react most strongly to globally induced climate change [30]. We found that the area of tundra & alpine steppe declined by approximately 88.0×10^4 km^2 from 1911 to 2000 globally. The south edge showed a persistent movement northward and eastward, and its current extent is likely to be encroached by grassland or forest [55]. Vegetation in semiarid areas, such as steppe, seemed to be replaced by vegetation accommodated to arid zones or sub-humid zones, e.g. semi-desert, savanna [4]. In Australia, shifts in rainfall patterns favored the establishment of woody vegetation e.g. woody savanna, and encroachment of warm desert [56,57].

According to the global NPP derived from the synthetic model, most of the productivities were attributed to tropical forest, savanna, and temperate forest, and nearly 60% was estimated to occur in the tropical regions [58,59]. In the period 1911–2000, the global NPP increased gradually by 2317.97 Tg DW yr^{-1}. Most of this increase resulted from the increasing trend of savanna (9.42%), tropical forest (0.99%), and temperate forest (5.49%). This trend was also captured by the research of Cramer et al. [60] in which six dynamic global vegetation models were used to evaluate global terrestrial ecosystems in response to CO_2 and climate change. Wang et al. [61] reported that the obvious climatic warming in the northern temperate regions had led to the general increase of temperate forest NPP from 1901 to 2009. In tropical regions, the NPP of tropical forest showed a slight increase of 0.70% [62]. The NPP of savanna increased by 8.99%. In

Africa, warm desert and savanna expanded in some arid and semi-arid regions which increased productivities. However, high temperature also resulted in strong evapotranspiration which led to NPP decrease in some regions [63,64]. Grace et al. [65] also reported similar results of tropical savanna in Australia, India, and South America. In northern frigid zones, rising temperature led to the decrease of the NPP of tundra & alpine steppe. In contrast, the expansion of steppe and temperate humid grassland resulted in an eventual increase in their NPP [66]. In temperate regions, temperate forest was the dominant ecosystem both in northern and southern temperate zones. However, we found a positive trend of NPP in northern temperate forest with 7.42%, whereas in southern temperate forest, NPP went up in the T2 period (3.59%) but went down in T3 period (6.02%) [67,68,69]. The reason for this may lie in the fact that during the 20th century, besides the obvious rising temperature in northern mid-latitude, precipitation in much of the northern hemisphere also displayed an increasing trend by 0.2% to 1% per decade. However, no similar variations were observed in the southern hemisphere [30]. All these variations may be a sign that the late 20th century is a critical turning point for the significant change of terrestrial ecosystems as a consequence of global climate change.

From a global perspective, the positive correlation between NPP and MAP was more obvious than that of MAT and BT, indicating that precipitation was the most important climatic factor determining the productivities of vegetation. This is in agreement with previous studies [48,70]. Recent global warming had extended vegetation growing seasons in most regions [71,72] which led to the increase of biological temperature for plant growth. Variations in MAT would exert a more significant influence on NPP at a regional scale [73]. Zhang et al. [52] reported that in the last 50 years NPP in Inner Mongolia showed a positive correlation ($R^2 = 0.64$) with temperature variations.

The overall conclusion from this analysis is that a range of impacts on global terrestrial ecosystems and NPP were observed under climate change from 1911 to 2000 and effects varied greatly among different ecosystems. Ecosystems in mid- and high latitudes, e.g. tundra & alpine steppe, were forced to adapt to a new habitat, whereas, the area of savanna, temperate forest, and tropical forest all increased. The past 90 years, particularly the latter half of last century, also witnessed an increasing trend of global NPP. Most of this increase was attributed to tropical forest, temperate forest, and savanna. NPP increase displayed a latitudinal distribution, particularly in tropical zones and northern temperate zones. In addition, there would be an increased chance for forest and grassland degradation from wild fires and invasive species as a result of global warming and reduced precipitation. In summary, although terrestrial productivities increased under climate change in the period 1911–2000, some ecosystems benefitted from it but others were negatively affected. Some regions were undergoing significant and irreversible change, and these effects may continue.

Supporting Information

Appendix S1 MAP in T1 period.

Appendix S2 MAP in T2 period.

Appendix S3 MAP in T3 period.

Appendix S4 MAT in T1 period.

Appendix S5 MAT in T2 period.

Appendix S6 MAT in T3 period.

Appendix S7 The simulated global vegetation in T1 period.

Appendix S8 The simulated global vegetation in T2 period.

Appendix S9 The simulated global vegetation in T3 period.

Appendix S10 Global NPP in T1 period.

Appendix S11 Global NPP in T2 period.

Appendix S12 Global NPP in T3 period.

Acknowledgments

We also thank Prof. Liu Bing from Leeds University and Prof. Liu Jiyuan from Institute of Geographic Sciences and Natural Resources Research, Chinese Academy of Sciences for their guidance to this work and data provided. We also appreciate the CRU and The Oak Ridge National Laboratory Distributed Active Archive Center (ORNL DAAC) for sharing datasets.

Author Contributions

Conceived and designed the experiments: CCG. Performed the experiments: WZ JLL. Analyzed the data: YZC SJM. Wrote the paper: CCG. Contributed to design of the experiments: WZ JLL. Collected the data: YZC SJM. Provided suggestions about improving the materials and methods: JZR. Provided suggestions about data analysis and helped with improving the writing of the paper: JMC PYG.

References

1. Parton WJ, Scurlock J, Ojima DS, Gilmanov TG, Scholes RJ, et al. (1993) Observations and modeling of biomass and soil organic matter dynamics for the grassland biome worldwide. Global Biogeochem Cy 7: 785–809.
2. Canadell JG, Pataki DE, Pitelka LF (2006) Terrestrial ecosystems in a changing world. Berlin: Springer.
3. Chinese Academy of Sciences (1994) Department of Life Sciences of State Commission of Chinese National Natural Science Foundation, Global changes and ecological systems. Shanghai: Shanghai science and technology press. pp62–95.
4. Shaver GR, Canadell J, Chapin FS III, Gurevitch J, Harte J, et al. (2000) Global warming and terrestrial ecosystems: a conceptual framework for analysis. Bioscience 50: 871–882.
5. Sykes MT, Prentice IC, Laarif F (1999) Quantifying the impact of global climate change on potential natural vegetation. Climatic Change 41: 37–52.
6. Zhang XS (1993) A vegetation-climate classification system for global change studies in China. Quaternary Sciences 2: 157–269.
7. Horion S, Cornet Y, Erpicum M, Tychon B (2013) Studying interactions between climate variability and vegetation dynamic using a phenology based approach. International Journal of Applied Earth Observation and Geoinformation 20: 20–32.
8. Foley JA, Levis S, Costa MH, Cramer W, Pollard D (2000) Incorporating dynamic vegetation cover within global climate models. Ecol Appl 10: 1620–1632.
9. Roy J, Saugier B, Mooney HA (2001) Terrestrial Global Productivity. San Diego: Academic Press.
10. Cao MK, Yu GR, Liu JY, Li KR (2005) Multi-scale observation and cross-scale mechanistic modeling on terrestrial ecosystem carbon cycle. Sci China Ser D 48: 17–32.
11. Holdridge LR (1947) Determination of world plant formations from simple climate data. Science 105: 367–368.
12. Woodward FI, Smith TM, Emanuel WR (1995) A global land primary productivity and phytogeography model. Global Biogeochem Cy 9: 471–490.
13. Lobo A, Rebollar JLG (2010) Model-based discriminate analysis of Iberian potential vegetation and bio-climatic indices. Phys Chem Earth 35: 52–56.
14. Hu ZZ, Gao CX (1995) Improvement of the comprehensive and sequential classification system of grasslands: I. Indices of grassland classes and index chart (in Chinese). Acta Pratac Sin 3: 1–7.
15. Ren JZ, Hu ZZ, Zhao J, Zhang DG, Hou FJ, et al. (2008) A grassland classification system and its application in China. Rangeland J 30: 199–209.
16. Lin HL, Feng QS, Liang TG, Rena JZ (2012) Modelling global-scale potential grassland changes in spatio-temporal patterns to global climate change. International Journal of Sustainable Development & World Ecology. doi:10.1080/13504509.2012.749815.
17. Li ZQ, Liu ZG, Chen ZZ, Yang ZG (2003) The effects of climate changes on the productivity in the Inner Mongolia steppe of China. Acta Prataculturae Sinica 12: 4–10.
18. Odum EP (1976) Fundamentals of Ecology. Philadelphia: W.B. Saunders.
19. Lieth H (1977) Modeling the primary productivity of the world. In: Lieth H, Whittaker RH, editors. Productivity of the Biosphere. New York: Springer. pp. 237–263.
20. McGuire AD, Melillo JM, Kicklighter DW, Joyce LA (1995) Equilibrium responses of soil carbon to climate change: Empirical and process-based estimates. J Biogeogr: 785–796.
21. Running SW, Hunt ER (1993) Generalization of a forest ecosystem process model for other biomes, BIOME-BGC, and an application for global-scale models. Scaling physiological processes: Leaf to globe: 141–158.
22. Potter CS, Randerson JT, Field CB, Matson PA, Vitousek PM, et al. (1993) Terrestrial ecosystem production- A process model based on global satellite and surface data. Global Biogeochem Cy 7: 811–841.
23. Prince SD (1991) A model of regional primary production for use with coarse resolution satellite data. Int J Remote Sens 12: 1313–1330.
24. Zhu WQ, Chen YH, Xu D, Li J (2005) Advances in terrestrial net primary productivity (NPP) estimation models. Chinese Journal of Ecology 24: 296–300.
25. Zhou GS, Zhang XS (1995) A natural vegetation NPP model (in Chinese). Acta Botanica Sinica 19: 193–200.
26. Nemani RR, Keeling CD, Hashimoto H, Jolly WM, Piper SC, et al. (2003) Climate-driven increases in global terrestrial net primary production from 1982 to 1999. Science 300: 1560–1563.
27. Harpole WS, Potts DL, Suding KN (2007) Ecosystem responses to water and nitrogen amendment in a California grassland. Global Change Biol 13: 2341–2348.
28. Dukes JS, Chiariello NR, Cleland EE, Moore LA, Shaw MR, et al. (2005) Responses of grassland production to single and multiple global environmental changes. Public Library of Science Biology 3: 1829–1837.

29. Mitchell TD, Jones PD (2005) An improved method of constructing a database of monthly climate observations and associated high-resolution grids. Int J Climatol 25: 693–712.

30. Intergovernmental Panel on Climate Change (IPCC) (2007) Climate Change 2007: The Scientific Basis. Cambridge: Cambridge University Press.

31. Zhou GS, Zheng YR, Chen SQ (1998) NPP model of natural vegetation and its application in China. Scientia Silvae Sinicae 34: 2–11.

32. Zhou GS, Zhang XS (1996) Study on NPP of natural vegetation in China under global climate change. Acta Phytoecological Sinica 20: 11–19.

33. Zhou GS, Wang YH, Jiang YL, Yang ZY (2002) Estimating biomass and net primary production from forest inventory data: a case study of China's Larix forests. Forest Ecol Manag 169: 149–157.

34. Gao QZ, Li Y, Wan YF, Qin XB, Jiangcun WZ, et al. (2009) Dynamics of alpine grassland NPP and its response to climate change in Northern Tibet. Climatic Change 97: 515–528.

35. Ito A (2011) A historical meta-analysis of global terrestrial net primary productivity: are estimates converging? Global Change Biol 17: 3161–3175.

36. Oak Ridge National Laboradtory, DAAC (2009) Net Primary Production (NPP) Project Web Page. Available: http://daac.ornl.gov/NPP/npp_home.shtml. Oak Ridge, Tennessee, USA. Accessed September 10, 2012.

37. Bailey RG (1989) Explanatory supplement to ecoregions map of the continents. Environ Conserv 16: 307–309.

38. Jager HI, Hargrove WW, Brandt CC, King AW, Olson RJ, et al. (2000) Constructive contrasts between modeled and measured climate responses over a regional scale. Ecosystems 3: 396–411.

39. Woodwell GM, Whittaker RH (1968) Primary production in terrestrial ecosystems. American Zoologist 8: 19–30.

40. Intergovernmental Panel on Climate Change (IPCC) (1990) Climate Change: The IPCC Scientific Assessment. Cambridge: Cambridge University Press.

41. Siegenthaler U, Sarmiento JL (1993) Atmospheric carbon dioxide and the ocean. Nature 365: 119–125.

42. Sundquist ET (1993) The global carbon dioxide budget. Science-New York, then Washington 259: 934.

43. Intergovernmental Panel on Climate Change (IPCC) (1996) Climate Change 1995: the Science of Climate Change. Cambridge: Cambridge University Press.

44. Schlesinger WH, Bernhardt ES (2013) Biogeochemistry: an analysis of global change. Waltham, MA: Academic Press.

45. Schimel DS, House JI, Hibbard KA, Bousquet P, Ciais P, et al. (2001) Recent patterns and mechanisms of carbon exchange by terrestrial ecosystems. Nature 414: 169–172.

46. Cramer W, Kicklighter DW, Bondeau A, Iii BM, Churkina G, et al. (1999) Comparing global models of terrestrial net primary productivity (NPP): overview and key results. Global Change Biol 5: 1–15.

47. Geider RJ, Delucia EH, Falkowski PG, Finzi AC, Grime JP, et al. (2001) Primary productivity of planet earth: biological determinants and physical constraints in terrestrial and aquatic habitats. Global Change Biol 7: 849–882.

48. Intergovernmental Panel on Climate Change (IPCC) (2001) Climate change 2001: The scientific basis. Cambridge: Cambridge University Press.

49. Gruber N, Friedlingstein P, Field CB, Valentini R, Heimann M, et al. (2004) The vulnerability of the carbon cycle in the 21st century: An assessment of carbon-climate-human interactions. Scope-Scientific committee on Problems of the environment international council of scientific union 62: 45–76.

50. Rohde R (2013) Comparison of Berkeley Earth, NASA GISS, and Hadley CRU averaging techniques on ideal synthetic data. Available: http://berkeleyearth.org/. Accessed 2013 Nov 3.

51. New M, Hulme M, Jones P (2000) Representing twentieth-century space-time climate variability. Part II: Development of 1901-96 monthly grids of terrestrial surface climate. J Climate 13: 2217–2238.

52. Zhang GG, Kang YM, Han GD, Sakurai K (2011) Effect of climate change over the past half century on the distribution, extent and NPP of ecosystems of Inner Mongolia. Global Change Biol 17: 377–389.

53. Zhang YJ, Zhou GS (2011) Exploring the effects of water on vegetation change and net primary productivity along the IGBP Northeast China Transect. Environmental Earth Science 62: 1481–1490.

54. Shiyatov SG, Terent'Ev MM, Fomin VV (2005) Spatiotemporal dynamics of forest-tundra communities in the polar urals. Russ J Ecol 36: 69–75.

55. Yue TX, Fan ZM, Liu JY (2005) Changes of major terrestrial ecosystems in China since 1960. Global Planet Change 48: 287–302.

56. Gifford RM, Howden M (2001) Vegetation thickening in an ecological perspective: significance to national greenhouse gas inventories. Environmental Science & Policy 4: 59–72.

57. Hughes L (2003) Climate change and Australia: Trends, projections and impacts. Austral Ecol 28: 423–443.

58. Melillo JM, Mcguire AD, Kicklighter DW, Moore B, Vorosmarty CJ, et al. (1993) Global climate change and terrestrial net primary production. Nature 363: 234–240.

59. Cao MK, Woodward FI (1998) Net primary and ecosystem production and carbon stocks of terrestrial ecosystems and their responses to climate change. Global Change Biol 4: 185–198.

60. Cramer W, Bondeau A, Woodward FI, Prentice IC, Betts RA, et al. (2001) Global response of terrestrial ecosystem structure and function to CO2 and climate change: results from six dynamic global vegetation models. Global Change Biol 7: 357–373.

61. Wang W, Hashimoto H, Ganguly S, Votava P, Nemani RR, et al. (2010) Characterizing uncertainties in recent trends of global terrestrial net primary production through ensemble modeling. In AGU Fall Meeting Abstracts. p. 3.

62. Boisvenue C, Running SW (2006) Impacts of climate change on natural forest productivity–evidence since the middle of the 20th century. Global Change Biol 12: 862–882.

63. Reich PF, Numbem ST, Almaraz RA, Eswaran H (2004) Land resource stresses and desertification in Africa. Agro-Science 2.

64. Muriuki GW, Njoka TJ, Reid RS, Nyariki DM (2005) Tsetse control and land-use change in Lambwe valley, south-western Kenya. Agr Ecosyst Environ 106: 99–107.

65. Grace J, José JS, Meir P, Miranda HS, Montes RA (2006) Productivity and carbon fluxes of tropical savannas. J Biogeogr 33: 387–400.

66. Bunn AG, Goetz SJ, Kimball JS, Zhang K (2007) Northern high-latitude ecosystems respond to climate change. Eos, Transactions American Geophysical Union 88: 333–335.

67. Field CB, Mortsch LD, Brklacich M, Forbes DL, Kovacs P, et al. (2007) North America. Climate Change 2007: Impacts, Adaptation and Vulnerability. Contribution of Working Group II to the Fourth Assessment Report of the Intergovernmental Panel on Climate Change. In: Parry ML, Canziani OF, Palutikof JP, Van Der Linden PJ, Hanson CE, editors. Cambridge, UK: Cambridge University Press. pp. 617–652.

68. Cruz RV, Harasawa H, Lal M, Wu S, Anokhin Y, et al. (2007) Asia. Climate Change 2007: Impacts, Adaptation and Vulnerability. Contribution of Working Group II to the Fourth Assessment Report of the Intergovernmental Panel on Climate Change. In: Parry ML, Canziani OF, Palutikof JP, Van Der Linden PJ, Hanson CE, editors. Cambridge, UK: Cambridge University Press. pp. 469–506.

69. Magrin G, García CG, Choque DC, Giménez JC, Moreno AR, et al. (2007) Latin America. Climate Change 2007: Impacts, Adaptation and Vulnerability. Contribution of Working Group II to the Fourth Assessment Report of the Intergovernmental Panel on Climate Change. In: Parry ML, Canziani OF, Palutikof JP, Van Der Linden PJ, Hanson CE, editors. Cambridge, UK: Cambridge University Press. pp. 581–615.

70. Del Grosso S, Parton W, Stohlgren T, Zheng DL, Bachelet D, et al. (2008) Global potential net primary production predicted from vegetation class, precipitation, and temperature. Ecology 89: 2117–2126.

71. Bonsal BR, Prowse TD (2003) Trends and variability in spring and autumn 0 C-isotherm dates over Canada. Climatic Change 57: 341–358.

72. Berthelot M, Friedlingstein P, Ciais P, Monfray P, Dufresne JL, et al. (2002) Global response of the terrestrial biosphere to CO2 and climate change using a coupled climate-carbon cycle model. Global Biogeochem Cy 16: 1084.

73. Piao SL, Fang JY, He JS (2006) Variations in vegetation net primary production in the Qinghai-Xizang Plateau, China, from 1982 to 1999. Climatic Change 74: 253–267.

4

Ectomycorrhizal-Dominated Boreal and Tropical Forests Have Distinct Fungal Communities, but Analogous Spatial Patterns across Soil Horizons

Krista L. McGuire[1]*, **Steven D. Allison**[2,3], **Noah Fierer**[4,5], **Kathleen K. Treseder**[2]

1 Department of Biology, Barnard College, Columbia University, New York, New York, United States of America, **2** Department of Ecology & Evolutionary Biology, University of California Irvine, Irvine, California, United States of America, **3** Department of Earth System Science, University of California Irvine, Irvine, California, United States of America, **4** Department of Ecology & Evolutionary Biology, University of Colorado, Boulder, Colorado, United States of America, **5** Cooperative Institute for Research in Environmental Sciences, University of Colorado, Boulder, Colorado, United States of America

Abstract

Fungi regulate key nutrient cycling processes in many forest ecosystems, but their diversity and distribution within and across ecosystems are poorly understood. Here, we examine the spatial distribution of fungi across a boreal and tropical ecosystem, focusing on ectomycorrhizal fungi. We analyzed fungal community composition across litter (organic horizons) and underlying soil horizons (0–20 cm) using 454 pyrosequencing and clone library sequencing. In both forests, we found significant clustering of fungal communities by site and soil horizons with analogous patterns detected by both sequencing technologies. Free-living saprotrophic fungi dominated the recently-shed leaf litter and ectomycorrhizal fungi dominated the underlying soil horizons. This vertical pattern of fungal segregation has also been found in temperate and European boreal forests, suggesting that these results apply broadly to ectomycorrhizal-dominated systems, including tropical rain forests. Since ectomycorrhizal and free-living saprotrophic fungi have different influences on soil carbon and nitrogen dynamics, information on the spatial distribution of these functional groups will improve our understanding of forest nutrient cycling.

Editor: Jean Thioulouse, CNRS - Université Lyon 1, France

Funding: Funding for this work came from the National Science Foundation grant to KKT, a NOAA Climate and Global Change Postdoctoral Fellowship to SDA, and Doctoral Dissertation Improvement Grant to KLM. The funders had no role in study design, data collection and analysis, decision to publish, or preparation of the manuscript.

Competing Interests: The authors have declared that no competing interests exist.

* E-mail: kmcguire@barnard.columbia.edu

Introduction

Ectomycorrhizal (ECM) and saprotrophic fungi are major contributors to nutrient cycling in forest ecosystems [1]. These functional groups are globally distributed and coexist in many forest ecosystems. Approximately 6000 tree species worldwide depend on ECM fungi for nutrient acquisition [2], and the distribution of ECM trees spans the globe ranging from northern boreal regions to tropical rain forests. Strikingly, a disproportionate number of the dominant trees in temperate, boreal and certain tropical forests form ECM associations [3–6], suggesting that ECM fungi are likely responsible for a significant quantity of C, N, and P cycling worldwide. In boreal forests, ECM fungi contribute up to 86% of total plant N [7]. Saprotrophic fungi are also critical to nutrient cycling, and are the major decomposers of complex, organic molecules such as lignin. Thus, understanding how ECM and saprotrophic fungi are distributed within and across ecosystems is critical for making inferences about nutrient cycling and related ecosystem functions in forest communities.

It is well established that mycorrhizal fungi interact with other soil organisms such as bacteria and invertebrates, but interactions among mycorrhizal and decomposer fungi have been more challenging to evaluate [1,8]. There is evidence from boreal and temperate forests that ECM and saprotrophic fungal taxa vertically segregate in soils [9–11], suggesting physiological specialization of fungi on organic substrates in various levels of decay [10]. However, there have been few studies of fungal spatial dynamics in tropical ECM forests, so it is unclear if the patterns detected in boreal and temperate forests are similar to those found in the tropics.

While the majority of trees in temperate and boreal forests form ECM associations, most species of trees in lowland tropical rain forests form arbuscular mycorrhizal (AM) associations. When tropical trees do form ECM symbioses, they are more likely to become locally dominant [6] or in some cases regionally dominant (e.g., the Dipterocarpaceae in Southeast Asia). At this point, we do not know if generalizations can be made about ECM forests at a global scale or if tropical ECM forests contain unique fungal communities that function differently from ECM fungi at higher latitudes. From the data that have been collected, it seems that tropical forests have lower ECM diversity than temperate and boreal ecosystems [12,13], although there is clearly a gap in our knowledge and a paucity of belowground studies in tropical ECM forests. Since tropical forests harbor 40% of all terrestrial biomass and are responsible for 32% of terrestrial net primary production [14,15], understanding the dynamics of fungal distribution and function in tropical forests is important for making inferences about global nutrient cycles.

In this study, we used sequence-based approaches to assess the distribution of fungal taxa in a tropical forest located in central Guyana and a boreal forest in Delta Junction, Alaska. The tropical forest site contained two types of rain forest: an ectomycorrhizal monodominant forest and a non-ectomycorrhizal mixed forest [16,17]. Our objectives were to: 1) examine the level of taxonomic similarity in fungal community composition across the two ECM forests in different biomes, 2) compare fungal community composition across organic and mineral soil horizons within each ecosystem, and 3) determine if patterns of functional group separation across soil horizons were analogous in the boreal and tropical forest. Since tropical ECM forest dynamics have been shown to be significantly different than non-EM forests within the same biome [6,18], we predicted that the boreal forest and tropical ECM forest would exhibit more similar fungal community patterns than the ECM and non-EM tropical forest.

Materials and Methods

Study Site and Sample Collection

Samples used for this study were collected from sites in Alaska, USA (63°55′N, 145°44′W) and Guyana (5°4′ N, 59°58 W) between 2007 and 2009. In Alaska, the site consisted of boreal spruce forest that has not burned in over 80 years [19].The forest canopy was dominated by *Picea mariana* (Mill.) Britton, Sterns & Poggenb. (Pinaceae), which forms ECM associations and comprises the vast majority of the canopy trees [19,20]. Likewise, the ECM forest site in Guyana consisted of mature forest dominated by the ECM tree *Dicymbe corymbosa* Spruce ex. Benth (Caesalpiniaceae), in which *D. corymbosa* comprises up to 90% of canopy trees [16,21]. *Dicymbe corymbosa* was also the only ECM host in the plots used for this study, thereby making it an ideal comparative site to the *Picea*-dominated boreal forest. As a non-ECM comparison, three plots from mixed forest in Guyana were also analyzed, which do not contain dominant ECM species [17]. Since the majority of boreal forest trees are ECM, we did not have a non-ECM forest comparison for the boreal biome. Permits for the field research in Guyana were granted by the Guyana Environmental Protection Agency and the Ministry of Amerindian Affairs. The sites were not located on private or protected land and did not involve endangered or protected species.

In each ecosystem, samples were separately collected from the litter and upper soil horizons (0–20 cm) from previously established plots at both sites. Plots in both sites were at least 100 m apart. In the boreal forest, a total of three plots were sampled, with each plot having dimensions of 10×10 m. One composite soil sample was derived from five soil cores taken from each plot. These plots were also used as control sites in a previous study [22]. In the tropical forest, ten composite soil samples were taken from three previously established forest plots (30×100m) in the *Dicymbe*-dominated forest [21]. At the same points of soil core sampling, we also collected litter samples from the forest floor. Plot sizes across sites were different, as these study sites were established independently without the original intention of comparative analyses. However, since samples were collected in a similar manner and DNA was extracted with the same protocol, the extracts were sequenced together for comparative analyses of vertical fungal separation, rather than comparisons of fungal species richness.

To evaluate litter fungi in a more controlled way so that only the dominant tree litter was used, we set out freshly fallen leaf litter in mesh bags on the forest floor at both sites. In the boreal forest, 4 g air-dried *Picea mariana* leaf litter was placed in litter bags composed of 2 mm mesh (window screen) lined with 0.5 mm mesh (bridal veil) to prevent loss of needle fragments. Leaf litter in the mesh bags from Guyana was composed of 10 g air-dried, freshly-fallen *D. corymbosa* leaves. After one year of incubation on the surface of the forest floor, decomposed litter from six bags in the boreal forest and ten bags in the tropical forest was transported to the laboratory, where it was frozen at –80°C until analysis. All samples remained frozen during transport, which was less than 10 h for both sites.

Molecular Analyses

To examine fungal community composition across ecosystems, we first analyzed environmental soil and litter samples from the six plots at each site using 454 pyrosequencing. To homogenize the soil samples, each composite sample was passed through a 2 mm sieve that had been sterilized with ethanol and 15 min of uv radiation. All homogenizations were accomplished in a sterile, benchtop PCR hood (AirClean Systems, Inc, Raleigh, NC). Litter was hand homogenized with sterile gloves. Since the litter was highly decomposed, mechanical grinding was not necessary. From each composite soil and litter sample for each plot, total DNA was extracted from three 0.25g subsamples to obtain a representative sample [23] using a Powersoil DNA extraction kit (MoBio, Carlsbad, CA) according to the manufacturer's instructions. These three DNA extracts were pooled to create one representative soil DNA extract. General fungal primers (SSU817f and SSU1196r) targeting a portion of the 18S rRNA gene were modified for 454 sequencing [24]. PCR amplifications were done as described previously [24–26] with 30 mM of each primer (0.25 µl), 22.5 µl Platinum PCR SuperMix (Invitrogen, Carlsbad, CA), and 3 µl of DNA template. Three PCR reactions per sample were pooled for analysis. PCR products were sequenced at the Environmental Genomics Core Facility at the University of South Carolina (Columbia, SC) on a Roche 454 Gene Sequencer with Titanium chemistry. Sequenced amplicons were quality checked, aligned, and grouped into operational taxonomic units (OTUs) at a 97% sequence similarity cutoff with the Quantitative Insights Into Microbial Ecology (QIIME) pipeline [27]. The centroid sequence from each OTU cluster was chosen and used to create a phylogenetic tree with the FastTree algorithm [28]. Taxonomic information for each OTU was determined using the BLAST algorithm [29] against identified sequences in both Genbank and the SILVA database [30]. Ultimately, we used an open-reference for OTU picking and sequences <400 bp were removed. The (phred) quality score cutoff was 25 and sequences containing ambiguous characters and those having an unreadable barcode were also removed. Non-fungal sequences were manually removed following taxonomic assignment. The average sequence length was ~450 bp. Fungal sequences have been deposited in the Sequence Read Archive of Genbank (Accession # SRP009079.1).

To gain more detailed taxonomic information about soil and litter fungi, we used clone library sequencing on the same soil samples analyzed in the pyrosequencing runs and on litter from incubated leaf litter bags to standardize for litter species and decomposition time. The same DNA extracts used in pyrosequencing were analyzed for the composite soil samples and DNA from litter bags was extracted with a PowerSoil DNA kit (Mo Bio Laboratories, Inc, CA) as described above. Three DNA extractions of each sample were again pooled for each site. Fungal DNA was selectively amplified from soil and litter DNA extractions using the ITS1-F forward primer [31] and the TW13 reverse primer [32]. These primers target ~600 bp of the ITS region and ~700bp of the 5′ portion of the 28S region. The reason for choosing these primers is that amplification of the 28S region allowed for alignment of amplicons and phylogenetic community analysis, whereas the hypervariable ITS region allowed for higher taxonomic resolution at the subgeneric level [31,33]. PCR reactions were carried out in 30 μL volumes with 200 mM Tris-HCl PCR buffer, 1.23 mM $MgSO_4$, 0.2 mM each dNTP, 0.5 μg μL^{-1} BSA, 0.1 μM each primer, and 0.01 U μL^{-1} Platinum Taq DNA Polymerase (Invitrogen, Carlsbad, CA), and 0.13 μL template DNA uL^{-1} reaction volume. PCR reactions were done in an iCycler thermocycler (BioRad) with the following program: 5 min initial denaturation at 95°C, followed by 35 cycles of 30 sec at 95°C, 45 sec of annealing at 50°C, 6 min of elongation at 72°C, and a final elongation for 10 min at 72°C. While these PCR conditions are frequently used in the literature, we acknowledge that the high cycle number may have skewed the mixed template amplifications in favor of more abundant groups [34,35].

PCR products were gel-purified by running each sample on a 1.5% agarose gel; target bands were cut from the gel and cleaned up with a Qiaquick gel extraction kit (Qiagen, Valencia, CA). Clone libraries were constructed with the gel-purified PCR products using the Topo TA Cloning Kit for Sequencing with PCR 4-TOPO vector (Invitrogen) following the manufacturer's instructions. This vector allowed for blue/white colony screening, such that only chemically competent E. coli cells that were white in color were selected for sequencing. We picked 96 colonies from each clone library, 384 sequences total for a total of 4 clone libraries; one for each organic soil fraction in each ecosystem to identify the dominant fungal taxa. Clones were bi-directionally sequenced at the Laboratory for Genomics and Bioinformatics at the University of Georgia (Athens, GA).

Raw DNA sequences were edited using using CodonCode Aligner version 2.0 (CodonCode Corporation, Dedham, MA) and Bioedit. Contiguous sequences were constructed for forward and reverse DNA sequences using Geneious version 3.7.0 (Biomatters Ltd., Auckland, New Zealand). Contiguous sequences have been deposited to Genbank (Accession #: JN889716 - JN890544). Alignments were made in ClustalW [36] using only the 28S portion of the DNA, as the ITS portions are too variable for alignment. Distance matrices were generated using the default parameters of Phylip DNADIST [37]. 28S sequences having ≥ 99% sequence similarity, as determined by DOTUR [38], were assigned the same Operational Taxonomic Unit (OTU). OTUs were assigned to taxa using the BLASTn algorithm against known sequences in GenBank [29] and the UNITE database [39]. For all BLAST searches the full consensus sequences were used, rather than just the 28S portion, for better identification resolution. A taxonomic name was assigned to an OTU only if the name occurred within the top ten best BLAST matches, query coverage was >95%, and the e-value was 0.0. If the top ten matches were all 'uncultured' or 'unidentified', then 'unknown' was assigned to the OTU. Chimeras were identified by separately BLAST searching the 28S and ITS regions; if the top five hits in GenBank did not match for both regions of DNA, the sequence was considered chimeric and discarded. Functional group assignments (saprobe, EM, pathogen, etc.) were given to OTUs with assigned identities only if the taxonomic affiliation could reliably be placed in a group where the majority of species are known to have that particular function. For a few groups (notably Amanitaceae, Entolomataceae, and Clavulinaceae), the ECM function was assigned since it is the dominant function of that family or if the OTUs aligned to genera known to be ECM, even though there are some cases of fungal taxa in those families that can be saprotrophic [40,41]. Ambiguous genera or families in which there was not a predominant function were listed as unknown functional groups.

Statistical Analyses

To determine differences in fungal community composition across soil and litter horizons in the forest plots, fungal sequences were rarified to 1000 sequences [42] and proportional counts of sequences per OTU group were then square-root transformed to minimize the influence of rare taxa. OTU abundance data were then analyzed by generating distance matrices with the Bray-Curtis coefficient followed by Analysis of Similarity [ANOSIM; 43] using Primer-version 6 software (Primer-E, Plymouth, UK). Nonmetric multidimensional scaling plots and dendrograms were used to visualize similarity in fungal community composition across sites and horizons. In the ECM forests at both sites, the relative proportions of the most abundant ECM fungal families were analyzed across sites and horizons (litter versus soil) using a multivariate general linear model. For the clone library data, a two-way ANOVA was used to assess differences in fungal taxonomic richness between soil horizons and recently-shed litter within and across ecosystems.

Results

We obtained 31,942 sequences from pyrosequencing with an average of 1330 sequences per sample and approximately 450 bp in length. Prior to downstream analyses, all non-fungal and unclassifiable sequences were removed, which represented approximately 7% of the sequences. Thus, a total of 29,837 sequences were used for downstream analyses. Of the sequences that could be identified as fungi from the 18S pyrosequencing data, an average of 327 unique operational taxonomic units (OTUs) were observed for each sample. Across all samples, 28% of sequences were Ascomycota, 55% were Basidiomycota, 9% were Chytridiomycota, 1% were Glomeromycota, 1% were basal fungal lineages, and 5% could not be assigned to a phylum. The inability to assign a phylum to these sequences may in part be due to the presence of deeply diverging fungal lineages that have not yet been characterized in Genbank [44].

Ordination of pyrosequencing data showed that fungal communities were distinct across tropical and boreal ecosystems and across horizons within site (Fig. 1A). These patterns were confirmed by ANOSIM for both site ($P = 0.02$) and horizon (litter versus soil) within site (Alaska $P < 0.01$; Guyana $P < 0.001$). When fungal communities in the tropical forests were analyzed separately from boreal samples, fungal taxa in the ECM forest were distinct from the non-EM forest across horizons in each forest type ($P < 0.001$; Fig. 1B).

When pyrosequencing-derived fungal OTUs from soil and litter samples were compared across tropical and boreal sites, the proportional abundances of fungal phyla were significantly different in litter samples ($P < 0.01$ for all comparisons; Fig. 2). However, the proportional abundance of fungal phyla in soil samples were not significantly different, with the exception of the Glomeromycota ($F_{(1,11)} = 7.4$, $P = 0.02$), which was more abundant in the tropical soils. The Ascomycota and Basidiomycota comprised 80–90% of fungal OTUs in both horizons at both sites. Thus, to determine if these phyla were differentially driving the observed biogeographical patterns of fungi, we separately analyzed OTUs assigned to each phylum. Cluster analysis showed that fungi in both phyla displayed similar patterns across sites and horizons (Fig. 3).

While pyrosequencing data provide limited taxonomic resolution as the sequences cannot be reliably identified beyond the family level, some fungal families are exclusively or mostly ECM and could be compared across ECM forests. In both boreal and tropical ECM sites, there were six predominantly ECM fungal

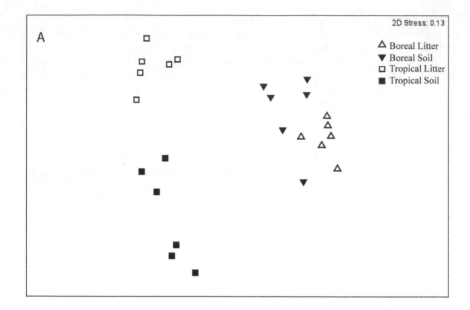

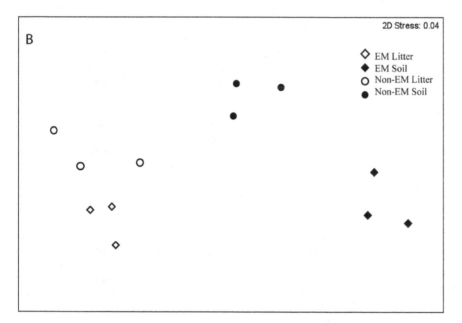

Figure 1. Non-metric multidimensional scaling plot for pyrosequencing OTUs in litter and soil horizons based on Bray-Curtis distance across sites (A) and across forest types (EM vs. non-EM) in the tropical ecosystem (B).

families that were among the 10 most abundant ECM families in both litter and soil horizons collected from the plots (in terms of sequence abundance), so these taxa were used to compare ECM communities across the boreal and tropical ECM forests (the non-EM tropical forest was excluded, as very few ECM taxa were detected). The relative abundance of these six predominantly ECM families was not significantly different across boreal and tropical forests with the exception of the Clavulinaceae, which had higher abundance in the tropical ECM forest (F $(1,8) = 237.0$; $P<0.001$; Fig. 4). Of these six ECM families, there were significantly more ECM taxa detected in soil compared to litter for all families except for the Clavulinaceae, which had a high relative abundance in the boreal litter.

Clone library sequencing generated a total of 119 unique OTUs out of 329 analyzed sequences; 56 of these OTUs were detected

from the boreal forest and 63 were detected from the tropical rain forest. From the original 384 sequences, a total of 55 sequences were discarded based on poor sequencing quality and chimera formation (5 chimeras detected). In both ecosystems there were more Basidiomycota than Ascomycota in the O_a (top mineral) horizons, but approximately equal representation of Ascomycota and Basidiomycota in the recently shed leaf litter. We found a total of 44 Ascomycota (22 in the boreal forest, 22 in the tropical forest), 73 Basidiomycota (34 in the boreal forest, 39 in the tropical forest) and 2 unknown taxa, both detected in the tropical forest.

Based on alignments unknown clone library sequences in GenBank using the BLASTn algorithm [29], the dominant fungal taxa were found to be distinct in each organic soil horizon. In the boreal forest, the fungal community from the upper soil horizons (0–20 cm) was dominated by the Agaricales, and particularly, the

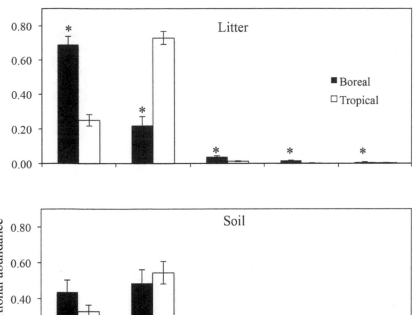

Figure 2. Proportional abundances of fungal taxa from pyrosequencing data assigned to each phylum across sites and horizons. Asterisks denote significance between boreal and tropical forests at $P<0.05$.

ECM genus *Cortinarius*. The next most abundant identified order in the boreal soil was the Helotiales, from sequences closely matched to pathogenic fungi. This order was also the second most abundant order of fungi in the boreal litter bag samples. The most abundant order on the boreal litter was the Microbotryomycetes (incertae sedis), specifically *Zymoxenogloea* sp., of which little ecological information is known. The Agaricales were the most common order of fungi found in both the soil and litter bag samples in the tropical rain forest.

Since ecological function could only be assigned to exclusively ECM families in the pyrosequencing data, we used the information from the clone library sequence identifications to determine how the dominant ECM and non-EM fungi were distributed. We were able to assign an ecological function (EM fungus, saprotroph, pathogen, etc.) to 96 of the 119 unique OTUs from clone library sequencing, as inferred from the GenBank sequence alignments. We found that saprotrophs occurred in recently-shed leaf litter and ECM fungi in underlying soil horizons (0–20 cm) in both the boreal and tropical ecosystem (Fig. 5).

Discussion

While numerous sporocarp surveys have been done in tropical forests [e.g., 45,46,47], our study provides some of the first molecular evidence that confirms biogeographical separation of fungal communities across a tropical and boreal forest, despite the occurrence of dominant trees that form ectomycorrhizae in both ecosystems. As has been found in other tropical ECM forests, the

major ECM fungal lineages reflect those already known to dominate temperate and boreal ecosystems [48,49]. Additionally, fungal communities were unique across soil and litter horizons within the same ecosystem, possibly due to fungal specialization on substrates in differing levels of decay [50]. Clone library sequencing and pyrosequencing showed analogous results in both ecosystems indicating that these patterns are robust to sequencing technology and gene region targeted, which has been a major concern among microbial ecologists [51]. While the clone library sequencing gave more reliable taxonomic information for the environmental DNA sequences (i.e., longer sequence reads), the OTUs generated from pyrosequencing aligned to similar taxa, probably as a result of incomplete coverage of fungal reference sequences in Genbank. However, because pyrosequencing allows for greater sequencing depth (for this study samples were rarified to 1000 sequences each), we can more reliably say that we have fully characterized the fungal community of a sample. Thus, the tandem use of these technologies provides strong support for our results in terms of fungal community characterization and taxonomic placement of environmental sequences. Another result supported by both pyrosequencing and clone library sequencing was that within the tropical and boreal ECM forests, ECM fungi were not prevalent in litter horizons from the forest floor, but rather occupied lower organic and mineral soil horizons.

Findings that ECM fungi were more abundant in deeper soil depths have also been observed in temperate [9] and Swedish boreal forests [11], indicating that vertical segregation of ECM

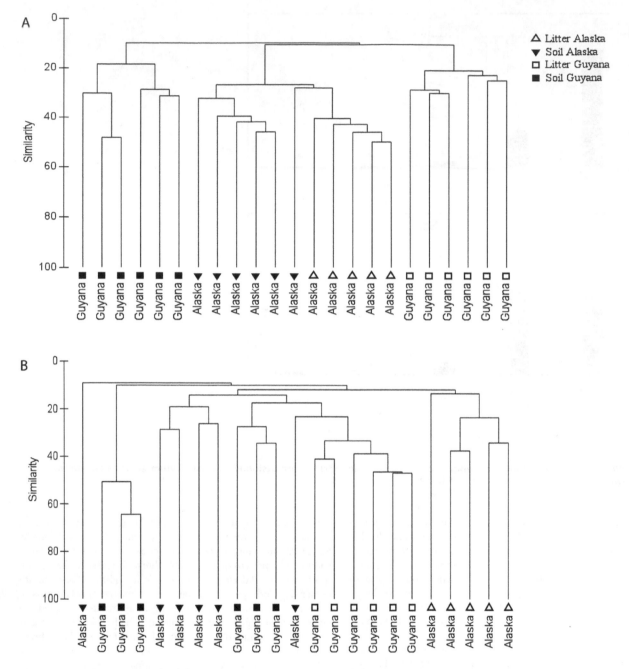

Figure 3. Dendrograms derived from cluster analyses are shown for pyrosequencing OTUs identified as Ascomycota (A) and Basidiomycota (B). Similar clustering patterns by site (Guyana versus Alaska) and horizon was observed for fungal communities in both phyla.

and saprotrophic fungi in soils may be a widespread phenomenon in ECM-dominated forest ecosystems. The reasons for spatial segregation of these fungal groups are likely due to the distribution of C and nutrients in litter versus soil. Since ECM fungi have decomposer abilities [52,53], but are not C-limited like other saprotrophs due to their access to plant photosynthate, ECM fungi may reside below the freshly-fallen litter layer in deeper horizons to target substrates richer in other nutrients [54,55]. An alternative, but not mutually exclusive explanation for the predominance of ECM fungi in the soil horizons may be due to antagonistic relationships between ECM and saprotrophic fungi [56–58]. Since these fungal groups compete for some of the same

resources, they may vertically segregate to avoid competitive exclusion [10].

Within the tropical ecosystem, pyrosequencing showed that soil fungal communities were distinct between the ECM and the diverse, non-ECM forests, indicating that at a local scale, the presence of an ECM tree can dramatically alter the general fungal community. The magnitude of differentiation in soil fungal communities across these tropical forests was almost as dramatic as the differentiation observed across biomes, and previous research in this site has shown that soil physicochemical properties are not responsible for determining these patterns [18]. Fungi detected in forest floor litter were also clustered by forest type in

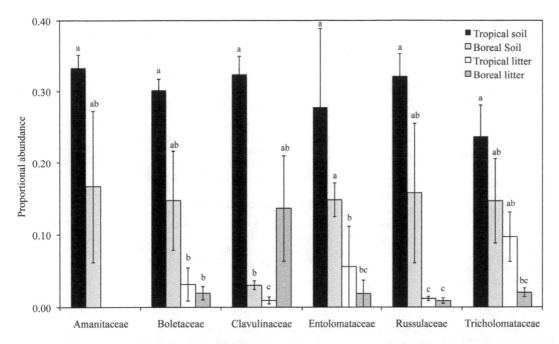

Figure 4. Proportional abundances of sequences derived from 454 pyrosequencing (calculated on a per-sample basis using total sequences as the denominator) of predominantly ectomycorrhizal fungal families found to be abundant in the boreal and tropical forests. Different letters indicate significance levels at $P<0.05$.

the tropical system, although the magnitude of difference was much less. This result may be due to the fact that the same species of non-ECM trees are present in both tropical forests [16,21], so the chemical composition of the leaf litter is somewhat similar [18]. However, there is an overwhelming abundance of litter in the ECM forest from the ECM, monodominant tree (*Dicymbe corymbosa*), which may explain the differences in the litter fungal communities across these forests. In another study, a reciprocal litter decomposition experiment has shown that leaf litter of *Dicymbe* and non-ECM trees decomposes slower in the ECM forest relative to the non-ECM forest [18], indicating that these differences in fungal communities may result in altered nutrient cycling.

In the boreal biome, the results of this study suggest that related fungal taxa may dominate the organic layers in boreal forest soils across different systems. For example, the genus *Cortinarius* was the most abundant in our boreal soil samples, and this genus also dominates Swedish boreal forest soil [11]. In an earlier study, Allison et al. [59] also found that the ECM genus *Cortinarius* was the dominant taxon from our Alaskan study site. Future work focusing on the function of *Cortinarius* in decomposition would be valuable, as it is a globally distributed genus and known to occupy litter at late stages of decomposition [60]. However, other than protease ability [61], its complete enzymatic capabilities are still unknown. We also found *Cortinarius* taxa in the tropical samples,

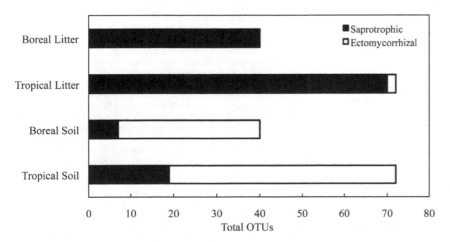

Figure 5. Distribution of clone library OTUs aligning with saprotrophic (black bars) and ectomycorrhizal fungi (white bars) across organic litter and underlying soil horizons (0–20 cm) in the boreal and tropical forest ecosystems. OTUs for this analysis were derived from clone library sequencing.

although our sequence analysis indicated that they are different genotypes than the boreal taxa.

Some of the ECM genera we detected from the Agaricales in the tropical samples are known to associate with the dominant ECM tree, *Dicymbe corymbosa*, as they have been described by mycologists working in that region [62,63]. However, some of the sequences we generated are likely undescribed taxa. This is probably true for the numerous *Clavulina* species we observed from the Cantharellales in the tropical soil, which was the second most abundant order. *Clavulina* diversity is known to be high in this region [64], which reflects what we detected in our environmental pyrosequencing data.

The finding that fungal communities are distinct in litter horizons also has implications for environmental sampling of fungal communities. For a comprehensive understanding of microbial community composition, sampling should incorporate both the organic and underlying soil horizons. In addition, environmental changes that affect one soil layer more than another may have disproportionate consequences for the two fungal groups. For instance, forest fires primarily burn the upper soil horizons (depending on severity), so direct effects of fire may be stronger on saprotrophic fungi than on ECM fungi. Making inferences about fungal communities from only mineral samples may, therefore, underestimate diversity and provide an incomplete picture of community composition.

Acknowledgments

We thank Jennifer Martiny, Ivan Edwards, and Terry Henkel for valuable intellectual contributions to this work. We are also grateful to the Patamona Amerindian tribe in Guyana for assisting with field work, Margaret and Malcolm Chana-Sue and Raquel Thomas for logistical support, and the Guyana EPA for granting permits.

Author Contributions

Conceived and designed the experiments: KLM SDA KKT. Performed the experiments: KLM SDA KKT. Analyzed the data: KLM SDA NF KKT. Contributed reagents/materials/analysis tools: KLM SDA NF KKT. Wrote the paper: KLM SDA NF KKT.

References

1. Cairney JWG, Meharg AA (2002) Interactions between ectomycorrhizal fungi and soil saprotrophs: implications for decomposition of organic matter in soils and degradation of organic pollutants in the rhizosphere. Canadian Journal of Botany-Revue Canadienne De Botanique 80: 803–809.

2. Brundrett MC (2009) Mycorrhizal associations and other means of nutrition of vascular plants: understanding the global diversity of host plants by resolving conflicting information and developing reliable means of diagnosis. Plant and Soil 320: 37–77.

3. Alexander IJ, Hogberg P (1986) Ectomycorrhizas of tropical angiospermous trees. New Phytologist 102: 541–549.

4. Connell JH, Lowman MD (1989) Low-diversity tropical rain forests: some possible mechanisms for their existence. American Naturalist 134: 88–119.

5. Read DJ (1991) Mycorrhizas in ecosystems. Experientia 47: 376–391.

6. Torti SD, Coley PD, Kursar TA (2001) Causes and consequences of monodominance in tropical lowland forests. The American Naturalist 157: 141–153.

7. Hobbie SE, Reich PB, Oleksyn J, Ogdahl M, Zytkowiak R, et al. (2006) Tree species effects on decomposition and forest floor dynamics in a common garden. Ecology 87: 2288–2297.

8. Bending GD, Aspray TJ, Whipps JM (2006) Significance of microbial interactions in the mycorrhizosphere. Advances in Applied Microbiology, Vol 60. 97–132.

9. O'Brien HE, Parrent JL, Jackson JA, Moncalvo JM, Vilgalys R (2005) Fungal community analysis by large-scale sequencing of environmental samples. Applied and Environmental Microbiology 71: 5544–5550.

10. Dickie IA, Xu B, Koide RT (2002) Vertical niche differentiation of ectomycorrhizal hyphae in soil as shown by T-RFLP analysis. New Phytologist 156: 527–535.

11. Lindahl BD, Ihrmark K, Boberg J, Trumbore SE, Hogberg P, et al. (2007) Spatial separation of litter decomposition and mycorrhizal nitrogen uptake in a boreal forest. New Phytologist 173: 611–620.

12. Tedersoo L, Bahram M, Toots M, Diédhiou AG, Henkel TW, et al. (2012) Towards global patterns in the diversity and community structure of ectomycorrhizal fungi. Molecular Ecology 21: 4160–4170.

13. Kennedy PG, Matheny PB, Ryberg KM, Henkel TW, Uehling JK, et al. (2012) Scaling up: examining the macroecology of ectomycorrhizal fungi. Molecular Ecology 21: 4151–4154.

14. Sala OE, Chapin FS, Armesto JJ, Berlow E, Bloomfield J, et al. (2000) Biodiversity - Global biodiversity scenarios for the year 2100. Science 287: 1770–1774.

15. Zimov SA, Schuur EAG, Chapin FS (2006) Permafrost and the global carbon budget. Science 312: 1612–1613.

16. Henkel TW (2003) Monodominance in the ectomycorrhizal Dicymbe corymbosa (Caesalpiniaceae) from Guyana. Journal of Tropical Ecology 19: 417–437.

17. McGuire KL, Henkel TW, Granzow de la Cerda I, Villa G, Edmund F, et al. (2008) Dual mycorrhizal colonization of forest-dominating tropical trees and the mycorrhizal status of non-dominant tree and liana species. Mycorrhiza 18: 217–222.

18. McGuire KL, Zak DR, Edwards IP, Blackwood CB, Upchurch R (2010) Slowed decomposition is biotically mediated in an ectomycorrhizal, tropical rain forest. Oecologia 164: 785–795.

19. Treseder KK, Mack MC, Cross A (2004) Relationships among fires, fungi, and soil dynamics in Alaskan Boreal Forests. Ecological Applications 14: 1826–1838.

20. Mack MC, Treseder KK, Manies KL, Harden JW, Schuur EAG, et al. (2008) Recovery of aboveground plant biomass and productivity after fire in mesic and dry black spruce forests of interior alaska. Ecosystems 11: 209–225.

21. McGuire KL (2008) Ectomycorrhizal Associations Function to Maintain Tropical Monodominance. In: Siddiqui ZA, Akhtar MS, Futai K, editors. Mycorrhizae: Sustainable Agriculture and Forestry. Netherlands: Springer. 287–302.

22. Allison SD, Hanson CA, Treseder KK (2007) Nitrogen fertilization reduces diversity and alters community structure of active fungi in boreal ecosystems. Soil Biology & Biochemistry 39: 1878–1887.

23. Feinstein LM, Sul WJ, Blackwood CB (2009) Assessment of Bias Associated with Incomplete Extraction of Microbial DNA from Soil. Applied and Environmental Microbiology 75: 5428–5433.

24. Rousk J, Baath E, Brookes PC, Lauber CL, Lozupone C, et al. (2010) Soil bacterial and fungal communities across a pH gradient in an arable soil. Isme Journal 4: 1340–1351.

25. Fierer N, Hamady M, Lauber CL, Knight R (2008) The influence of sex, handedness, and washing on the diversity of hand surface bacteria. Proceedings of the National Academy of Sciences of the United States of America 105: 17994–17999.

26. Lauber CL, Hamady M, Knight R, Fierer N (2009) Pyrosequencing-Based Assessment of Soil pH as a Predictor of Soil Bacterial Community Structure at the Continental Scale. Applied and Environmental Microbiology 75: 5111–5120.

27. Caporaso JG, Kuczynski J, Stombaugh J, Bittinger K, Bushman FD, et al. (2010) QIIME allows analysis of high-throughput community sequencing data. Nature Methods 7: 335–336.

28. Price MN, Dehal PS, Arkin AP (2009) FastTree: Computing Large Minimum Evolution Trees with Profiles instead of a Distance Matrix. Molecular Biology and Evolution 26: 1641–1650.

29. Altschul SF, Madden TL, Schaffer AA, Zhang JH, Zhang Z, et al. (1997) Gapped BLAST and PSI-BLAST: a new generation of protein database search programs. Nucleic Acids Research 25: 3389–3402.

30. Pruesse E, Quast C, Knittel K, Fuchs BM, Ludwig WG, et al. (2007) SILVA: a comprehensive online resource for quality checked and aligned ribosomal RNA sequence data compatible with ARB. Nucleic Acids Research 35: 7188–7196.

31. Gardes M, Bruns TD (1993) ITS primers with enhanced specificity for basidiomycetes - application to the identification of mycorrhizae and rusts. Molecular Ecology 2: 113–118.

32. Taylor DL, Bruns TD (1999) Community structure of ectomycorrhizal fungi in a Pinus muricata forest: minimal overlap between the mature forest and resistant propagule communities. Molecular Ecology 8: 1837–1850.

33. Nilsson RH, Kristiansson E, Ryberg M, Hallenberg N, Larsson KH (2008) Intraspecific ITS variability in the kingdom fungi as expressed in the international sequence databases and its implications for molecular species identification. 4: 193–201.

34. Suzuki MT, Giovannoni SJ (1996) Bias caused by template annealing in the amplification of mixtures of 16S rRNA genes by PCR. Applied Microbiology and Biotechnology 80: 99–106.

35. Polz MF, Cavanaugh CM (1998) Bias in template-to-product ratios in multitemplate PCR. Applied and environmental Microbiology 64: 3624–3730.

36. Thompson JD, Higgins DG, Gibson TJ (1994) Clustal-W - Improving the sensitivity of progressive multiple sequence alignment through sequence

weighting, position-specific gap penalties and weight matrix choice. Nucleic Acids Research 22: 4673–4680.

37. Felsenstein J (2005) PHYLIP (Phylogeny Inference Package). 3.6 ed. Seattle: Department of Genome Sciences, University of Washington.

38. Schloss PD, Handelsman J (2005) Introducing DOTUR, a computer program for defining operational taxonomic units and estimating species richness. Applied and Environmental Microbiology 71: 1501–1506.

39. Kõljalg U, Larsson KH, Abarenkov K, Nilsson RH, Alexander IJ, et al. (2005) UNITE: a database providing web-based methods for the molecular identification of ectomycorrhizal fungi. 166: 1063–1068.

40. Wolfe BE, Kuo M, Pringle A (2012) Amanita thiersii is a saprotrophic fungus expanding its range in the United States. Mycologia 104: 22–33.

41. Miller J, Orson K (1983) Ectomycorrhizae in the Agaricales and Gasteromycetes. Canadian Journal of Botany 61: 909–916.

42. Gihring TM, Green SJ, Schadt CW (2012) Massively parallel rRNA gene sequencing exacerbates the potential for biased community diversity comparisons due to variable library sizes. Environmental microbiology 14: 285–290.

43. Clarke KR, Warwick RM (2001) Change in marine communities: an approach to statistical analysis and interpretation. Plymouth, UK: Primer-E.

44. Hibbett DS, Ohman A, Glotzer D, Nuhn M, Kirk P, et al. (2011) Progress in molecular and morphological taxon discovery in Fungi and options for formal classification of environmental sequences. Fungal Biology Reviews 25: 38–47.

45. Watling R (2001) The relationships and possible distributional patterns of boletes in South-East Asia. Mycological Research 105: 1440–1448.

46. Watling R, Taylor A, See LS, Sims K, Alexander I (1995) A rain forest *Pisolithus* - its taxonomy and ecology. Nova Hedwigia 61: 417–429.

47. Henkel TW, Aime MC, Chin MML, Miller SL, Vilgalys R, et al. (2011) Ectomycorrhizal fungal sporocarp diversity and discovery of new taxa in Dicymbe monodominant forests of the Guiana Shield. Biodiversity and Conservation 21: 2195–2220.

48. Tedersoo L, May TW, Smith ME (2009) Ectomycorrhizal lifestyle in fungi: global diversity, distribution, and evolution of phylogenetic lineages. Mycorrhiza 20: 217–263.

49. Bâ AM, Duponnois R, Moyersoen B, Diédhiou AG (2012) Ectomycorrhizal symbiosis of tropical African trees. Mycorrhiza.

50. McGuire KL, Bent E, Borneman J, Majumder A, Allison SD, et al. (2010) Functional diversity in resource use by fungi. Ecology 91: 2324–2332.

51. Tedersoo L, Nilsson RH, Abarenkov K, Jairus T, Sadam A, et al. (2010) 454 Pyrosequencing and Sanger sequencing of tropical mycorrhizal fungi provide

similar results but reveal substantial methodological biases. New Phytologist 188: 291–301.

52. Read DJ, Perez-Moreno J (2003) Mycorrhizas and nutrient cycling in ecosystems - a journey towards relevance? New Phytologist 157: 475–492.

53. Smith JE, Read D (2008) Mycorrhizal Symbiosis. San Diego (CA): Academic Press. 787 p.

54. Colpaert JV, vanTichelen KK (1996) Decomposition, nitrogen and phosphorus mineralization from beech leaf litter colonized by ectomycorrhizal or litter-decomposing basidiomycetes. New Phytologist 134: 123–132.

55. Osono T, Takeda H (2002) Comparison of litter decomposing ability among diverse fungi in a cool temperate deciduous forest in Japan. Mycologia 94: 421–427.

56. Gadgil RL, Gadgil GD (1971) Mycorrhiza and litter decomposition. Nature 233: 133.

57. Gadgil RL, Gadgil PD (1975) Suppression of litter decomposition by mycorrhizal roots of *Pinus radiata*. New Zealand Journal of Forest Science 5: 35–41.

58. Lindahl B, Stenlid J, Olsson S, Finlay R (1999) Translocation of P-32 between interacting mycelia of a wood-decomposing fungus and ectomycorrhizal fungi in microcosm systems. New Phytologist 144: 183–193.

59. Allison SD, Czimczik CI, Treseder KK (2008) Microbial activity and soil respiration under nitrogen addition in Alaskan boreal forest. Global Change Biology 14: 1156–1168.

60. Visser S (1995) Ectomycorrhizal Fungal Succession in Jack Pine Stands Following Wildfire. New Phytologist 129: 389–401.

61. Lilleskov EA, Hobbie EA, Fahey TJ (2002) Ectomycorrhizal fungal taxa differing in response to nitrogen deposition also differ in pure culture organic nitrogen use and natural abundance of nitrogen isotopes. New Phytologist 154: 219–231.

62. Matheny PB, Aime MC, Henkel TW (2003) New species of Inocybe from Dicymbe forests of Guyana. Mycological Research 107: 495–505.

63. Henkel TW, Terborgh J, Vilgalys RJ (2002) Ectomycorrhizal fungi and their leguminous hosts in the Pakaraima Mountains of Guyana. Mycological Research 106: 515–531.

64. Smith ME, Henkel TW, Aime MC, Fremier AK, Vilgalys R (2011) Ectomycorrhizal fungal diversity and community structure on three co-occurring leguminous canopy tree species in a Neotropical rainforest. New Phytologist 192: 699–712.

5

Community Turnover of Wood-Inhabiting Fungi across Hierarchical Spatial Scales

Nerea Abrego[1,2]*, Gonzalo García-Baquero[2], Panu Halme[1,3], Otso Ovaskainen[4], Isabel Salcedo[2]

1 Department of Biological and Environmental Science, University of Jyväskylä, Jyväskylä, Finland, 2 Department of Plant Biology and Ecology, University of the Basque Country (UPV/EHU), Bilbao, Spain, 3 Natural History Museum, University of Jyväskylä, Jyväskylä, Finland, 4 Department of Biosciences, University of Helsinki, Helsinki, Finland

Abstract

For efficient use of conservation resources it is important to determine how species diversity changes across spatial scales. In many poorly known species groups little is known about at which spatial scales the conservation efforts should be focused. Here we examined how the community turnover of wood-inhabiting fungi is realised at three hierarchical levels, and how much of community variation is explained by variation in resource composition and spatial proximity. The hierarchical study design consisted of management type (fixed factor), forest site (random factor, nested within management type) and study plots (randomly placed plots within each study site). To examine how species richness varied across the three hierarchical scales, randomized species accumulation curves and additive partitioning of species richness were applied. To analyse variation in wood-inhabiting species and dead wood composition at each scale, linear and Permanova modelling approaches were used. Wood-inhabiting fungal communities were dominated by rare and infrequent species. The similarity of fungal communities was higher within sites and within management categories than among sites or between the two management categories, and it decreased with increasing distance among the sampling plots and with decreasing similarity of dead wood resources. However, only a small part of community variation could be explained by these factors. The species present in managed forests were in a large extent a subset of those species present in natural forests. Our results suggest that in particular the protection of rare species requires a large total area. As managed forests have only little additional value complementing the diversity of natural forests, the conservation of natural forests is the key to ecologically effective conservation. As the dissimilarity of fungal communities increases with distance, the conserved natural forest sites should be broadly distributed in space, yet the individual conserved areas should be large enough to ensure local persistence.

Editor: Rick Edward Paul, Institut Pasteur, France

Funding: This study was partially funded by a PhD student fellowship to NA by the University of the Basque Country (UPV/EHU) (PIF10/2010/PIF10008), by Maj and Tor Nessling Foundation (a grant to PH), and by the Academy of Finland (grant no. 250444 to OO). The funders had no role in study design, data collection and analysis, decision to publish, or preparation of the manuscript.

Competing Interests: The authors have declared that no competing interests exist.

* Email: nerea.abrego@ehu.es

Introduction

Biodiversity is decreasing faster than ever in the history of the earth, largely as a result of human activity [1]. Consequently, conservation and management actions aiming at sustainable use of ecosystems and maintenance of biological diversity have become a priority worldwide. The most commonly applied conservation measures in terrestrial ecosystems have focused on preserving key habitats, species or communities of conservation interest. To date, most conservation strategies have focused on designating protected set aside areas or increasing the areas of existing protected areas. For example, regarding temperate forests, a considerable proportion of the remaining old-growth and semi-natural forests are already conserved, and the protected area is expected to continue increasing [2]. To make conservation efforts cost-effective in an evidence based manner, protection efforts need to be supported by research about the effects of habitat protection [2]. One factor potentially compromising the cost-effectiveness of biodiversity conservation plans is that the patterns and drivers of distribution and turnover of biodiversity across different spatial scales are still poorly understood for many organism groups [3].

A successful design of conservation area networks requires an understanding of the processes determining species composition and persistence at different spatial and temporal scales [4]. For example, if local processes are the most important factor determining the persistence and diversity of the focal species group (as has been observed for several taxa; [5–7]), it may be more effective to aim management actions or site selection procedures towards maintaining the heterogeneity and variability of habitats at the local scale, instead of e.g. increasing the connectivity among protected areas [5]. The influences of local versus regional processes in community assembly often interact with the traits of the species. For example, in a study of forest-dwelling beetles, Gering et al. [8] found that the spatial turnover of rare species was realised across large spatial scales (ecoregions), whereas the spatial turnover of common species was realised across smaller spatial scales. The most commonly used concept in measuring spatial species turnover is beta-diversity. However, this concept has been defined and used in multiple ways, making it difficult to synthesize studies based on this concept, and leading to a debate on its correct use [9,10].

Wood-inhabiting fungi are a diverse and functionally important group of organisms in forest ecosystems. In addition to their central role in nutrient cycling [11], they play a key role on the creation of new habitats and resources for other wood-dwelling species [12]. Furthermore, their dependence on dead wood makes them especially vulnerable to the changes caused by forest management (e.g. [13,14]). At the resource unit level, most species show preferences to a specific substrate type in terms of host species, log type (snag, log, stump…), decay stage and log size [15]. Therefore, the reduction of particular dead wood types (especially of coarse woody debris) in managed forests has directly led to declines and extinctions of certain species [16,17]. At the stand level, fungal communities are highly influenced by microclimatic conditions, which modify e.g. the water content of dead wood and thus can inhibit or facilitate the colonization and fructification of species [18,19]. Further, the size and isolation of forest stands can significantly influence fungal communities: specialist species have been found to be negatively affected by isolation while generalist species can even benefit from fragmented surroundings [15]. At the broadest scale, wood-inhabiting fungi are influenced by regional factors, related e.g. to climatic variation and management history [20].

Only a few studies have examined the importance of environmental factors at different spatial scales in determining the diversity of wood-inhabiting fungi [21,22]. Berglund et al. [23] applied a hierarchical study design and a hierarchical data analysis to show that habitat suitability determines the presence of polypore species at a small spatial scale. Bässler et al. [24] quantified the relative importance of the plot and forest management levels, showing that in spruce forests, species turnover is higher among plots than among management categories. Based on these previous studies, wood-inhabiting fungal species diversity is much affected by small scale processes, though also regional-scale processes seem important especially for specialized and threatened species. However, we are not aware of any previous studies in which the variation in species richness and community composition have been explicitly partitioned to different spatial scales. Such information would be critically needed to design effective conservation areas.

The overall objective of this study is to partition the variation in the assembly of wood-inhabiting fungal communities into different spatial levels. More specifically, our aim is to examine how community turnover of wood-inhabiting fungi is realised at the levels of forest management types, forest sites and plots, and how much of community variability is explained by variation in resource composition and spatial proximity. We utilize the conclusions of the present study to derive recommendations for the design of protection area networks for the conservation of wood-inhabiting fungi in temperate broadleaved forests.

Materials and Methods

Study permits were provided by the Government of Navarre (the local official institution for forest management) and the field studies did not involve endangered or protected species.

1. Study area

The study was carried out in the northern part of Navarre in northern Spain. The study area covers two biogeographical regions with temperate climate: the Atlantic region in the Northwest and the Alpine region in the Northeast. The predominant forest type is beech-dominated forest (with total area of ca. 61,000 ha, covering ca. 14% of the landscape), which also represents the most productive forest type for commercial harvesting in the area.

The northern part of Navarre consists of a valley-dominated landscape with a mountainous topography. Towns and farms are generally located in the lower parts of the valleys and the forests grow in the hillsides. The major part of the forested area has been intensively managed by thinning and selective cutting, but there is also a network of permanently protected areas (see Files S1 and S2). In this study, we examined the difference between two management types, called henceforth managed and natural forests. As natural forests we considered such protected areas in which logging activities have been prohibited, allowing the development of old-growth beech forests. As managed forests we considered such unprotected sites in which commercial harvesting has been continuously conducted (File S1).

2. Study design

2.1. Sampling design. We assembled a list of natural and managed forest sites based on the above definitions. From this list, eight managed and eight natural forest sites (File S1) were randomly selected. Within each site, fungal data (species abundances, defined as number of logs in which each species was observed) were recorded in five randomly placed 100 m^2 sampling plots (File S2). The sampling plots were separated from each other by tens of meters to hundreds of meters, whereas the study sites were separated by hundreds of meters to kilometres (the minimum, maximum and mean distances between plots were 0.07, 114 and 44 km respectively). The sites belonging to the two management types were distributed in space without a strong spatial autocorrelation, i.e. the managed and natural sites did not form their own clusters. The nested design has thus two factors: management (considered as fixed factor; with two levels) and forest site (considered as random factor; with eight levels, nested within management). Variation among study sites represents spatial variation at a regional scale, whereas the residual variation among the sampling plots represents spatial variation at a local scale.

2.2. Data collection. The fieldwork was carried out in the year 2011 during the main period of fungal fruit body production (from late September to early November). All woody debris (called henceforth resource units) with minimum diameter of 1 cm were measured for diameter and decay class, and examined for the presence of fungal species. The resource units were classified according to their diameter (three classes) and decay stage (five classes), thus forming altogether 15 classes. The diameter classes correspond to those used by Abrego & Salcedo [17]: very fine woody debris (VFWD), including resource units with diameter in the range 1–5 cm; fine woody debris (FWD), with 5 cm<diameter ≤10 cm; and coarse woody debris (CWD), with diameter >10 cm. The decay stages were classified into 5 categories based on a modified version of the classification by Renvall [25].

The presence of all saproxylic macromycetes (fungi with visible fruit bodies) was registered at the level of the resource units. Microscopic identification was carried out in the laboratory when necessary. The literature consulted for identifying fungal fruit bodies is summarized in File S3. Each species found on one resource unit was considered as one record, regardless of the number of fruit bodies. Thus, the abundances of the species correspond to the number of woody debris pieces in which each species was found in each sampling plot.

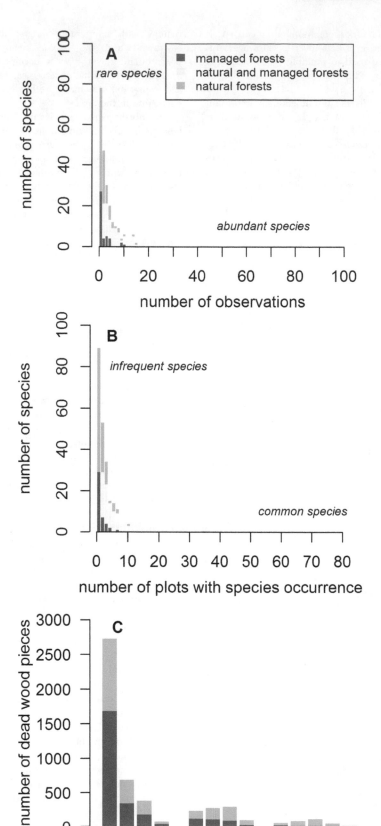

Figure 1. The abundance distributions of species and dead wood types. The total number of occurrences of each species (Panel A), the number of plots in which each species was recorded (Panel B) and the number of occurrences of each dead wood type (Panel C). The species and dead wood pieces that were recorded in managed forests are represented with blue colour, those that were recorded in natural forests with yellow colour, and those species recorded in both forest types with green colour.

3. Statistical analyses

We first conducted a descriptive analysis examining how species richness in fungal communities varies across the three hierarchical scales included in this study. To do so, we constructed accumulation curves in which the total number of species was computed as a function of the number of surveyed plots, assuming that the plots were selected randomly either a) within the same randomly selected managed forest site, b) within the same randomly selected natural forest site, c) among all sites in managed forests, d) among all sites in natural forests, or e) among all sites. We estimated the mean number of species and the interquartile (25%–75%) range for each of these five cases and for each number of sampling plots using 1000 random replicates.

We supplemented the accumulation curves by an additive partitioning in species richness at different hierarchical scales [26,27]. In this methodology, the total diversity γ (here, species

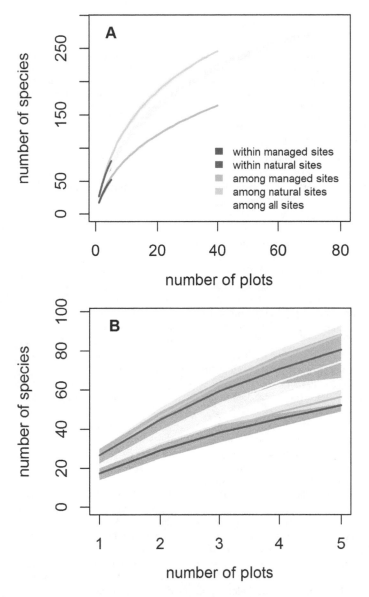

Figure 2. Species accumulation curves showing the total number of species in a given number of plots. The thick lines show means and the shaded regions the interquartile (25%–75%) ranges. The different lines correspond to the five cases which assume that the plots were selected randomly either within the same randomly selected managed forest site (red), the same randomly selected natural forest site (dark blue), among all sites in managed forests (green), among all sites in natural forests (light blue), or among all sites (yellow). Note that the maximum number of plots for which the number of species could be computed is either 5, 40 or 80, depending on the category. Panel A shows the full data, panel B a zoom, in which the number of plots varies from 1 to 5.

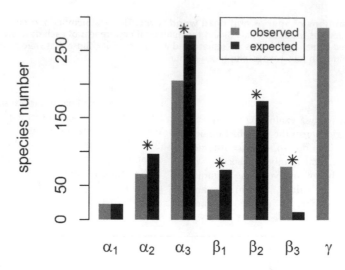

Figure 3. Total species richness (γ) explained by alpha and beta components of diversity at three spatial scales: the mean species number per plot (α₁), the mean species number per site (α₂), the mean species number per management category (α₃), the average addition of species from plots to sites (β₁), the average addition of species from sites to management categories (β₂), the average addition of species from management categories to the total species number (β₃) and the total species number (γ). The grey bars correspond to the observed values and the black bars to the expected values based on a null model. The asterisks indicate the levels of statistical significance (* corresponding to $p<0.001$) for a difference between observed and expected values.

richness) is modelled as a sum of the average diversities within sampling units (α) and among sampling units (β), the latter across the scales included in the hierarchical design. In our case, the partitioning of species richness γ was thus expressed as $\gamma = \alpha_{1(plots)} + \beta_{1(plots)} + \beta_{2(sites)} + \beta_{3(management)}$, where α₁ is the mean species number in the sampling plots, β₁ is the turnover (beta-diversity) between sampling plots, β₂ is the turnover between forest sites, and β₃ is the turnover between managed and natural forests. The additive partitioning and the related significance tests were performed under 999 permutations of the null model r0 of the function *adipart* in the R-package *vegan* [28]. The null model r0 is suited to binary (presence/absence) data, and it randomizes the community matrix in a way that preserves the number of species observed in each site.

We then analysed variation in community composition at the different levels of our hierarchical study design by calculating Bray-Curtis similarity measures [29] for all pairs of sampling plots. We constructed similarity matrices both for the fungal data (square root transformed species abundances, organised as a matrix for 80 sampling plots by 285 species) and for the dead wood data (square root transformed counts of resource unit types, organised as a matrix of 80 sampling plots by 15 resource unit types). We examined how much of the similarity in dead wood composition (SDC) was explained by the two hierarchical factors of management and site, as well as spatial distance (D). We examined how much of the similarity in fungal community composition (SFC) was explained by management and site, as well as by similarity in dead wood composition (SDC) and spatial distance (D). To do so, we constructed matrices measuring similarity in site (SS; with value 1 if the plots belonged to the same site and 0 otherwise), similarity in management (SM; with value 0 if the plots belonged to a different management category, and values 1 or 2 if both plots belonged to the managed or to the natural category, respectively) and log-transformed spatial distance (D) between the sampling plots. We treated SS and SM as fixed factors and SDC and D as continuous covariates. In the models (SDC = SS+SM+D and SFC = SS+SM+SDC+D) we considered also all two-way interactions (except for D). We used backward variable selection to include in the final

model only the statistically significant variables. To account for the non-independence of data points that relate to pairs of plots rather than to individual plots, we tested for the statistical significance by comparing the *F-values* related to each factor to the null distribution of *F-values*. To compute the null distribution, we constructed 999 permutations of the species community, in each of which we permuted randomly the identity numbers of the sampling plots. As a complementary approach to quantify the variation on species and resource similarity explained by each scale, we carried out a permutational multivariate analysis of variance (PERMANOVA; [30]). The *pseudo-F* values were used to evaluate which components of the models explained more variability than expected by a null model. This analysis was performed using 999 permutations of residuals under a reduced model with the function *nested.npmanova* of the R-package *BiodiversityR*.

All statistical analyses were performed using R software [31], the code and the original data (the fungal species data matrix, UTM locations of all sampling plots and resource type data matrix) being available in the Supporting Information section (Files S4–S7).

Results

1. Variation in species richness and resource types

We recorded in total 3241 occurrences of 285 species in 2724 dead wood pieces. The species abundance distribution was dominated by rare species (Fig. 1A), with 27% of the recorded species (2.4% of the occurrences) being represented only once in our data. The five most abundant species were simultaneously the five species with highest prevalence at the plot level, comprising over 25% of all occurrences. In general, the species with high abundances were found in both (managed and natural) forest types (Fig. 1, A and B; black bars). Nevertheless, note that in the tail of Fig. 1 (A) there are also some species that were found only in managed forests. These species are *Steccherinum fimbriatum* and *Hyphodermella rosae*, which were recorded in 7 managed plots, 18 and 11 times respectively.

Table 1. Linear models for similarity in fungal and dead wood composition.

Response variable: Similarity in fungal composition (SFC)						
Source of variation	df	Estimate	SSq	MS	F	p
Management (SM)	2	MM: 0 MU: −0.060 UU: −0.047	1.93	0.96	152	<0.001
Site (SS)	1	Different: 0 Same: 0.019	0.71	0.71	112	<0.001
Similarity in dead wood composition (SDC)	1	0.11	1.11	1.11	175	<0.001
Spatial distance (D)	1	−0.02	0.46	0.46	73	<0.001
Residual	3154		20.02	0.01		
Response variable: Similarity in dead wood composition (SDC)						
Source of variation	df	Estimate	SSq	MS	F	p
Management (SM)	2	MM: 0 MU: −0.067 UU: −0.016	3.07	1.54	192	<0.001
Site (SS)	1	Different: 0 Same: 0.019	0.05	0.05	6.58	<0.001
Residual	3156		25.26	0.01		

df = degrees of freedom, SSq = sum of squares, MS = mean squares, MM = pairs of plots from managed forests, MU = pairs of plots belonging to a different management types, UU = pairs of plots from natural (unmanaged) forests. Statistical significance (p) based on permutation test.

The dead wood profile was numerically strongly dominated by small diameter classes in early decay stages (Fig. 1C). The coarser dead wood classes in the latest decay stages were the most uncommon, and were mostly found in natural forests.

The species accumulation curves indicate that natural forests were substantially more species rich than managed forests (Fig. 2A). Further, the number of species increased much faster if the sampling plots were located only in natural forests (Fig. 2A; light blue line) than if they were located both in natural and managed forests (Fig. 2A; yellow line), suggesting that the species assemblages in managed forests were largely a subset of those present in natural forests. The total number of species increased only marginally faster if the sampling plots were spread across forest sites than if they were located within a forest site, both in managed (Fig. 2B; green and red lines) and natural (Fig. 2B; light blue and dark blue lines).

The additive partitioning of total species richness (Fig. 3) was in line with the patterns visible in the species accumulation curves. The increase in species number when moving from the plot level to the forest site level (β_1) was lower than would expected by chance, and consequently so was also the number of species within a site, averaged across the sites (α_2). Similarly, the increase in species number when moving from the site level to the management level (β_2) was lower than would expected by chance, and consequently so was also the mean number of species within a management category, averaged across the two management categories (α_3). Both of these results reflect the fact that the most important part of species diversity was realized among the management categories (β_3). This is consistent with Fig. 2, which shows that natural forests were much more species rich than management forests, and that managed forests largely contained a subset of the species found from natural forests. Note that α_3 corresponds to the mean of the light blue and green lines in Fig. 2, which indeed is below the yellow line, which corresponds to a null expectation.

2. Variation in community and resource similarities

After applying the model selection procedure, the model for similarity in fungal community composition (SFC) included as explanatory variables management type similarity (SM), forest site similarity (SS), similarity in dead wood composition (SDC) and spatial distance (D) (Table 1). The model for similarity in dead wood composition (SDC) included management type similarity (SM) and forest site similarity (SS) but not spatial distance (Table 1). In both models none of the two-way interactions were significant (Permutation comparison of F-values, $p > 0.05$). The variability explained by both models was low, as only the 17% of the total variability was explained in the SFC model and 11% in the SDC model. According to the predicted values of these models, the plots within managed forests showed higher fungal and dead wood similarities than the plots within natural forests, and the lowest fungal and dead wood similarities were found among pairs of plots out which one belonged to a natural and the other one to a managed forest (Fig. 4A). The effect of forest site similarity (SS) on the similarity of both fungal composition (SFC) and dead wood was small but yet significant (Fig. 4, Table 1). Fungal community similarity decreased with increasing spatial distance (Fig. 4B).

The results from the Permanova analysis (Table 2) were in line with the results of the linear model described above: for both fungal and dead wood species composition, both forest management type and forest site level explained a significant amount of variation. For dead wood composition, the variation explained by forest management type was much higher than that explained by forest site level.

Discussion

We found that the assembly of fungal communities was influenced by all scales that we considered in this study. Indeed, the similarity of fungal communities was higher within sites and within management categories than among sites or between the two management categories; it decreased with increasing distance

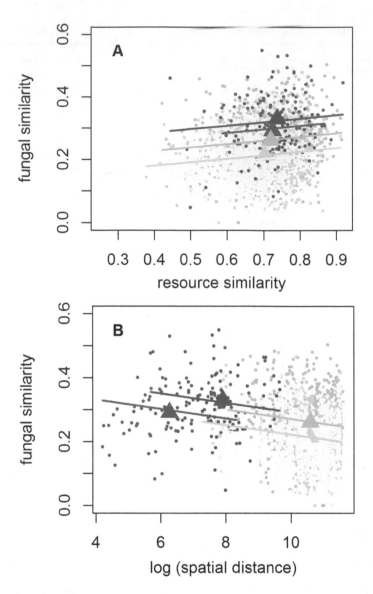

Figure 4. Predicted similarity of the fungal composition (SFC), among sampling plots. When two sampling plots are located either within the same managed forest site are represented by red colour, when located within the same natural forest site are represented by dark blue colour, when located in two different sites in managed forests are represented by green colour, when located in two different sites in natural forests are represented by light blue colour, and when plots are located in sites from different management types are represented by yellow colour. The dots show raw data, the triangles the mean values and the lines the model predictions. When constructing the model prediction as a function of resource similarity (panel A), we normalized spatial distance to its mean value within each category. When constructing the model prediction as a function of spatial distance (panel B), we normalized resource similarity to its mean value within each category.

among the sampling plots and increased with increasing similarity of dead wood resources. However, the majority of the variation remained unexplained, suggesting that biological interactions, neutral mechanisms such as inherent randomness in dispersal, establishment and fruiting [32], as well as environmental constraints not considered here are likely to play an important role in determining local assemblies of wood-inhabiting fungi. This result partly reflects the fact that the wood-inhabiting fungal communities studied here were dominated by locally rare and infrequent species, which directly leads to a high species turnover from one plot to another. This result is in line with a previous study in a coniferous forest where a high variability of wood inhabiting fungal species at small scales was also detected [24].

Bässler et al. [24] partitioned wood-inhabiting fungal richness between plots and management categories, with the result that community turnover was higher between study plots than between management categories. Thus Bässler et al. [24] concluded that the stand level is the most relevant scale for the conservation the wood-inhabiting diversity. In contrast, we found that fungal communities were significantly influenced by all scales that we considered, and in particular by the management category, as natural forests were significantly more diverse than managed forests. The difference between our results and those of Bässler et al. [24] can be at least partially attributed to a difference in the study design, as the study of Bässler et al. [24] involved only a single forest site for each management category.

Table 2. Output from the Permanova analysis for fungal species and dead wood composition models.

Source of variation	df	Response variable: Similarity in fungal composition (SFC)				Response variable: Similarity in dead wood composition (SDC)			
		SS	MS	Pseudo-F	p	SSq	MS	Pseudo-F	p
Management	1	17737	17737	3.53	0.001	8549.2	8549.2	12.45	0.001
Site	14	70343	5024.5	2.08	0.001	9612.1	686.58	1.74	0.002
Residual	64	154670	2416.7			25288	395.13		

df = degrees of freedom, SSq = sum of squares, MS = mean squares.

In general, wood-inhabiting fungal species show preferences for certain resource types in terms of tree species, size and decay class [15], so from a niche differentiation perspective [33], variation in resource type is expected to explain a major part of species variation at a small scale. In our study, although increasing similarity in resource composition (as measured by dead wood decay class and size) significantly increased the similarity of wood-inhabiting fungal communities, the trend was weak. Hence, at a small spatial scale, wood-inhabiting fungal communities seem to be largely influenced by stochastic processes related to species' idiosyncrasy and spatial proximity, rather than habitat factors. Some hierarchically designed studies have reported similar results, and showed that the variation in the occurrence of species among different forest stands [23] or study plots [34] remains largely unexplained even when the most important explanatory variables were considered.

The dispersal capabilities of species can be considered as a species-specific factor limiting potentially their occurrence on suitable habitats [35,36]. One of the covariates that significantly influenced fungal species similarity in our study was the spatial distance, with the result that similarity in fungal composition decreased with increasing distance between the study plots. This result suggests that dispersal limitation is one major factor shaping the local assembly of fungal communities.

Other factors creating unexplained variation between study units are related to the low detectability of wood-inhabiting fungi [37], partly due to their fruiting phenology and its dependence on microclimatic variation [38]. Many fungal species have ephemeral, short living fruit bodies, which makes it difficult to assess fungal diversity by a single fruit body based survey [37]. Further, a large proportion of fungal species can never be detected as fruit bodies [39]. With presence-absence data, low detectability generates a larger amount of random variation on small spatial scales because on larger scale the increasing number of potentially detectable individuals increases the likelihood that at least one is detected [40]. Thus, also in our study, the low general detectability of fungi as well as the climatic dependence of their phenology may have inflated the variation realized at the smallest spatial scale.

Due to the overall high number of species and the large proportion of rare species, the species accumulation curves we constructed were far from saturated. The only way to achieve saturated accumulation curves for hyperdiverse organism groups consists of sampling intensively over a limited area during several years [8,41]. Nevertheless, in spatially extensive studies even unsaturated accumulation curves provide a suitable method for the assessment of the heterogeneity of species diversity [8].

It is well documented that wood-inhabiting fungal communities in natural forests are significantly more species rich and different from those in managed forests (e.g. [16,42,43]), including the communities in beech forests [17]. The main factors behind this difference are related to resource availability and connectivity at forest stand level [15,42]. Our previous study [17] showed that species specialised in utilizing certain dead wood types were absent from managed forests, due to the absence of suitable resource type. Based on the previous [17] and present results, the community differences between different management categories are not related to qualitative differences in species compositions, but to quantitative differences, in the sense that the species from managed forests are a subset from those in natural ones (because their required resource has disappeared). Therefore we cannot say that the natural and managed forests have different species compositions, but rather that natural forests have more species. Indeed, the general trend reported in the literature is that the reduction of particular dead wood pieces caused by forestry practices in managed forests causes extinctions of habitat specialist saproxylic species (e.g. [44–46]). Thus, forest management significantly affects resource composition, and managed forests are more homogeneous in their dead wood and fungal species compositions. Nevertheless we found some species to be more abundant in managed forests than in natural ones, in particular *Steccherinum fimbriatum* and *Hyphodermella rosae*. Consistently with our results, earlier studies have shown that there are species adapted to disturbances, such as those caused by forest management [15,47].

Conservation implications and conclusions

Identifying the spatial scales at which species vary in their diversity and composition is an important research goal in conservation ecology [4,48]. Additive partitioning of diversity is one of the most commonly applied statistical approaches for quantifying the fraction of diversity explained by different scales and for concluding which scales are critical for species conservation (e.g. [5–8]). We found that fungal communities show a high level of turnover at all the studied scales, and that only a small part of the turnover can be explained by factors such as spatial proximity or resource type similarity. The large number of random variation and high number of rare species suggests that a large total area of protected forests is needed to conserve a substantial part of fungal diversity. As managed forests contributed only little to the total number of species, an ecologically effective way to conserve fungal communities necessarily requires conserving natural forests. Nevertheless, leaving dead wood in managed forests can be seen an additional supplementary method, partly because some rare species only appear in managed forests, and partly because increasing the quality of the matrix (here, managed forests) can be essential for ensuring the long-term viability of populations in fragmented landscapes [49]. We found the similarity of fungal communities to decrease with increasing spatial distance, suggesting that protected areas should be distributed thorough the landscape. However, the individual

protected areas or local networks of protected areas must simultaneously be kept large enough to ensure the persistence of local populations, which process we did not however consider in the present study.

Supporting Information

File S1 Principal characteristics of the sampled beech forest sites.

File S2 A map of the sampled forest sites and plots in Navarre.

File S3 Literature consulted for the identification of wood-inhabiting species.

File S4 Code for performing the analyses in R.

File S5 Fungal species data.

File S6 Geographical UTM coordinates.

File S7 Dead wood composition data.

Acknowledgments

We are grateful to the Government of Navarre for giving the sampling permissions. We also thank A. Chiarucci and an anonymous reviewer for their valuable comments on the manuscript.

Author Contributions

Conceived and designed the experiments: NA GG-B IS. Performed the experiments: NA. Analyzed the data: NA GG-B OO. Contributed to the writing of the manuscript: NA GG-B PH OO IS.

References

1. Barnosky AD, Matzke N, Tomiya S, Wogan GOU, Swartz B, et al. (2011) Has the earth/'s sixth mass extinction already arrived? Nature 471: 51–57. Available: http://dx.doi.org/10.1038/nature09678.
2. Götmark F. (2013) Habitat management alternatives for conservation forests in the temperate zone: Review, synthesis, and implications. For Ecol Manage 306: 292–307. Available: http://dx.doi.org/10.1016/j.foreco.2013.06.014.
3. Chiarucci A, Bacaro G, Rocchini D. (2008) Quantifying plant species diversity in a natura 2000 network: Old ideas and new proposals. Biol Conserv 141: 2608–2618. Available: http://dx.doi.org/10.1016/j.biocon.2008.07.024.
4. Summerville KS, Boulware MJ, Veech JA, Crist TO. (2003) Spatial variation in species diversity and composition of forest lepidoptera in eastern deciduous forests of north america. Conserv Biol 17: 1045–1057. 10.1046/j.1523-1739.2003.02059.x.
5. Wagner H, Wildi O, Ewald K. (2000) Additive partitioning of plant species diversity in an agricultural mosaic landscape. Landscape Ecol 15: 219–227. Available: http://dx.doi.org/10.1023/A%3A1008114117913.
6. Summerville KS, Crist TO. (2003) Determinants of lepidopteran community composition and species diversity in eastern deciduous forests: Roles of season, ecoregion and patch size. Oikos 100: 134–148. 10.1034/j.1600-0706.2003.11992.x.
7. Müller J, Goßner MM. (2010) Three-dimensional partitioning of diversity informs state-wide strategies for the conservation of saproxylic beetles. Biol Conserv 143: 625–633. Available: http://dx.doi.org/10.1016/j.biocon.2009.11.027.
8. Gering JC, Crist TO, Veech JA. (2003) Additive partitioning of species diversity across multiple spatial scales: Implications for regional conservation of biodiversity. Conserv Biol 17: 488–499. 10.1046/j.1523-1739.2003.01465.x.
9. Tuomisto H. (2010) A consistent terminology for quantifying species diversity? yes, it does exist. Oecologia 164: 853–860. Available: http://dx.doi.org/10.1007/s00442-010-1812-0.
10. Anderson MJ, Crist TO, Chase JM, Vellend M, Inouye BD, et al. (2011) Navigating the multiple meanings of β diversity: A roadmap for the practicing ecologist. Ecol Lett 14: 19–28. 10.1111/j.1461-0248.2010.01552.x.
11. Boddy L, Frankland JC, van West P. (2008) Ecology of saprotrophic basidiomycetes. London, United Kingdom: Elsevier Academic Press. Available: http://dx.doi.org/10.1016/S0275-0287(08)80001-X.
12. Stokland JN, Siitonen J, Jonsson BG. (2012) Biodiversity in dead wood. Cambridge: Cambridge University Press.
13. Penttilä R, Siitonen J, Kuusinen M. (2004) Polypore diversity in managed and old-growth boreal picea abies forests in southern finland. Biol Conserv 117: 271–283. 10.1016/j.biocon.2003.12.007.
14. Hottola J, Ovaskainen O, Hanski I. (2009) A unified measure of the number, volume and diversity of dead trees and the response of fungal communities. J Ecol 97: 1320–1328. 10.1111/j.1365-2745.2009.01583.x.
15. Nordén J, Penttilä R, Siitonen J, Tomppo E, Ovaskainen O. (2013) Specialist species of wood-inhabiting fungi struggle while generalists thrive in fragmented boreal forests. J Ecol 101: 701–712. 10.1111/1365-2745.12085.
16. Küffer N, Senn-Irlet B. (2005) Influence of forest management on the species richness and composition of wood-inhabiting basidiomycetes in swiss forests. Biodivers Conserv 14: 2419–2435. 10.1007/s10531-004-0151-z.
17. Abrego N, Salcedo I. (2013) Variety of woody debris as the factor influencing wood-inhabiting fungal richness and assemblages: Is it a question of quantity or quality? For Ecol Manage 291: 377–385. 10.1016/j.foreco.2012.11.025.
18. Heilmann-Clausen J. (2001) A gradient analysis of communities of macrofungi and slime moulds on decaying beech logs. Mycol Res 105: 575–596. Available: http://dx.doi.org/10.1017/S0953756201003665.
19. Siitonen P, Lehtinen A, Siitonen M. (2005) Effects of forest edges on the distribution, abundance, and regional persistence of wood-rotting fungi. Conserv Biol 19: 250–260. 10.1111/j.1523-1739.2005.00232.x.
20. Ódor P, Heilmann-Clausen J, Christensen M, Aude E, Dort K, et al. (2006) Diversity of dead wood inhabiting fungi an bryophytes in semi-natural beech forest in Europe. Biol Conserv. 131: 58–71. Available: http://dx.doi.org/10.1016/j.biocon.2006.02.004.
21. Junninen K, Komonen A. (2011) Conservation ecology of boreal polypores: A review. Biol Conserv 144: 11–20. 10.1016/j.biocon.2010.07.010.

22. Sverdrup-Thygeson A, Gustafsson L, Kouki J. (2014) Spatial and temporal scales relevant for conservation of dead-wood associated species: Current status and perspectives. Available: http://dx.doi.org/10.1007/s10531-014-0628-3.

23. Berglund H, Hottola J, Penttilä R, Siitonen J. (2011) Linking substrate and habitat requirements of wood-inhabiting fungi to their regional extinction vulnerability. Ecography 34: 864–875. 10.1111/j.1600-0587.2010.06141.x.

24. Bässler C, Müller J, Svoboda M, Lepšová A, Hahn C, et al. (2012) Diversity of wood-decaying fungi under different disturbance regimes–a case study from spruce mountain forests. Biodivers Conserv 21: 33–49. 10.1007/s10531-011-0159-0.

25. Renvall P. (1995) Community structure and dynamics of wood-rotting Basidiomycetes on decomposing conifer trunks in northern Finland. Karstenia 35: 1–51.

26. Lande R. (1996) Statistics and partitioning of species diversity, and similarity among multiple communities. Oikos 76: 5–13. Available: http://www.jstor.org/stable/3545743.

27. Crist TO, Veech JA, Gering JC, Summerville KS. (2003) Partitioning species diversity across landscapes and regions: A hierarchical analysis of α, β, and γ diversity. Am Nat 162: 734–743. Available: http://www.jstor.org/stable/10.1086/378901.

28. Oksanen J, Blanchet FG, Roeland R, Legendre P, Minchin PR, et al. (2013) Community ecology package "vegan", R package version 2.0–7.

29. Bray JR, Curtis JT. (1957) An ordination of the upland forest communities of southern wisconsin. Ecol Monogr 27: 325–349.

30. Anderson MJ. (2001) A new method for non-parametric multivariate analysis of variance. Austral Ecol 26: 32–46. 10.1111/j.1442-9993.2001.01070.pp.x.

31. R Development Core Team. (2013) R: A language and environment for statistical computing. Vienna, Austria: R Foundation for Statistical Computing.

32. Hubbell SP. (2001) The unified neutral theory of biodiversity and biogeography. Princeton, N. J.: Princeton Univ. Press.

33. Chase JM, Leibold MA. (2003) Ecological niches: Linking classical and contemporary approaches. Chicago, Illinois, USA: University of Chicago Press.

34. Abrego N, Salcedo I. (2014) Response of wood-inhabiting fungal community to fragmentation in a beech forest landscape. Fungal Ecol 8: 18–27. Available: http://dx.doi.org/10.1016/j.funeco.2013.12.007.

35. Edman M, Gustafsson M, Stenlid J, Ericson L. (2004) Abundance and viability of fungal spores along a forestry gradient - responses to habitat loss and isolation? Oikos 104: 35–42. 10.1111/j.0030-1299.2004.12454.x.

36. Norros V, Penttilä R, Suominen M, Ovaskainen O. (2012) Dispersal may limit the occurrence of specialist wood decay fungi already at small spatial scales. Oikos 121: 961–974. 10.1111/j.1600-0706.2012.20052.x.

37. Halme P, Kotiaho JS. (2012) The importance of timing and number of surveys in fungal biodiversity research. Biodivers Conserv 21: 205–219. 10.1007/s10531-011-0176-z.

38. Moore D, Gange AC, Gange EG, Boddy L. Chapter 5 fruit bodies: Their production and development in relation to environment. In: Anonymous British Mycological Society Symposia Series: Academic Press. pp.79–103. Available: http://dx.doi.org/10.1016/S0275-0287(08)80007-0.

39. Halme P, Heilmann-Clausen J, Rämä T, Kosonen T, Kunttu P. (2012) Monitoring fungal biodiversity – towards an integrated approach. Fungal Ecol 5: 750–758. Available: http://dx.doi.org/10.1016/j.funeco.2012.05.009.

40. MacKenzie DI, Nichols JD, Royle JA, Pollock KH, Bailey LL, et al. (2006) Occupancy estimation and modeling: Inferring patterns and dynamics of species occurrence. San Diego, USA: Elsevier.

41. Novotný V, Basset Y. (2000) Rare species in communities of tropical insect herbivores: Pondering the mystery of singletons. Oikos 89: 564–572. 10.1034/j.1600-0706.2000.890316.x.

42. Penttilä R, Lindgren M, Miettinen O, Rita H, Hanski I. (2006) Consequences of forest fragmentation for polyporous fungi at two spatial scales. Oikos 114: 225–240. 10.1111/j.2006.0030-1299.14349.x.

43. Juutilainen K, Mönkkönen M, Kotiranta H, Halme P. (2014) The effects of forest management on wood-inhabiting fungi occupying dead wood of different diameter fractions. For Ecol Manage 313: 283–291. Available: http://dx.doi.org/10.1016/j.foreco.2013.11.019.

44. Ódor P, Standovár T. (2001) Richness of bryophyte vegetation in near-natural and managed beech stands: The effects of management-induced differences in dead wood. Ecol Bull: 219–229. Available: http://www.jstor.org/stable/20113278.

45. Siitonen J. (2001) Forest management, coarse woody debris and saproxylic organisms: Fennoscandian boreal forests as an example. Ecol Bull: 11–41. Available: http://www.jstor.org/stable/20113262.

46. Stokland J, Kauserud H. (2004) Phellinus nigrolimitatus–a wood-decomposing fungus highly influenced by forestry. For Ecol Manage 187: 333–343. Available: http://dx.doi.org/10.1016/j.foreco.2003.07.004.

47. Stokland JN, Larsson K. (2011) Legacies from natural forest dynamics: Different effects of forest management on wood-inhabiting fungi in pine and spruce forests. For Ecol Manage 261: 1707–1721. 10.1016/j.foreco.2011.01.003.

48. Dray S, Pélissier R, Couteron P, Fortin M-, Legendre P, et al. (2012) Community ecology in the age of multivariate multiscale spatial analysis. Ecol Monogr 82: 257–275. 10.1890/11-1183.1. Available: http://dx.doi.org/10.1890/11-1183.1.

49. Jules ES, Shahani P. (2003) A broader ecological context to habitat fragmentation: Why matrix habitat is more important than we thought. J Veg Sci 14: 459–464. 10.1111/j.1654-1103.2003.tb02172.x.

6

Predicting Impacts of Climate Change on the Aboveground Carbon Sequestration Rate of a Temperate Forest in Northeastern China

Jun Ma[1,2], Yuanman Hu[1]*, Rencang Bu[1], Yu Chang[1], Huawei Deng[3], Qin Qin[1,2]

1 State Key Laboratory of Forest and Soil Ecology, Institute of Applied Ecology, Chinese Academy of Sciences, Shenyang, People's Republic of China, 2 University of Chinese Academy of Sciences, Beijing, People's Republic of China, 3 Shengli Oilfield's Shengli Engineering co., LTD, Dongying, People's Republic of China

Abstract

The aboveground carbon sequestration rate (ACSR) reflects the influence of climate change on forest dynamics. To reveal the long-term effects of climate change on forest succession and carbon sequestration, a forest landscape succession and disturbance model (LANDIS Pro7.0) was used to simulate the ACSR of a temperate forest at the community and species levels in northeastern China based on both current and predicted climatic data. On the community level, the ACSR of mixed Korean pine hardwood forests and mixed larch hardwood forests, fluctuated during the entire simulation, while a large decline of ACSR emerged in interim of simulation in spruce-fir forest and aspen-white birch forests, respectively. On the species level, the ACSR of all conifers declined greatly around 2070s except for Korean pine. The ACSR of dominant hardwoods in the Lesser Khingan Mountains area, such as Manchurian ash, Amur cork, black elm, and ribbed birch fluctuated with broad ranges, respectively. Pioneer species experienced a sharp decline around 2080s, and they would finally disappear in the simulation. The differences of the ACSR among various climates were mainly identified in mixed Korean pine hardwood forests, in all conifers, and in a few hardwoods in the last quarter of simulation. These results indicate that climate warming can influence the ACSR in the Lesser Khingan Mountains area, and the largest impact commonly emerged in the A2 scenario. The ACSR of coniferous species experienced higher impact by climate change than that of deciduous species.

Editor: Gil Bohrer, The Ohio State University, United States of America

Funding: This research was co-funded by the "Strategic Priority Research Program–Climate Change: Carbon Budget and Related Issues" of the Chinese Academy of Sciences (XDA05050201) and National Natural Science Foundation of China (30870441,31070422). An additional funding resource is National Natural Science Foundation of China (41371198). The funders had no role in study design, data collection and analysis, decision to publish, or preparation of the manuscript.

Competing Interests: One of the authors is employed by a commercial company (Shengli Oilfield's Shengli Engineering Co., Ltd). This author was a student of our research department, and he was employed by a company after graduation. This study was completed before he graduated. The authors made sure that the Shengli Oilfield's Shengli Engineering Co., Ltd had nothing to do with this study. No competing interests exist between this company and the authors' research group.

* E-mail: huym@iae.ac.cn

Introduction

Forests store the most carbon of any unit of the terrestrial ecosystem [1,2], and the majority of the carbon sequestrated is held in woody biomass [3]. Forests play a vital role in climate change mitigation [4] and water conservation [5,6]. Also, forests can avoid soil erosion [7], although the impact of vegetation cover on soil erosion is not straight forward [8,9]. They provide many ecological services including biodiversity protection, a supply of wood and fiber, and functions related to tourism and recreation [10]. The potential capacity of forests to sequester carbon will obviously influence the future balance of global carbon flux; however, this potential is largely determined by the rate of carbon sequestration occurring in forests [11]. The aboveground carbon sequestration rate (ACSR) of forests is an important index reflecting the usefulness of forest ecosystems to humans. The future dynamics of ACSR has aroused many concerns, especially when one considers the impacts of climate change.

Human activity has altered the concentration of atmospheric carbon dioxide in a way that will create serious consequences such as warmer climates and irregular patterns of precipitation [12–14].

Emissions of greenhouse gas that continue at or above current levels are likely to cause additional climatic warming in the future, and this will transform some processes related to forest carbon sequestration such as the productivity, species distribution, and large alterations in nature disturbance regimes [15]. Also, the extension of the growing season and increased rates of photosynthesis which are caused by climate change will enhance forest growth rates [16]. Climate change can change tree species migration patterns [17], which can further affect forest carbon sequestration [18]. The net primary productivity (NPP) of tree species as well as their competitiveness can be changed by the alteration of climatic conditions such as temperature, precipitation, and solar irradiation [19,20], and the forest ACSR will experience parallel impacts of higher temperatures [21]. In addition, the changing climate will lead to changes in disturbance regimes that will diminish the process of carbon sequestration to a large extent. This makes exploring the rate and potential capacity of forest carbon sequestration necessary [22,23].

Many studies focus on the process of forest carbon sequestration under climate change, and most of which have researched the climate change impact on forest composition and carbon accumulation [24–27] and the responses of forest growth to altered climates [21,28,29]. However, relatively little research has explored the speed of forest carbon sequestration, which is important to those developing forest management policies.

Globally, temperate forest is a widely distributed forest type, and its carbon flux has been significantly altered by the changing climate [30–32]. The Lesser Khingan Mountains lie in a transitional region between a cold temperate and a moderate zone and these mountains are covered by typical temperate forests. A variety of vegetation and forest communities can be found in this area (**Fig. 1**), including coniferous forests, mixed broad-leaved conifer forests, and deciduous broad-leaved forests. In the past decade, many research studies related to the impact of climate change on forest ecosystems have been conducted in this area [33–35]. However, the dynamics of forest ACSR under climate change scenarios is still unclear. Rational forest management policies, which are designed to maintain sustainable productivity in the future, need to carefully explore the dynamics of forest ACSR. By assessing and understanding the future status of forest carbon sequestration under different climate change scenarios, foresters can make well-designed policies related to forest management.

Quantifying the complex effects on forests caused by global climate and land use changes has proved difficult [26], but we can use ecological models to simulate the forests dynamics. Many previous studies have proved that ecological models have an ability to estimate historical as well as future forest carbon pool dynamics [36,37]. Forest landscape succession and disturbance (LANDIS) model is a spatially explicit forest landscape model capable of simulating forest succession under multiple natural and anthropogenic disturbance regimes based on the current species distribution, age cohorts, and individuals [38,39]. Many studies [17,26,39–41] about forest dynamics under different conditions in North America and China have proved that it has the ability to detect the effects of alternative future climate scenarios on forests [26].

In this study, we couple projected meteorological data using Earth System Models, LANDIS Pro7.0 model and logistics model to simulate the response of forest ACSR to climate change. The objectives of this study were to (1) explore the dynamics of the ACSR in four main communities and fourteen tree species of a temperate forest in the Lesser Khingan Mountains for 200 years starting from 2000, (2) analyze statistical discrepancies found in the different effects of various climate change scenarios on forest ACSR, and (3) provide useful suggestions on how to carry out forest management in a changing climate.

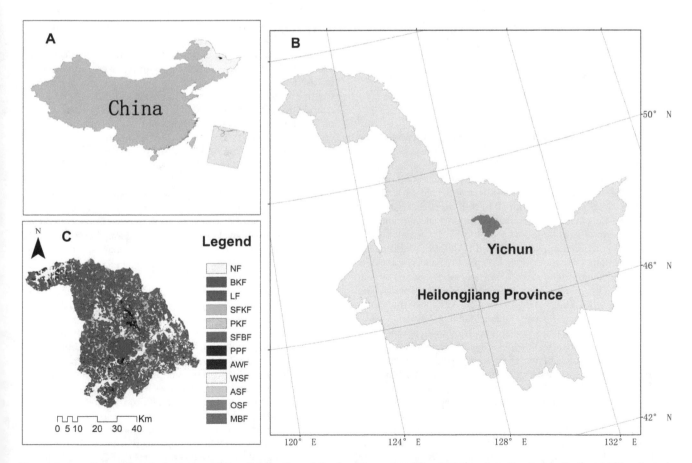

Figure 1. Location of the study area and the distribution of the forest communities. A: Heilongjiang Province in Northeast China; B: study area in Heilongjiang Province; C: The distribution of forest communities in Lesser Khingan Mountains. NF: no forest zone, BKF: Broad-leaved Korean pine forest, LF: Larch conifer forest, SFKF: Spruce-fir Korean pine forest, PKF: Pinus sylvestris-Korean pine conifer forest, SFBF: Spruce-fir broad-leaved forest, PPF: Planted pinus sylvestris forest, AWF: Aspen-white birch forest, WSF: White birch softwood forest, ASF: Aspen softwood forest, OSF: Oak softwood forest, MBF: Mixed broad-leaved forest.

Methods

Study Area

Our study area (**Fig. 1**) is located in the northern part of the Lesser Khingan Mountain region near the city of Yi Chun. This area extends across 47.85°–48.05°N, 128.43°–129.62°E, is covered by three forestry bureaus, includes a nature reserve zone, and has total area of about 315,000 ha. Dark brown soil, homogeneously distributed in this area, constitutes the typical soil of this region. The elevation ranges between 400 m and 600 m. The temperate continental monsoon climate experiences cold, long winters (mean January temperature, –25°C) while summers are warm and transitory (mean July temperature, 21°C). The average annual precipitation (550–700 mm) mostly falls from June to August.

The Lesser Khingan Mountains lie in a transitional zone between a cold and a moderate temperate zone. Several conifers and soft-hard woods coexist in this typical temperate forest. Common species include Korean pine (*Pinus koraiensis*), spruce (*Picea koraiensis* and *P. jezoensis*), Khingan fir (*Abies nephrolepis*), larch (*Larix gmelinii*), Manchurian ash (*Fraxinus mandshurica*), Amur cork (*Phellodendron amurense*), Mongolia oak (*Quercus mongolica*), black elm (*Ulmus japonica*), mono maple (*Acer mono*), ribbed birch (*Betula costata*), black birch (*Betula davurica*), Amur linden (*Tilia amurensis*), white birch (*Betula platyphylla*), aspen (*Populus davidiana*). Korean pine is the regionally dominant species, while spruce and Khingan fir are dominant only in high elevation areas.

Four representative communities occur in the study area: mixed Korean pine hardwood forests, spruce-fir forests, mixed larch hardwood forests, and aspen-white birch forests. These four communities are common vegetation types in the Lesser Khingan Mountain area. The entire region has suffered severe deforestation except in the Fenglin Natural Reserve.

LANDIS Model

LANDIS is a dynamic forest landscape model simulating forest succession, seed dispersal, species establishment, and various types of disturbance such as wind, fire, and timber harvesting [39,42,43]. LANDIS Pro 7.0 is derived from an earlier version of LANDIS, in which the landscape is represented as a grid of cells. The cell size can be set from 10 m × 10 m to as large as 500 m × 500 m. This model can simulate changes over long temporal (e.g. >100 years) and at large spatial scales (e.g. >107 ha). Several species were contained in every cell, and LANDIS Pro 7.0 model grouped trees into different species age cohorts. Besides, the model keeps track of the number of trees for each species age cohorts in all cells [44]. The modeled landscape was divided into several land types according to altitude, slope, climatic conditions and other environmental factors. Species establishment coefficients (SEC) (ranging from 0 to 1.0), which are an important input parameter in a LANDIS model, quantify whether a specific land type favors or works against the establishment of a selected species. Similar SECs would be developed in modeling of the same land type [43].

LANDIS Pro 7.0 simulates seedling establishment, growth, death, regeneration, random mortality, and vegetative reproduction on the basis of SEC and species vital attributes (**Table 1**) at the scale of a single modeled cell [45]. A detailed description of these species' physiological process can be found and consulted in the LANDIS Pro 7.0 user's guide [44]. An important advance of the latest LANDIS version is that the number of tree for each species age cohort was added in the model. This makes it easy for us to calculate the aboveground biomass of each tree as well as the total biomass of the forest. At a landscape scale, LANDIS simulates spatial processes such as seed dispersal and seedling

establishment [17]. Seed dispersal simulates the seed travel process based on a species' effective and maximum seed dispersal distance. A seedling establishment algorithm starts to work when seeds reached a particular site to decide whether a particular seed can become established based on consideration of other species that occur on the site and the shade tolerance rank of the seeding species relative to the species occupying the site [17,45]. If a site is occupied by species with a higher shade tolerance (e.g. Korean pine), species with lower shade tolerance (e.g. aspen, white birch) cannot spread into this site. A uniform random number from 0 to 1 will be set to compare with the SEC to decide if seeds can become established. Only when a species' SEC is greater than the random number can the species become successfully established. That means species with a high SEC will obtain high probabilities of establishment [43].

The species' growth curve is another essential input parameter used to calculate species biomass in the LANDIS model. In any outputting year, the model reads the corresponding diameter at breast height (DBH) and then outputs biomass by applying allometric growth equations.

Model Parameterizations

Species' vital attributes are driving factors of succession and dispersal in LANDIS [45]. Some other input data such as disturbance and management parameters, species composition maps, land type maps, and the species establishment coefficients for each land type are also included in LANDIS [17,45,46]. Species' growth curves and the average number of tree individuals are specifically used to calculate the biomass of each species in the biomass module. Growth of fourteen of the main tree species (four conifer and ten broad-leaf species) in our study area are simulated in LANDIS. The values of these input data were mainly compiled from previous LANDIS parameterization, plot investigation, and consultations with local experts [17,41,45,47,48]. With the consent of Fenglin Nature Reserve administration, we also investigated some plant plots (Table S1) in the reserve to test whether the parameters of the LANDIS Pro7.0 model were generally reasonable; our field experiments did no harm to the animals and tree species.

We generated an initial species composition map including species as well as age information from a forest stand inventory map and database produced in 2000 (provided by the Forestry Planning and Design Bureau of Heilongjiang Province, 2003). Useful information such as stand boundaries, the relative abundance of canopy species, and the average age of dominant canopy species were obtained from the database. The grid format stand maps with a resolution of 90 m × 90 m were converted from vector format with the goal of reducing computational loads during model simulations. Many previous studies in this area revealed that single species stands occur only during early successional stages [49–51]. On most occasions, multiple species occur in a 90 m × 90 m sized pixel. The LANDIS Pro 7.0 model traces species age cohorts in all pixels during the process of succession. The simulated processes of dispersal, establishment, growth, and death were all recorded. So, the model can output species' aboveground biomass at the end of every simulation time step (we set the time step at 10 years) by calculating biomass using tree biomass equations that are embedded in the LANDIS model.

Logistics models were used to project the SEC for each species, and this model's input parameters are mainly environmental variables, including slope, aspect, elevation, annual average temperature and precipitation, topographical position index and compound topographical index [17,52]. We used logistics models to simulate the probability of each species' occurrence in all cells in

Table 1. Species vital attributes in the Lesser Khingan mountains area, Northeastern China.

Species	LONG	MTR	ST	FT	ESD	MSD	VP	MVP	MD	CCC
Korean pine (*P. koraiensis*)	320	80	4	3	200	600	0	0	130	0.457
Spruce (*P. koraiensis and jezoensis*)	300	30	4	3	80	200	0	0	100	0.447
Khingan fir (*A. nephrolepis*)	300	30	4	3	80	200	0	0	60	0.440
Larch (*L. gmelinii*)	300	20	3	4	80	200	0	0	100	0.454
Manchuria ash (*Fraxinusmandshurica*)	250	40	3	5	400	1000	0.9	50	110	0.433
Amur cork (*P. amurense*)	250	15	3	4	60	300	0.8	60	90	0.443
Mongolia oak (*Q. mongolica*)	320	20	3	5	50	200	1	50	100	0.429
Black elm (*Ulmus japonica*)	250	10	3	3	200	1000	0.5	60	100	0.465
Mono maple (*A. mono*)	200	10	3	3	500	1000	0.5	50	60	0.420
Ribbed birch (*B. costata*)	250	15	3	3	500	4000	0.9	40	90	0.448
Black birch (*B. davurica*)	150	15	3	5	500	4000	0.9	30	50	0.433
Amur linden (*T. amurensis*)	300	15	3	2	80	250	0.8	30	80	0.448
White birch (*B. platyphylla*)	150	15	1	2	500	4000	0.8	50	60	0.451
Aspen (*P. davidiana*)	150	10	1	1	600	5000	0.9	10	60	0.433

LONG: longevity (years); MTR: age of maturity (years); ST: shade tolerance (1–5); FT: fire tolerance (1–5); ESD: effective seeding distance (m); MSD: maximum seeding distance (m); VP: vegetative production probability (0–1); MVP: minimum age of vegetative reproduction (years); MD: maximum diameter at breast height (cm); CCC: carbon content coefficient (0–1).

the land type every 10 years with the current and future climatic data and output the resulting map. The mean value of the probability of a species appearing in the cells of a specific land type is the final SEC of this species in that land type. We modeled the SECs of 14 species under different climatic conditions from 2000 to 2100 with 10-year increments; the initial SECs in 2000 were generated under current climatic conditions. The SECs after 2100 were assumed to remain stable.

The development of forest and the dynamics of carbon flux were affected by some disturbances such as land use change, CO_2 fertilization, and outbreak of insect [53,54]. However, they were not included in full research area, and they are not the main disturbance factors in this region. Therefore, in our study, we only considered climate change and ignored other disturbance or management factors, and we assumed that no fire or other events occurred in this simulation. Three types of climate change scenarios (B1, A1B, and A2) as well as current climatic condition were taken into account, and we attempted to compare different impacts of climatic conditions on the forest ACSR for the next 200 years (2000–2200).

Climate Data

The current meteorological data were collected from the Northeastern Institute of Weather in China and were compiled for 1961–2005 from 78 weather stations. Regression models were built between spatial position and temperature as well as precipitation. We calculated mean annual temperature and precipitation based on the daily temperature and precipitation. Climate projections generated by the third version of the Canadian Global Coupled Model (CGCM3) were used in this study. Three different scenarios simulating different levels of carbon emissions (B1, A1B, and A2) were adopted to produce future climatic data. The B1 emission scenario represents the lower emission scenario while A1B and A2 scenarios represent the median and higher emission scenario, respectively.

We interpolated the mean annual temperature and precipitation from 78 weather stations distributed throughout Northeastern China into grids with 90 m × 90 m resolution, indicating the distribution of current temperature and precipitation. According to CGCM3, the mean annual temperatures and precipitations of all climate scenarios would increase in first 100 years (2000–2100), and some studies believed the climates would enter into a stable state after 2100 and fluctuate around the level in 2100 [17,55]. We calculated the annual temperature differences between the future warming and current climate in 2000, as projected by CGCM3, using equation (1):

$$\Delta T_{i,j} = T_{\omega i,j} - T_{2000} \tag{1}$$

where T_ω represents temperature under warming climate, T_{2000} represents temperature in 2000 projected by CGCM3, i represents the year ($2000 < i < 2100$), and j is the decade ($2010 < j < 2100$ with a 10-year increment); therefore, $\Delta T_{i,j}$ is the climate change for year i and decade j. The projected mean annual temperature and precipitation of each decade from 2000 to 2100 were obtained when $\Delta T_{i,j}$ was added into the initial grids. **Table 2** shows the increment of mean annual precipitation and temperature in 2100. The current mean precipitation and temperature were 555.1±24.1 mm, and –0.54±0.48°C, respectively. The final results of climatic data were adopted by the logistics models to output SECs.

Modeling and Analyzing Approaches

This study coupled the LANDIS model and a logistics model to simulate the forest ACSR in the Lesser Khingan Mountains region. The logistics model was used to simulate the species' physiological response to current and changing climate conditions, and it outputted SECs for each time step corresponding with those of the LANDIS model. Furthermore, the LANDIS model was used to simulate species establishment, succession, and the process of forest carbon sequestration using the SECs obtained from the logistics model under the different climate scenarios. LANDIS Pro 7.0 can read SECs for every simulated time step, so the results output for species' biomass could reflect the effect of climate change on forests. The current climate and three warming climate scenarios were adopted in this study, and these results were compared.

Carbon content coefficient (CCC) as a vital factor was applied to convert biomass to carbon content. Results from an existing study [56] using CCCs of several main tree species in China's northeastern forest region were adopted in this study (**Table 1**). We used the variation of carbon content every ten years to represent the forest ACSR; the ACSR of fourteen species as well as four forest communities were analyzed. Equation (2) displayed the formula used to calculate the species' carbon sequestration rate:

$$v_{i,j} = c_i \cdot \left(B_{i,j} - B_{i,j-10} \right) \tag{2}$$

where i represents tree species and j is the decade ($2010 < j < 2200$ with a 10-year increment). c_i represents the CCC of species i. $B_{i,j}$ is biomass of species i in decade j, and $v_{i,j}$ is the average carbon sequestration rate of species i in decade j. For example, $v_{aspen,2050}$ is the average forest carbon sequestration rate of aspen in the period from 2040 to 2050. Five replicas were simulated for each climate scenario to assess model stochasticity, and we took the average values as the final carbon sequestration rate. We divided the total simulation into four time periods (2000–2050, 2060–2100, 2110–2150, and 2160–2200) and used one-way ANOVA to test the hypothesis that differences of ACSR existed among various climate scenarios. The LSD multiple comparison method was used to detect the difference of ACSR under various climate conditions. Furthermore, we conducted T-tests among the 157 measured biomass data from the plots in a natural reserve and the value of each time step's total biomass output from 2100 to 2200 in our simulation. All statistics were conducted using open resource statistical graphics and computing environment R with $P<0.05$ used as a test of significance [57].

Results

ACSR at Community Level

Parallel dynamics of the ACSR under current climate conditions and three climate warming scenarios during the simulation were detected in all four forest communities. The ACSR of mixed Korean pine hardwood forests and mixed larch hardwood forests fluctuated during the entire simulation (**Fig. 2A, 2B**). The range of variation of ACSR in the two communities were 0.84–1.87 t ha^{-1} 10 a^{-1} and 1.03–4.78 t ha^{-1} 10 a^{-1}, respectively. The ACSR of the two communities under all warming scenarios was always higher than that under current climate.

The ACSR in the spruce-fir forest and aspen-white birch forest communities initially fell, followed by a rising trend (**Fig. 2C, 2D**). However, different beginning time of falling and falling ranges existed in the ACSR of the two communities. The ACSR of spruce-fir forests experienced a short rise before 2030. The minimum values of the ACSR of these two communities were –

Table 2. The variation values of average annual precipitation and temperature in 2100 modeled by CGCM3.

Climate scenarios	ΔP (mm)	ΔT (°C)
B1	65.03±16.75	1.00±0.07
A1B	60.22±7.32	2.00±0.02
A2	240.38±17.30	3.21±0.41

ΔP:Increment of precipitation in 2100 modeled by CGCM3; ΔT: Increment of temperature in 2100 modeled by CGCM3.

0.88 t ha^{-1} 10 a^{-1} in the 2070s and −18.02 t ha^{-1} 10 a^{-1} in the 2140s, respectively. In the spruce-fir community, the ACSR under the warming scenarios generally was lower than that under the current climate, especially after the year of 2140. The ACSR of the aspen-white birch community almost tended to zero at the end of the simulation.

ACSR at Species Level

In conifers, the ACSR of the species adapted to a warm-climate, e.g. Korean pine (**Fig. 3A**), fluctuated in the first half of the simulation under current and warming climates and varied in the range of 0.61–0.86 t ha^{-1} 10 a^{-1}. The ACSR of spruce, Khingan fir, and larch (**Fig. 3**) had similar dynamics, which displayed a rising trend after an initial falling trend and finally fluctuated after 2140, and they varied in the range of 0.71–1.35 t ha^{-1} 10 a^{-1}, 0.26–0.79 t ha^{-1} 10 a^{-1}, and 0.44–0.78 t ha^{-1} 10 a^{-1}, respectively. Most of them reach a minimum ACSR in the period from 2070 to 2090. The ACSR of the three conifers under warming scenarios were relative lower than that under current climatic conditions.

The ACSR of broad-leaved tree species (**Fig. 4**) demonstrated complex dynamics during the simulation. The ACSR of the Manchurian ash, Amur cork tree, black elm, and ribbed birch initially fluctuated and then rose rapidly and finally fell. The ranges

of the variations in these four species were 0.02–0.05 t ha^{-1} 10 a^{-1}, 0.01–0.03 t ha^{-1} 10 a^{-1}, 0.06–0.23 t ha^{-1} 10 a^{-1}, and 0.81–1.63 t ha^{-1} 10 a^{-1}, respectively. Black elm was the only specie that the ACSR under warming climate lower than that under the current climate. The dynamics of ACSR of Mongolia oak and Amur linden showed a trend of rising before a sharp decline and then recovering later in the simulation. The ACSR of these two species reached a minimum in the 2080s and 2090s, respectively. Two unique patterns of ACSR fluctuations existed for mono maple and black birch in the initial and final periods of the simulation. Initially, the ACSR of these two broad-leaved trees varied in the ranges of 0.05–0.13 t ha^{-1} 10 a^{-1} and 0.06–0.15 t ha^{-1} 10 a^{-1}, while the ranges of variation in the last period were −0.47– −0.31 t ha^{-1} 10 a^{-1} and −0.05–0.12 t ha^{-1} 10 a^{-1}, respectively. The dynamics of the ACSR of these pioneer species, aspen and white birch, tended to rise after an initial fall and fluctuated around zero after 2160, and the ranges of variation of them before 2080 were 0.10–0.92 t ha^{-1} 10 a^{-1} and 0.96–4.08 t ha^{-1} 10 a^{-1}, respectively. The ACSR of these pioneer species reached a minimum in the 2140s simultaneously, and the curve of the ACSR of the two species under all climates nearly overlapped.

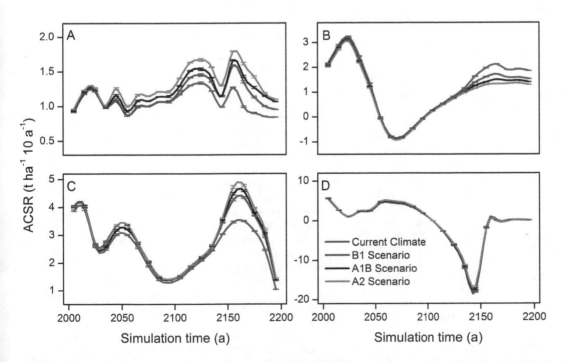

Figure 2. Forest aboveground carbon sequestration rates of different forest communities. A: Mixed Korean pine hardwood forests, B: Spruce-fir forests, C: Mixed larch hardwood forests, and D: Aspen-white birch forests.

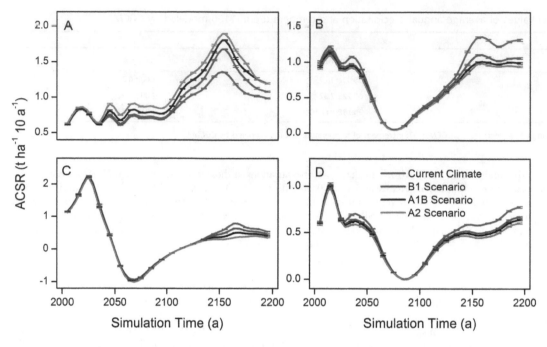

Figure 3. Forest aboveground carbon sequestration rates of four conifers. A: Korean pine, B: Spruce, C: Khingan fir, and D: Larch.

Differences of ACSR among Climate Scenarios

According to the results of variance analysis (**Table 3**), significant differences ($P<0.05$) of ACSR existed among the various climate scenarios in mixed Korean pine hardwood forests at 2060–2100 and 2110–2150, and in spruce-fir forests at 2160–2200. No significant difference of the ACSR among those climates was identified in the other two communities. In those two periods, the ACSR of the mixed Korean pine hardwood forests under scenario A2 was significantly higher than that under other climates (**Fig. 5A, 5B**). In the spruce-fir forest community, the ACSR under all warming scenarios were significantly higher than that under current climate scenario, and significant differences existed between any two warming scenarios (**Fig. 5C**).

In the species level, only in the periods of 2060–2100 and 2160–2200, significant differences of ACSR existed among various climate scenarios (**Table 3**). The ACSR of all the conifers among all climate scenarios had significant differences ($P<0.05$) in the last quarter (2160–2200) of the simulation, while these differences could be only detected in three hardwoods (**Fig. 6F–J**) (Amur cork tree, Mongolia oak and black elm) in the same period. From 2060 to 2100, the ACSR of Korean pine and Amur cork under the warming scenarios were significantly different from that under the current climate scenario. From 2160 to 2200, the ACSR of the species, with significant differences among climates, under the A1B and A2 scenarios were significantly differ from that under the current climate scenario (**Fig. 6**). Unique differences of ACSR between scenario A1B and A2 was detected in spruce (**Fig. 6C**), while no difference of ACSR was identified between scenario B1 and A1B in all species except spruce and Khingan fir (**Fig. 6D**). Differences of ACSR between the current climate and the scenario B1 were found in spruce, Khingan fir, and Amur cork.

Spatial Distribution of Forest Biomass

The forest total biomass under current climate and three warming scenarios presented similar spatial patterns in 2000, 2050, 2100, and 2200 (**Fig. 7**). The total forest biomass of this area is generally low except in the Fenglin Natural Reserve. Total biomass constantly accumulated during the simulation, and it becomes hard to detect differences in biomass between the forest reserve zone and other locations. However, large changes in total forest biomass accumulation can be identified during the second half of the simulation. A decrease of simulated total biomass was observed in most areas while it remained at a high level in *Fenglin* Forest Reserve. At the end of the simulation, the total biomass accumulation of the reserve zone was no longer higher than in other zones. The geographic center of the spatial distribution of total biomass tended to move north during the simulation.

Discussions

The LANDIS Pro 7.0 model is a spatially explicit model, and plays an important role in simulating the processes of germination, seed dispersal and establishment, and growth for various species under various disturbance regimes. The output of species' biomass makes it possible to explore the dynamics of the forest ACSR.

The trends in the variation of ACSR of all communities and species mostly occurred simultaneously among all the climate scenarios, though some relatively large differences appeared.

At the community level, the effect of multiple climate change scenarios appeared to involve complex processes affecting changes in forest ACSR. The possible explanation of this phenomenon is that usually a small increase in temperature has little impact on the growth of Korean pine, larch, and some other broadleaf species that are dominant species in these communities. This also corresponds with some previous studies [58,59], which have reported that the forest growth rate will not change until a temperature threshold appears. The temperatures under scenario A2 are likely to exceed this threshold while temperatures under other scenarios are not, and this can lead to ACSR under scenario A2 usually significantly higher than that under others. Moreover, the mixed Korean pine hardwood community is the climax community in this area, and it appears to be well-adapted to the warming climates of the past decades. The general rising trend of

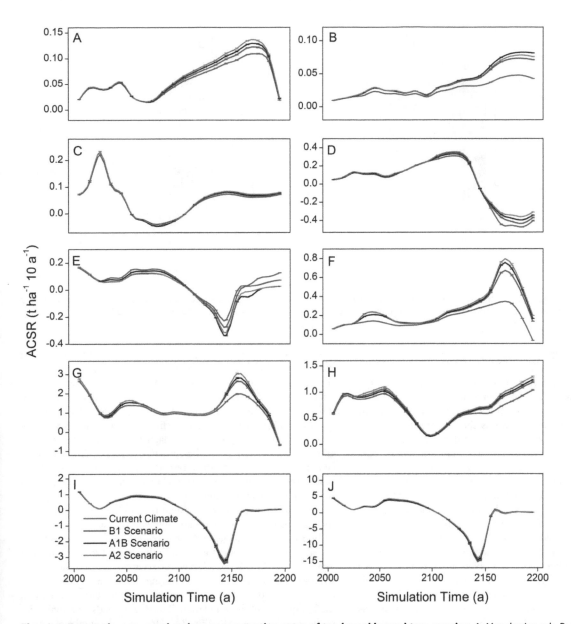

Figure 4. Forest aboveground carbon sequestration rates of ten broad leaved tree species. A: Manchurian ash, B: Amur cork, C: Mongolia oak, D: Mono maple, E: Black birch, F: Black elm, G: Ribbed birch, H: Amur linden, I: Aspen, and J: White birch.

ACSR may follow that adaption. This possible inference is in agreement with the local reality of forest distribution.

Significant differences existed between any two climate scenarios in the last quarters of the simulation in the spruce-fir community, and this community was the only one where the ACSR decreased when climate changed from current scenario to warming scenarios. Most likely, this community is vulnerable if the climate becomes warmer, and much greater impact on the ACSR may be caused by higher temperatures. The warming is not good for the growth of Khingan fir. Nonetheless, one study found the biomass of fir increase notably during a 6-year warming trend [60]. However, the increase of biomass is not equal to the increase of ACSR; the biomass will continue to increase as long as the ACSR is greater than 0. Similar patterns in mixed larch hardwood forests in the later period of the simulation were also remarkable; however, the difference was that warming scenarios brought about higher ACSR values. This probably occurred because climate

warming created a sharp increase in the abundance of broadleaf species in this community. Broadleaf trees can adapt to higher temperatures better than conifers and this may lead to a northward shift of broadleaf forest [58]. This corresponds well with the northward movement in the spatial distribution of the forest total biomass.

The final decrease of ACSR at the end of the simulation may suggest that the mixed larch hardwood community is a transitional community and will eventually be replaced by hardwood communities. Aspen and white birch are both pioneer species in the aspen-white birch community and they occupy an extensive range in this study area. The sharp decrease of the ACSR in the middle of the simulation may be caused by natural mortality of these species. According to the species vital attributes (**Table 1**), aspen and white birch both live about 150 years; most individuals died around 2150 in the simulations. The fluctuation ACSR

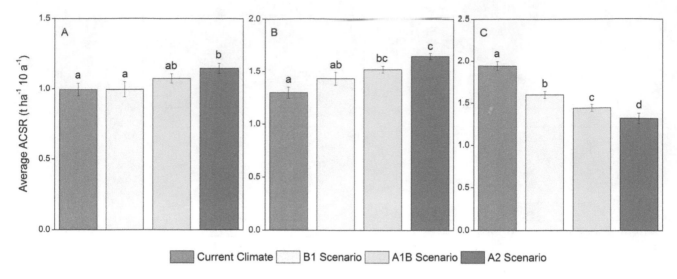

Figure 5. Results of multiple comparisons of the influences on communities' aboveground carbon sequestration rate. A: Mixed Korean pine hardwood forests in 2060–2100, B: Mixed Korean pine hardwood forests in 2110–2150, and C: Spruce-fir forests in 2160–2200.

around zero in the last decades of the simulation suggests that this community will eventually disappear.

At the species level, the dynamics of the ACSR display a more complex process. The ACSR of most conifers (except Korean pine) in the simulations demonstrated V-shaped alterations, and differences of the ACSR among various climate scenarios emerge late in the simulation. This may denote that the growth process of some conifers, including spruce, Khingan fir, and larch, may essentially stop during the simulation. It has already been proven that many conifers, including spruce and fir, become more vulnerable when climates become warmer [61]. The ACSR of Khingan fir once even dropped below zero, indicating that Khingan fir is not adapted to warm temperature, and the growth would be restrained while climate continues to warm. Global climate has changed as a result of hundreds of years of industrial activity, and some other disturbances such as fire and over exploitation have caused great destruction of forests [62]. This may be the reason why the responses of forests, which have already experienced climate change, appear insensitive to various climate change scenarios. As for Korean pine, increasing temperatures tend to accelerate the increase in ACSR, and its position as a dominant species will undoubtedly be enhanced under warming climatic conditions.

The ACSR of most hardwood species have experienced a rising trend, although some of them (Manchurian ash, black elm, and ribbed birch) decreased at the end of the simulation. This suggests that broadleaf trees have better adaption to warm conditions and grow at a relatively high speed. The final decrease in the ACSR of some broadleaf species likely occurred because they approached their natural longevity at the later time of simulation and began to decline. The smaller initial distribution and lower competitive ability to capture nutrients can well explain why Mono maple has a declining trend of its ACSR [63]. The decline of the ACSR of aspen and white birch is in agreement with the successional nature of aspen-white birch community.

We adopted three climate change scenarios (B1, A1B, and A2) as well as modeled the current climate conditions to detect the different influences of these scenarios on the ACSR. However, significant differences in the ACSR among various scenarios can be only discovered in a small number of communities and species,

most of which existed in the last quarter (2160–2200) of the simulation. These facts demonstrate that the effects of various climate change scenarios on the ACSR have no effect in a relatively short time (150 years). This does not mean that no action should be taken to control the current high levels of greenhouse gas emissions, because different climate change scenarios may manifest their effects on the ACSR later than in only 150 years. In other words, a hysteresis phenomenon exists in the process of climate change effects on the forest ACSR in this temperate forest. A slow response time of forests to climate change may be the real reason of this phenomenon.

The stability of a forest ecosystem is expressed in many aspects, such as species composition and biomass accumulation [64]. The carbon sequestration rate and capacity are vital factors reflecting a forest's ability to sequester carbon. In this study, the distribution of total biomass (**Fig. 7**) on a landscape scale reflects that the carbon sequestration of forest fluctuates during the modeled time period, and the forest displayed an extremely unstable state especially in the last quarter of the simulation. This corresponds with dynamics of the total forest ACSR (**Fig. 8A**). However, the total capacity of the forest for aboveground carbon sequestration maintained at a relative stable level in this landscape (**Fig. 8B**). This indicates that a forest actually reaches a dynamic equilibrium based on the stable state of a forest ecosystem when its biomass accumulation reaching a dynamic equilibrium. The species composition and the spatial distribution of biomass can reflect the status of a forest ecosystem at a large time scale; however, difficulties persist in interpreting the instantaneous status of forest growth. Reflection of the dynamics of forest carbon flux, especially the complex temperate forest, needs indicators that can represent the development process of forest succession. However, traditional indexes such as stand structure, species composition, and biomass are all static indicators. They only reflect the state of forest carbon sequestration at some point. ACSR displays the speed of forest carbon accumulation and demonstrate the ability of forest carbon sequestration in a certain period. Therefore, ACSR can express the instantaneous status of forest growth. The ACSR supplies an indication of the proper method to solve this problem.

Developing longtime successive forest inventory data under a warming climate in the future is impossible, and this makes it

Table 3. ANOVA results of differences among various climate scenarios effect on forest aboveground carbon sequestration rate.

Community with species	2010–2050		2060–2100		2110–2150		2160–2200	
	F	P	F	P	F	P	F	P
Mixed Korean pine hardwood forests	0.250	0.860	3.675	**0.035**	6.098	**0.006**	2.813	0.073
Korean pine (*P. koraiensis*)	0.317	0.813	9.618	**0.001**	1.359	0.291	3.210	**0.046**
Manchurian ash (*Fraxinus mandshurica*)	0.019	0.996	0.071	0.975	0.839	0.492	0.180	0.908
Amur corktree (*P. amurense*)	0.163	0.920	3.627	**0.013**	2.451	0.101	20.382	**<0.001**
Mongolia oak (*Q. mongolica*)	0.012	0.998	0.097	0.961	0.045	0.987	4.623	**0.016**
Black elm (*Ulmus japonica*)	0.388	0.763	2.002	0.154	0.753	0.536	2.419	0. 104
Mono maple (*A. mono*)	0.079	0.970	0.005	0.999	0.046	0.986	1.841	0.180
Spruce-fir forests	0.058	0.981	0.033	0.992	0.143	0.933	83.712	**<0.001**
Spruce (*P. koraiensis and jezoensis*)	0.919	0.454	0.021	0.996	0.266	0.849	140.216	**<0.001**
Khingan fir (*A. nephrolepis*)	0.009	0.999	0.030	0.993	0.047	0.986	16.843	**<0.001**
Mixed larch hardwood forests	0.081	0.969	0.067	0.977	0.098	0.960	0.506	0.684
Larch (*L. gmelinii*)	0.120	0.947	0.015	0.997	0.196	0.898	3.713	**0.034**
Ribbed birch (*B. costata*)	0.035	0.991	0.255	0.856	0.181	0.908	0.137	0.936
Black birch (*B. davurica*)	0.044	0.987	5.366	**0.016**	0.115	0.950	3.354	**0.045**
Amur linden (*T. amurensis*)	0.150	0.928	0.040	0.989	0.104	0.957	1.369	0.288
Aspen-white birch forests	0.010	0.999	0.094	0.962	0.007	0.999	0.032	0.992
White birch (*B. platyphylla*)	0.011	0.998	0.093	0.963	0.007	0.999	0.053	0.983
Aspen (*P. davidiana*)	0.008	0.999	0.098	0.960	0.009	0.999	0.015	0.997

df = 4; Bold P values mean the effect of treatment is significant ($\alpha = 0.05$).

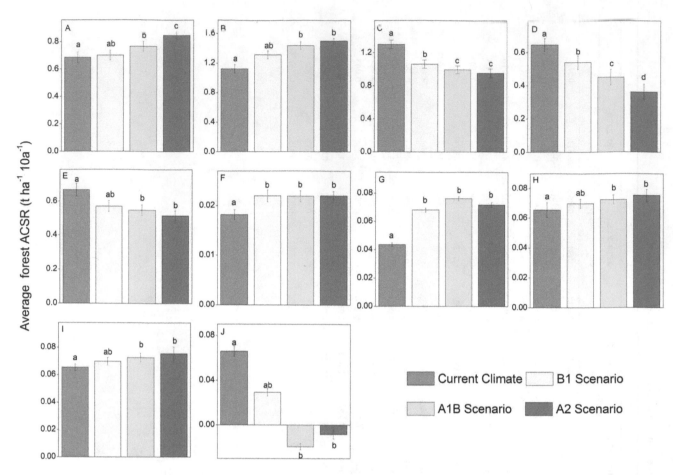

Figure 6. Results of multiple comparisons of the influences on species aboveground carbon sequestration rate. A: Korean pine in 2060–2100, B: Korean pine in 2160–2200, C: Spruce in 21260–2200, D: Khingan fir in 2160–2200, E: Larch in 2160–2200, F: Amur cork in 2060–2100, G: Amur cork in 2160–2200, H: Mongolia oak in 2160–2200, I: Black birch in 2060–2100, and J: Black birch in 2160–2200.

difficult for landscape models to predict forests dynamics precisely [65,66]. Applying forest inventory data collected in the field over a short time period with a limited number of data points used to verify the results of our 200-year simulation cannot allow for the creation of an entirely precise model. Comparing simulation results with other studies, including results from other models and experiments, may provide a feasible way to verify the precision of the LANDIS model; this method is sure to enhance the model's confidence [67].

The analysis of the uncertainty and sensitivity of LANDIS model has been carried out in some studies [26,43,68]. The results show that the output of this model is quite stable. The simulated biomass result in this study was also unanimous, and the mean proportional error of the biomass was less than 0.01 (**Fig. 2–4**). This indicates that the uncertainty of LANDIS model is very low.

One study [69] indicates that the range of biomass accumulation of mixed Korean pine hardwood in the Lesser Khingan Mountains is 199–371 t ha^{-1}. Yan *et al.* [70] used the NEWCOP model to simulate the biomass of natural forests in the Lesser Khingan Mountains and found that the value is about 250 t ha^{-1}. Our simulation provided a total biomass output of about 250 t ha^{-1} which is in agreement with those previous studies. Moreover, *Fenglin* natural reserve is the largest and oldest reserve in the Lesser Khingan Mountains area. Forests in this reserve have barely obtained damage, and most of the forest communities are in climax state. Therefore, forests in this reserve are considered to be

the future state of the forests which are distributed outside the reserve, and the biomass accumulation in reserve is also regarded as the upper limit of forest in the Lesser Khingan Mountains area. The general state of forest in the Lesser Khingan Mountains area 200 years later simulated by LANDIS Pro7.0 model also has been regarded as the climax. We have conducted T-tests among measured biomass data for 157 plots in *Fenglin* natural reserve and calculated each time step's output of total biomass from 2100 to 2200 in our simulation; the result revealed no significant differences (P>0.05). These facts show that the results of biomass output of the LANDIS model conform to field reality at the landscape scale, and the accuracy of the LANDIS model's biomass output can be trusted. The LANDIS model was used to evaluate species distribution under possible warming climates in northeast China [17]. The results showed that Korean pine thrives better than other species under warming climatic conditions, and the northern boundary of its range would shift northward while larch decreased under warming climates. The hardwoods were the dominant species in warmer conditions while conifers favored colder circumstances. Our results are consistent with other studies although coniferous species increased in abundance late in our simulation. This simulated effect may occur because the restoration and growth process was slowed by warming climates. Increasing temperature and precipitation would boost the growth rate of sugar maple and white spruce as has been reported in the literature, while the growth rate of balsam fir decreased under the

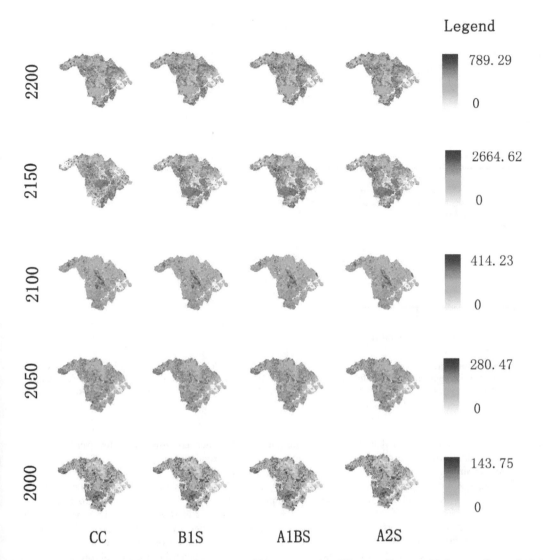

Figure 7. The spatial distribution of forest total biomass under different climates. CC: Current climate, B1S: B1 scenario, A1B: A1B scenario, A2: A2 scenario. Unit: t ha^{-1}.

same conditions [71]. Sabedi & Sharma [72] pointed out that the trend of diameter growth in jack pine and black spruce increased and decreased under warming conditions, respectively. Though these two species do not exist in our study area, they provide useful information in evaluating the ACSR of conifers.

In addition, effective comparisons of our results with previous studies such as those discussed above prove the LANDIS model has the ability to simulate forest growth accurately; also, the various results in comparisons provide us with clues for future analysis and modification of the model. The prediction of trends and the alteration of the range of forest ACSR under various climate scenarios in this study tend to agree with previous results.

Great uncertainty exists in climate change predictions of GCMs. Our simulation adopted the climate predictions of CGCM3 because of its applicability in northeastern China that has been proven by many studies [73,74]. We used only annual mean temperature and precipitation every 10-years and neglected the climate alteration in adjacent years and for a given year; however, this variability could affect the species' growth, morality, and establishment [75,76]. Nevertheless, every 10-year average annual temperature and precipitation could reflect the trend of climate

change in this area, and the effect of climate change on the forest carbon sequestration rate can also be interpreted appropriately by 10-year increment meteorological data.

In the LANDIS model, species establishment coefficients (SECs) play an important role in deciding the process of forest succession. Future SECs are simulated by the logistics model during different climate change scenarios, and the LANDIS model read the SEC parameters in every time-step to reflect the successive effects of climate change. We care more about forest landscape processes, and some processes do not get enough attention on a more detailed level. However, these processes, such as physiological processes, that are influenced by climate conditions are overlooked while we only use SECs to reflect climate changes indirectly; these physiological processes may influence the biomass accumulation of species. Moreover, biomass equations were used to calculate species' biomass; however, instability existed in these equations while applying them in different regions.

In this study, species growth curves, as input parameters in LANDIS Pro7.0 model, were obtained from field investigation and previous published references. These results were all based on the historical climate, and the effect of climate condition then had

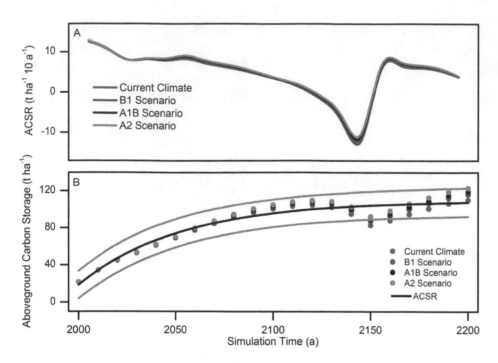

Figure 8. Total forest carbon sequestrations and its rate under various climate scenarios in simulation time. A: Forest total ACSR of different future climates. B: Forest aboveground carbon storage of different future climates. The black line means the average aboveground carbon storage of all possible future climates, and the grey lines indicate the forecasting interval of aboveground carbon storage with the confidence level of 95%.

been coupled in the results. Many previous studies [77–79] had explored the relationship between tree radial growth and climate conditions; however, we did not conduct similar research, and it might affect the final results of modelling at some extent. In future research, making input parameters more accurate can enhance the credibility of the simulated results.

Furthermore, an important vegetation driver, the fire, was not considered in this study. The regime of fire disturbance would be altered under climate change [80,81], and it would finally influence forest carbon sequestration. Nonetheless, the purpose of this study was to analysis forest ACSR under different climate change scenarios, and insight was obtained from this research on how forests response to possible climate conditions.

Model simulation is a vital tool in the research of forest dynamics under climate change at long-time and large-scale conditions. Although the model itself has many shortcomings, by appropriately analyzing and verifying the model, user may greatly improve the accuracy of the model. In addition, indigenization of model parameters will produce excellent prospects for successful and accurate application of the model.

Conclusions

The LANDIS Pro7.0 model is capable of modeling forest dynamics, especially biomass change, under changing climatic conditions. Based on the simulated results, several conclusions can be drawn as follows. Climate warming can influence the ACSR in the Lesser Khingan Mountains area. Mixed Korean pine hardwood forest is the climax community in the Lesser Khingan Mountains area, and it has higher adaption to climate warming

when compared with other communities. However, spruce-fir forest has a decline trend while climate becomes warmer. Differences of ACSR of various communities almost emerge around 2140, and this expresses that a hysteresis phenomenon exists in the process of climate change affecting temperate forests ACSR. The ACSR of coniferous species is more strongly influenced by climate change than is the ACSR of deciduous species. The biomass composition of climax communities is not stable, and ACSR can also reflect the response of forest to climate change. The differences of ACSR among various climate change scenarios are complex, and generally, climate change causes the largest impact in the scenario A2.

Acknowledgments

We thank Zaiping Xiong, Linlin Jiao, Xiaomei Liu for their field work. We also thank Marcus Higgs form University of the Third Age Kingborough Inc. for his pre-submission review.

Author Contributions

Conceived and designed the experiments: YH RB YC JM HD. Performed the experiments: HD JM RB. Analyzed the data: JM QQ. Contributed reagents/materials/analysis tools: JM RB. Wrote the paper: JM YC RB.

References

1. Iverson LR, Prasad AM (1998) Predicting abundance of 80 tree species following climate change in the eastern United States. Ecological Monographs 68: 465–485.

2. Pan Y, Birdsey RA, Fang J, Houghton R, Kauppi PE, et al. (2011) A Large and Persistent Carbon Sink in the World's Forests. Science 333: 988–993.

3. Scott NA, Rodrigues CA, Hughes H, Lee JT, Davidson EA, et al. (2004) Changes in carbon storage and net carbon exchange one year after an initial shelterwood harvest at Howland Forest, ME. Environmental Management 33: S9–S22.

4. Millar CI, Stephenson NL, Stephens SL (2007) Climate change and forests of the future: Managing in the face of uncertainty. Ecological Applications 17: 2145–2151.

5. Bent GC (2001) Effects of forest-management activities on runoff components and ground-water recharge to Quabbin Reservoir, central Massachusetts. Forest Ecology and Management 143: 115–129.

6. Alexander HD, Arthur MA (2010) Implications of a predicted shift from upland oaks to red maple on forest hydrology and nutrient availability. Canadian Journal of Forest Research-Revue Canadienne De Recherche Forestiere 40: 716–726.

7. Ammer U, Breitsameter J, Zander J (1995) Contribution of mountain forests towards the prevention of surface runoff and soil-erosion. Forstwissenschaftliches Centralblatt 114: 232–249.

8. Nanko K, Onda Y, Ito A, Moriwaki H (2008) Effect of canopy thickness and canopy saturation on the amount and kinetic energy of throughfall: An experimental approach. Geophysical Research Letters 35.

9. Nanko K, Mizugaki S, Onda Y (2008) Estimation of soil splash detachment rates on the forest floor of an unmanaged Japanese cypress plantation based on field measurements of throughfall drop sizes and velocities. Catena 72: 348–361.

10. Fiorese G, Guariso G (2013) Modeling the role of forests in a regional carbon mitigation plan. Renewable Energy 52: 175–182.

11. Chen QQ, Xu WQ, Li SG, Fu SL, Yan JH (2013) Aboveground biomass and corresponding carbon sequestration ability of four major forest types in south China. Chinese Science Bulletin 58: 1551–1557.

12. Forster PMD, Shine KP, Stuber N (2007) It is premature to include non-CO2 effects of aviation in emission trading schemes (vol 40, pg 1117, 2006). Atmospheric Environment 41: 3941–3941.

13. Rosenzweig C, Karoly D, Vicarelli M, Neofotis P, Wu QG, et al. (2008) Attributing physical and biological impacts to anthropogenic climate change. Nature 453: 353–U320.

14. Solomon D, Lehmann J, Kinyangi J, Amelung W, Lobe I, et al. (2007) Long-term impacts of anthropogenic perturbations on dynamics and speciation of organic carbon in tropical forest and subtropical grassland ecosystems. Global Change Biology 13: 511–530.

15. IPCC (2007) the phisical science basis. Contribution of Working Group I to the Fourth Assessment Report of the Intergovermental Panel on Climate Change. S.Solomon, D,Qin, M.Manning, Z.Chen, Marquis, K.B.Averyt, M.Tignor, and H.L.Miller, editors. Cambridge University Press, Cambridge, UK.

16. Lindner M, Maroschek M, Netherer S, Kremer A, Barbati A, et al. (2010) Climate change impacts, adaptive capacity, and vulnerability of European forest ecosystems. Forest Ecology and Management 259: 698–709.

17. Bu RC, He HS, Hu YM, Chang Y, Larsen DR (2008) Using the LANDIS model to evaluate forest harvesting and planting strategies under possible warming climates in Northeastern China. Forest Ecology and Management 254: 407–419.

18. Bunker DE, DeClerck F, Bradford JC, Colwell RK, Perfecto I, et al. (2005) Species loss and aboveground carbon storage in a tropical forest. Science 310: 1029–1031.

19. Peng CH, Apps MJ (1999) Modelling the response of net primary productivity (NPP) of boreal forest ecosystems to changes in climate and fire disturbance regimes. Ecological Modelling 122: 175–193.

20. Pan Y, Birdsey R, Hom J, McCullough K, Clark K (2006) Improved estimates of net primary productivity from MODIS satellite data at regional and local scales. Ecological Applications 16: 125–132.

21. Xu CG, Gertner GZ, Scheller RM (2007) Potential effects of interaction between CO2 and temperature on forest landscape response to global warming. Global Change Biology 13: 1469–1483.

22. Omeja PA, Obua J, Rwetsiba A, Chapman CA (2012) Biomass accumulation in tropical lands with different disturbance histories: Contrasts within one landscape and across regions. Forest Ecology and Management 269: 293–300.

23. Heimann M, Reichstein M (2008) Terrestrial ecosystem carbon dynamics and climate feedbacks. Nature 451: 289–292.

24. Nystrom M, Holmgren J, Olsson H (2012) Prediction of tree biomass in the forest-tundra ecotone using airborne laser scanning. Remote Sensing of Environment 123: 271–279.

25. Kasischke ES, Christensen NL, Stocks BJ (1995) Fire, global warming, and the carbon balance of boreal forests. Ecological Applications 5: 437–451.

26. Gustafson EJ, Shvidenko AZ, Sturtevant BR, Scheller RM (2010) Predicting global change effects on forest biomass and composition in south-central Siberia. Ecological Applications 20: 700–715.

27. Xu CG, Gertner GZ, Scheller RM (2012) Importance of colonization and competition in forest landscape response to global climatic change. Climatic Change 110: 53–83.

28. Sato H, Ise T (2012) Effect of plant dynamic processes on African vegetation responses to climate change: Analysis using the spatially explicit individual-based dynamic global vegetation model (SEIB-DGVM). Journal of Geophysical Research-Biogeosciences 117.

29. Manusch C, Bugmann H, Heiri C, Wolf A (2012) Tree mortality in dynamic vegetation models - A key feature for accurately simulating forest properties. Ecological Modelling 243: 101–111.

30. Dai LM, Jia J, Yu DP, Lewis BJ, Zhou L, et al. (2013) Effects of climate change on biomass carbon sequestration in old-growth forest ecosystems on Changbai Mountain in Northeast China. Forest Ecology and Management 300: 106–116.

31. Bonan GB (2008) Forests and climate change: Forcings, feedbacks, and the climate benefits of forests. Science 320: 1444–1449.

32. Hyvonen R, Agren GI, Linder S, Persson T, Cotrufo MF, et al. (2007) The likely impact of elevated CO2, nitrogen deposition, increased temperature and management on carbon sequestration in temperate and boreal forest ecosystems: a literature review. New Phytologist 173: 463–480.

33. Deng HP, Wu ZF, Zhou DW (2000) Response of broadleaved Pinus koraiensis forests in Xiaoxinganling Mt. to global climate change-a dynamic modeling. Chinese Journal of Applied Ecology 11: 4 (in Chinese).

34. Guo R, Bu RC, Hu YM, Chang Y, He HS, et al. (2010) Simulation of timber-harvesting area in Xiao Xing' anling Mountains under climate change. Chinese Journal of Applied Ecology 21: 8 (in Chinese).

35. Yan XD, Fu CB, Shugart HH (2000) Simulating the effecs of climate changes on Xiaoxing'an mountain forests. Acta Phytoeco Sinica 24: 8 (in Chinese).

36. Scheller RM, Van Tuyl S, Clark KL, Hom J, La Puma I (2011) Carbon Sequestration in the New Jersey Pine Barrens Under Different Scenarios of Fire Management. Ecosystems 14: 987–1004.

37. Bernier PY, Guindon L, Kurz WA, Stinson G (2010) Reconstructing and modelling 71 years of forest growth in a Canadian boreal landscape: a test of the CBM-CFS3 carbon accounting model. Canadian Journal of Forest Research-Revue Canadienne De Recherche Forestiere 40: 109–118.

38. He HS (2008) Forest landscape models: Definitions, characterization, and classification. Forest Ecology and Management 254: 484–498.

39. He HS, Mladenoff DJ, Crow TR (1999) Linking an ecosystem model and a landscape model to study forest species response to climate warming. Ecological Modelling 114: 213–233.

40. Xu CG, Guneralp B, Gertner GZ, Scheller RM (2010) Elasticity and loop analyses: tools for understanding forest landscape response to climatic change in spatial dynamic models. Landscape Ecology 25: 855–871.

41. He HS, Mladenoff DJ, Gustafson EJ (2002) Study of landscape change under forest harvesting and climate warming-induced fire disturbance. Forest Ecology and Management 155: 257–270.

42. Gustafson EJ, Shifley SR, Mladenoff DJ, Nimerfro KK, He HS (2000) Spatial simulation of forest succession and timber harvesting using LANDIS. Canadian Journal of Forest Research-Revue Canadienne De Recherche Forestiere 30: 32–43.

43. Mladenoff DJ, He HS (1999) Design and behavior of LANDIS, an objectoriented model of forest landscape disturbance and succession. In: Spatial Modeling of Forest Landscape Change: Approaches and Applications. Cambridge University Press, Cambridge, UK 125–162.

44. He SH, Wang WJ, Shifley SR, Fraser JS, Larsen DR, et al. (2011) LANDIS: a spatially explicit model of forest landscape disturbance, management, and succession, Landis Pro7.0 users guide.

45. He HS, Hao ZQ, Mladenoff DJ, Shao GF, Hu YM, et al. (2005) Simulating forest ecosystem response to climate warming incorporating spatial effects in north-eastern China. Journal of Biogeography 32: 2043–2056.

46. Mladenoff DJ (2004) LANDIS and forest landscape models. Ecological Modelling 180: 7–19.

47. Yan X, Zhao S (1996) Simulating sensitivities of Changbai Mountain forests to potential climate change. Journal of Environments science 8: 354–366.

48. Yan X, Fu CB, Shugart HH (2000) Simulating the effect of climate changes on Xiaoxing'an Mountain forests. Acta Phytoeco Sinica 24: 312–319 (in Chinese).

49. Xu H (2001) Natural Forest of Pinus koraiensis in China. Beijing: Forestry Press of China. 242.

50. Zhou Y (1994) Vegetation in the Small Khingan Mountains of China. Beijing: Science Press. 337.

51. Ge J, Guo H., Cheng D (1990) Study on age structure and spatial pattern of old-growth Korean pine forest in lesser Xingan Mountain. Journal of Northeast Forestry University 18: 6 (in Chinese).

52. Bu RC (2006) The response of main tree species to climate warming in the small Khingan Mountains. PhD Dissertation: 30 (in Chinese).

53. Albani M, Medvigy D, Hurtt GC, Moorcroft PR (2006) The contributions of land-use change, CO2 fertilization, and climate variability to the Eastern US carbon sink. Global Change Biology 12: 2370–2390.

54. Medvigy D, Clark KL, Skowronski NS, Schafer KVR (2012) Simulated impacts of insect defoliation on forest carbon dynamics. Environmental Research Letters 7.

55. Flato GM, Boer GJ (2001) Warming asymmetry in climate change simulations. Geophysical Research Letters 28: 195–198.

56. Yu Y, Fan WY, Li MZ (2012) Forest carbon rates at different scales in Northeast China forest area. Chinese Journal of Applied Ecology 23: 341–346 (in Chinese).

57. R Development Core Team (2011) R: A Language and Environment for Statistical Computing. R Foundation for Statistical Computing. Vienna.

58. Ni J (2011) Impacts of climate change on Chinese ecosystems: key vulnerable regions and potential thresholds. Regional Environmental Change 11: S49–S64.

59. Chen XW (2002) Study of the ecological safety of the forest in northeast China under climate change. International Journal of Sustainable Development and World Ecology 9: 49–58.

60. Wang JC, Duan BL, Zhang YB (2012) Effects of experimental warming on growth, biomass allocation, and needle chemistry of Abies faxoniana in even-aged monospecific stands. Plant Ecology 213: 47–55.

61. Coops NC, Waring RH (2011) Estimating the vulnerability of fifteen tree species under changing climate in Northwest North America. Ecological Modelling 222: 2119–2129.

62. Duchesne L, Prevost M (2013) Canopy disturbance and intertree competition: implications for tree growth and recruitment in two yellow birch-conifer stands in Quebec, Canada. Journal of Forest Research 18: 168–178.

63. Yao XQ, Liu Q (2009) Responses in Some Growth and Mineral Elements of Mono Maple Seedlings to Enhanced Ultraviolet-B and to Nitrogen Supply. Journal of Plant Nutrition 32: 772–784.

64. Reiners WA, Driese KL, Fahey TJ, Gerow KG (2012) Effects of Three Years of Regrowth Inhibition on the Resilience of a Clear-cut Northern Hardwood Forest. Ecosystems 15: 1351–1362.

65. Gardner RH, Urban DL (2003) Model validation and testing: past lessons, present concerns, future prospects. In: Canham CD, Cole, J.J, Lauenroth, W.K (Eds.), editor. Princeton, NJ, USA: Princeton University Press.

66. Rastetter EB (1996) Validating models of ecosystem response to global change. Bioscience 46: 190–198.

67. Bugmann H (2001) A review of forest gap models. Climatic Change 51: 259–305.

68. He HS, Mladenoff DJ (1999) Spatially explicit and stochastic simulation of forest-landscape fire disturbance and succession. Ecology 80: 81–99.

69. Feng ZW (1993) Biomass and production of forest ecosystems in China: Science Press (in Chinese).

70. Yan XD, Zhao SD, Fu CB, Shugart HH, Yu ZL (1999) A simulation study of natural forests in Lesser Khingan Mountain under cliamte change. Journal of Natural Resources 14: 372 (in Chinese).

71. Goldblum D, Rigg LS (2005) Tree growth response to climate change at the deciduous-boreal forest ecotone, Ontario, Canada. Canadian Journal of Forest Research-Revue Canadienne De Recherche Forestiere 35: 2709–2718.

72. Subedi N, Sharma M (2013) Climate-diameter growth relationships of black spruce and jack pine trees in boreal Ontario, Canada. Global Change Biology 19: 505–516.

73. Liu ZH, Yang J, Chang Y, Weisberg PJ, He HS (2012) Spatial patterns and drivers of fire occurrence and its future trend under climate change in a boreal forest of Northeast China. Global Change Biology 18: 2041–2056.

74. Sun F, Yang S, Chen P (2005) Climatic warming-drying trend in Northeastern China during the last 44 years and its effects. Chinese Journal of Ecology 24: 751.

75. Adams HD, Kolb TE (2004) Drought responses of conifers in ecotone forests of northern Arizona: tree ring growth and leaf sigma(13)C. Oecologia 140: 217–225.

76. Guarin A, Taylor AH (2005) Drought triggered tree mortality in mixed conifer forests in Yosemite National Park, California, USA. Forest Ecology and Management 218: 229–244.

77. Pretzsch H (1999) Changes in forest growth. Forstwissenschaftliches Centralblatt 118: 228–250.

78. Solberg S, Dobbertin M, Reinds GJ, Lange H, Andreassen K, et al. (2009) Analyses of the impact of changes in atmospheric deposition and climate on forest growth in European monitoring plots: A stand growth approach. Forest Ecology and Management 258: 1735–1750.

79. Worbes M (1999) Annual growth rings, rainfall-dependent growth and long-term growth patterns of tropical trees from the Caparo Forest Reserve in Venezuela. Journal of Ecology 87: 391–403.

80. Sitch S, Smith B, Prentice IC, Arneth A, Bondeau A, et al. (2003) Evaluation of ecosystem dynamics, plant geography and terrestrial carbon cycling in the LPJ dynamic global vegetation model. Global Change Biology 9: 161–185.

81. Westerling AL, Hidalgo HG, Cayan DR, Swetnam TW (2006) Warming and earlier spring increase western US forest wildfire activity. Science 313: 940–943.

7

Estimating Carbon Flux Phenology with Satellite-Derived Land Surface Phenology and Climate Drivers for Different Biomes: A Synthesis of AmeriFlux Observations

Wenquan Zhu[1,2]*, Guangsheng Chen[3], Nan Jiang[2], Jianhong Liu[2], Minjie Mou[2]

1 State Key Laboratory of Earth Surface Processes and Resource Ecology, Beijing Normal University, Beijing, China, **2** College of Resources Science and Technology, Beijing Normal University, Beijing, China, **3** Environmental Sciences Division, Oak Ridge National Laboratory, Oak Ridge, Tennessee, United States of America

Abstract

Carbon Flux Phenology (CFP) can affect the interannual variation in Net Ecosystem Exchange (NEE) of carbon between terrestrial ecosystems and the atmosphere. In this study, we proposed a methodology to estimate CFP metrics with satellite-derived Land Surface Phenology (LSP) metrics and climate drivers for 4 biomes (i.e., deciduous broadleaf forest, evergreen needleleaf forest, grasslands and croplands), using 159 site-years of NEE and climate data from 32 AmeriFlux sites and MODIS vegetation index time-series data. LSP metrics combined with optimal climate drivers can explain the variability in Start of Carbon Uptake (SCU) by more than 70% and End of Carbon Uptake (ECU) by more than 60%. The Root Mean Square Error (RMSE) of the estimations was within 8.5 days for both SCU and ECU. The estimation performance for this methodology was primarily dependent on the optimal combination of the LSP retrieval methods, the explanatory climate drivers, the biome types, and the specific CFP metric. This methodology has a potential for allowing extrapolation of CFP metrics for biomes with a distinct and detectable seasonal cycle over large areas, based on synoptic multi-temporal optical satellite data and climate data.

Editor: Bruno Hérault, Cirad, France

Funding: This work was supported by the National Natural Science Foundation of China (Grant No. 41371389), the State Key Laboratory of Earth Surface Processes and Resource Ecology (Grant No. 2013-ZY-14), and the Fundamental Research Funds for the Central University. The funders had no role in study design, data collection and analysis, decision to publish, or preparation of the manuscript.

Competing Interests: The authors have declared that no competing interests exist.

* E-mail: zhuwq75@bnu.edu.cn

Introduction

Vegetation phenology plays an important role in adjusting the annual Net Ecosystem Exchange (NEE) (see Acronym S1 in supporting information for a list of acronyms and definitions used in this paper) of carbon between terrestrial ecosystems and the atmosphere [1–5]. The interannual variation in ecosystem productivity caused by vegetation phenology shifts was widely investigated by field studies [6–9] and ecosystem models [10–14]. An earlier start or/and a later end of vegetation growing season can extend the period of photosynthesis, and thus increased primary productivity is expected. Indeed, some previous studies have shown a positive effect of Growing Season Length (GSL) on net productivity (e.g., 5.9 g $C \cdot m^{-2} \cdot d^{-1}$ in a deciduous forest [15] and around 4 g $C \cdot m^{-2} \cdot d^{-1}$ in a subtropical forest stand [16]). Moreover, the length of Carbon Uptake Period (CUP) has much predictive power about the spatial variation of annual NEE. For example, the length of CUP can explain 80% of the spatial variance in annual NEE for deciduous forests across a latitudinal and continental gradient [17].

There are currently numerous data sources available for estimating the timing of recurrent vegetation phenology transitions, such as the ground-, satellite- and eddy covariance flux-based data sources [18]. Land Surface Phenology (LSP) is defined as the study of the timing of recurring seasonal pattern of variation in vegetated land surfaces observed from synoptic sensors [19,20].

Satellite-based LSP is characterized by the Start (SOS) and End (EOS) of growing Season, which are closely related to vegetation growth or photosynthesis. Carbon Flux Phenology (CFP) is defined as the detrended zero-crossing timing of NEE from a source to a sink in spring and *vice versa* in autumn [3,4,18,19]. CFP is characterized by the Start (SCU) and End (ECU) of Carbon Uptake, which are closely related to the difference between growth and respiration. LSP allows the determination of GSL or the duration of canopy coverage from the difference between EOS and SOS, while CFP allows the determination of CUP from the difference between ECU and SCU. The CUP is controlled by GSL, but is not identical because growth will typically commence and terminate some time before and after the NEE changes sign in spring and autumn, respectively [19,21]. White & Nemani [13] found that there was a strong relationship between NEE and CUP, but a very weak relationship between NEE and GSL for deciduous forests. Thus, CUP is a potentially useful indicator of annual carbon sequestration [3]. However, the application of CUP is hindered by the limited number of flux towers and the distribution and footprint of these flux towers [3,19,21]. Although more than 500 tower sites from approximately 30 regional networks across 5 continents are currently operating on a long-term basis, these globally distributed eddy flux sites sample only a small subset of the Earth's biomes, disturbance regimes, and land management systems. Thus, estimation of CUP over large areas remains challenging [19,21,22].

Some limited attempts have been made to estimate CFP dates beyond the footprints of flux towers [18,19,21,22]. Using over 30 site-years of data from 12 eddy flux sites, Baldocchi *et al.* [22] found that 64% of variance in SCU can be explained by the date when soil temperature matched the mean annual air temperature. Remote sensing provides spatially comprehensive measures of ecosystem activity and therefore is a potentially powerful tool to allow extrapolation of CUP over large areas. To test the capabilities of remote observations in estimating CUP, Churkina *et al.* [21] related the GSL from remotely sensed data to the CUP from eddy flux tower measurements and found a strong relationship between them. However, a comparison of multiple phenology data sources indicated that no single source of phenological data was able to accurately describe annual patterns of flux phenology [18]. Therefore, Gonsamo et al. [19] combined LSP dates with the mean monthly surface temperature derived from remote sensing observations to predict CUP. Their results indicated that remote sensing-derived multiple surface variables can explain CUP variability by more than 70% in spring and autumn. However, this CUP determination approach is just based on four selected temperate and boreal deciduous forest CO_2 flux tower sites. A more comprehensive analysis, based on multi-year data from eddy flux sites across large areas for various biome types, is still expected. Moreover, improved estimation of LSP dates combined with optimal climate drivers may further enhance the CUP estimation performance.

Using data from a large number of AmeriFlux sites, this study aims to estimate CFP metrics with satellite-derived LSP metrics and climate drivers for different biomes, including deciduous broadleaf forest, evergreen needleleaf forest, grasslands and croplands. We first evaluated different LSP retrieval methods and Vegetation Index (VI) products based on the observed CFP dates and selected the best performing method and VI product to retrieve LSP dates as the explanatory variables in estimating both SCU and ECU. Then, we carried out a sensitive analysis to search the optimal explanatory climate drivers for the estimation of SCU and ECU. Finally, the estimated LSP dates and the selected optimal explanatory climate drivers were combined to estimate CFP dates, and a comprehensive discussion was given to highlight the limitations and potentials of the proposed methodology.

Data and methods

Data and pre-processing

Site carbon flux and meteorological data. The daily NEE (g C•m^{-2}•d^{-1}), air temperature (°C) and precipitation (mm) data used in this study were derived from the post-processed Level 4 product (available at: http://daac.ornl.gov/FLUXNET/fluxnet. shtml) of the AmeriFlux sites (Figure 1, Dataset S1). The covered period for the product was generally from 1995 to 2007 but depending on the specific site. For example, the acquired NEE and meteorological data were from 1998 to 2007 for the Niwot Ridge site, while they were from 1995 to 1999 for the Walker Branch site in the United States. The same years for having both NEE and Moderate Resolution Imaging Spectroradiometer (MODIS) VI data were used for analysis. Therefore, our analysis only focused on the period of 2000–2007 since the overlay time period for both data sets only covered from February 2000 (Start date for MODIS VI data) to December 2007 (End date for available NEE data). For each biome type, we first excluded the sites with more than 60 days deviations from the average SCU and ECU. We regarded each year for each flux tower site as one site-year and excluded the

site-years whose daily NEE values were missing for the carbon source-sink or sink-source transition period. Moreover, only the biomes with at least 10 site-years were included for analysis in order to get robust estimations for CFP dates. Therefore, we got 32 eddy flux sites, which covered 4 biome types according to the International Geosphere-Biosphere Program (IGBP) classification system (Figure 1). There were totally 73 site-years involved in the spring carbon source-sink transition period and 86 site-years involved in the autumn carbon sink-source transition period.

Remotely sensed data. The Terra's MODIS 250 m 16-day composited VI products (MOD13Q1, V005) for the 32 flux tower sites were used in this analysis (available at: http://daac.ornl.gov/ MODIS/). The first VI product was the standard Normalized Difference Vegetation Index (NDVI), and the second was the Enhance Vegetation Index (EVI). The VI time series for the pixel located at the center of the flux tower was used to retrieve land surface phenological metrics. The covered period for the VI data was the same as the NEE data for a given flux site. Much noise existed in the VI time series because of cloud contamination, atmospheric variability and sun-sensor-surface viewing geometries [23,24]. A filtering process was needed before using VI to retrieve phenological metrics [25]. We used the Savitzky-Golay filter method to remove the noise in the VI time series [26].

Methods

Retrieving CFP dates from NEE data. The SCU and ECU were retrieved based on the method proposed by Baldocchi *et al.* [22]. The original method is based on visual interpretation of the daily NEE time series. We developed this method to retrieve SCU and ECU automatically through fitting a regression equation between the daily NEE and the Julian Day of Year (DOY), using subsets of NEE data from spring source-sink and autumn sink-source transition periods, respectively (Figure 2). Specifically, SCU and ECU were automatically retrieved by the following three steps: (1) the original daily NEE was smoothed with a moving average of a 15-day width; (2) based on the smoothed daily NEE, a 10-day width window with the first 5 elements greater than zero and the last 5 elements less than zero was selected in the spring/ summer period to predict SCU, and another 10-day width window with the first 5 elements less than zero and the last 5 elements greater than zero was selected in the autumn/winter period to predict ECU; and (3) the smoothed daily NEE in the two selected windows was linearly regressed to predict SCU and ECU at the zero intersection.

Retrieving LSP dates from remotely sensed data. A number of methods have been developed to retrieve land surface phenology metrics using satellite VI time series [27–29]. These methods can be classified into 3 types: the threshold method (i.e., a global absolute threshold value or a local relative threshold value defined as a fraction of the annual amplitude) [30–33], the autoregressive moving average method [27,34] and the function fitting method [28,35–39]. Almost all the methods mentioned above have been proven to be consistent with their given references (e.g., ground observed phenology events, model simulated vegetation phenology or eddy covariance flux tower-derived phenological metrics), but it was very difficult to give the ordinal rank of SOS methods because they varied geographically [27]. Therefore, this study first investigated these 3 types of satellite methods (including 6 specific retrieval methods) based on the first MODIS VI product (i.e., NDVI), and selected the one with the best performance to retrieve LSP dates. Then, a comparison between the two MODIS VI products (i.e., NDVI and EVI) was carried out based on the best-performing LSP retrieval method, and the more suitable VI product was selected to

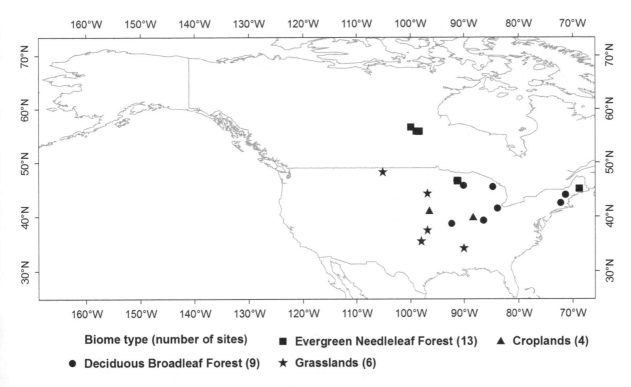

Biome type (number of sites)   ■ **Evergreen Needleleaf Forest (13)**   ▲ **Croplands (4)**

● **Deciduous Broadleaf Forest (9)**   ★ **Grasslands (6)**

Figure 1. Distribution of eddy flux towers and their corresponding biome types.

retrieve LSP dates as the explanatory variables in estimating CFP dates. The detailed descriptions about these 6 retrieval methods and the evaluation process were given in Text S1.

Identifying the explanatory climate drivers. The LSP dates derived from the more suitable VI product with the best satellite retrieval method, the cumulative daily air temperature (above 0°C) and total precipitation were used to identify the explanatory climate drivers. Previous studies indicated that SCU usually occurs 0–20 days later than SOS while ECU usually occurs 0–60 days earlier than EOS [4,19,21]. Therefore, we restricted the impact period of climate drivers on SCU/ECU in the range from 60 days before SOS/EOS to 20 days after SOS/EOS. To identify the optimal impact period for each climate driver, different impact periods were tested according to the distance (in days) from SOS/EOS, 10-day after SOS/EOS and 20-day after SOS/EOS with a step of 10 days (Figure 3). Therefore, we got 18 candidate impact periods for each climate driver (i.e., cumulative daily air temperature (above 0°C) or total precipitation). The coefficient of determination (R^2) between observed SCU/ECU and each climate driver with different candidate impact periods was used to select the best explanatory climate drivers. Only the climate driver in a given impact period with the highest R^2 in its group (i.e., 18 candidate cumulative temperature or total precipitation data for each biome type and each phenological metric) and with a statistical significance at the 0.05 level will be selected to estimate SCU/ECU.

Estimating CFP dates. Using the least-squares linear regression model, the CFP dates (i.e., SCU and ECU) can be estimated with the LSP dates (i.e., SOS and EOS) and the selected explanatory climate drivers. The estimating performance of the linear regression models was evaluated with coefficient of determination (R^2), Root Mean Square Error (RMSE) and the leave-one-out cross-validation approach [19,40]. Significance test

for these linear regression models was conducted by F-test with the standard 0.05 cutoff indicating statistical significance (i.e., $P<0.05$).

Results

The relationship between CFP and LSP dates

Our evaluation about the different LSP retrieval methods and MODIS VI products indicated that the NDVI-derived LSP dates with the local mean midpoint threshold method were more consistent with the observed CFP dates (see details in Text S2). Table 1 showed the relationship between CFP and LSP dates. SOS explained the SCU variance by 43.1%–78.4% for different biomes. The RMSE between SCU and SOS ranged from 2.7 to 7.6 days, which was far smaller than the temporal resolution of the satellite data (~16 days). This indicated that the SCU can be estimated with SOS to the accuracy that was comparable to the 16-day composited temporal resolution of satellite sensor. Comparing with the SCU, lower performance was found for estimating ECU based on EOS, with relatively lower explanatory variances and higher RMSE for different biomes. However, this RMSE was still comparable with the 16-day composited temporal sampling resolution of satellite data.

Different biomes showed distinctive CFP dates as estimated based on the LSP dates (Table 1). For example, evergreen needleleaf forest had the highest explanatory variance (78.4%) in estimating SCU based on SOS while grasslands had the lowest (43.1%). On the contrary, evergreen needleleaf forest showed the poorest performance in estimating ECU based on EOS while grasslands showed the best performance (67.1%). In general, the performance in estimating CFP dates for a single biome was better than multiple biomes.

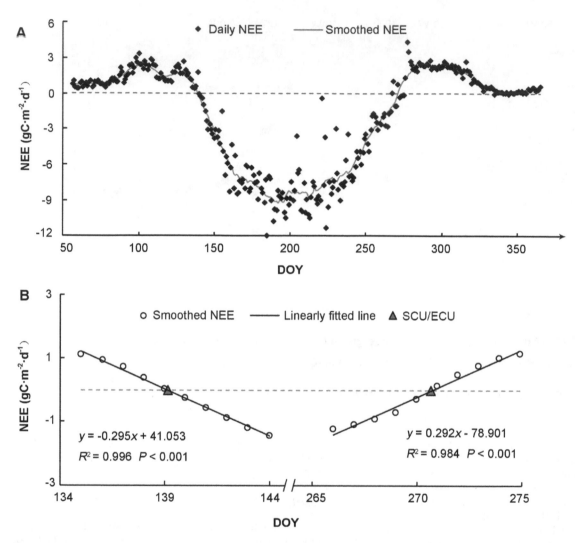

Figure 2. A schematic demonstration of the retrieval method for carbon flux phenology dates. A) The original and smoothed 15-day mean Net Ecosystem Exchange (NEE) of carbon, and B) the two selected transition periods for spring source-sink and autumn sink-source for identifying linear regressions between NEE and the Julian Day of Year (DOY). Start/End of Carbon Uptake (SCU/ECU) is estimated at the zero intersection.

The relationship between CFP dates and climate drivers

Figure 4 showed the coefficient of determination (R^2) between CFP dates and climate drivers during different impact periods. For SCU explanatory variances by the cumulative daily air temperature (above 0°C) (Figure 4A), evergreen needleleaf forest demonstrated better performance than other 3 biomes with consistently the highest R^2 values (ranging from 88.6% to 95.7% among the 18 tested impact periods). Considering its high explained variances, we selected the cumulative daily air temperature above 0°C during 20 days before NDVI-derived SOS as one of the explanatory variables in estimating SCU. Similarly, the total precipitation during the period of 30-day before SOS and 10-day after SOS was also selected for evergreen needleleaf forest in estimating SCU (Figure 4B). All the total precipitation variables during different impact periods were not significant at the 0.05 level for both deciduous broadleaf forest and croplands (Figure 4B). Therefore, no precipitation variables were selected for these two biomes to estimate SCU. In summary, the optimal impact periods for different climate drivers (i.e., cumulative air temperature and total precipitation) and different CFP

metrics (i.e., SCU and ECU) were marked with stars in Figure 4. Only the climate drivers being significant during their impact periods were selected as the explanatory variables to estimate CFP dates.

The sensitivity of CFP metrics to climate drivers varied among different biomes (Figure 4). For SCU, evergreen needleleaf forest showed higher sensitivities to both cumulative temperature (above 0°C) and total precipitation, while deciduous broadleaf forest and cropland were only sensitive to cumulative temperature (Figure 4 A, B). Grassland had a higher sensitivity to total precipitation during the period of 20-day before SOS and 10-day after SOS but a lower sensitivity to cumulative temperature for the 6 selected grassland sites. For ECU, herbaceous biomes (i.e., grasslands and croplands) showed a higher sensitivity to cumulative temperature than woody biomes (e.g., deciduous broadleaf and evergreen needleleaf forest) (Figure 4 C). Deciduous broadleaf forest showed higher explained variances by total precipitation variables in different impact periods for ECU, while evergreen needleleaf forest demonstrated lower explained variances (Figure 4 D).

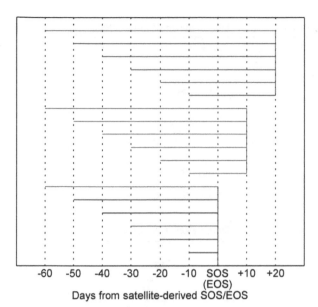

-60　-50　-40　-30　-20　-10　SOS　+10　+20
(EOS)
Days from satellite-derived SOS/EOS

Figure 3. A schematic representation of the different impact periods for climate drivers. The different impact periods of climate drivers on carbon flux phenology dates were determined in terms of the distance (in days) from satellite-derived Start/End of Season (SOS/EOS), 10-day after SOS/EOS and 20-day after SOS/EOS with a step of 10 days. There were totally 18 candidate impact periods for each climate driver. Negative values indicate the days before SOS/EOS and positive values indicates the days after SOS/EOS.

Estimation of CFP dates with LSP dates and climate drivers

Figure 5 showed the relationships between observed and estimated CFP dates based on the linear regression (Figure 5 A, C, E, G) and the leave-one-out cross-validation (Figure 5 B, D, F, H) approaches in terms of the best performing explanatory variables. All of the estimation performances were statistically significant ($P<0.05$). The explained variance for SCU ranged from 71% to 97%, and the RMSE for SCU ranged from 2.6 to 5.2 days (Figure 5 A, C, E, G). Comparing with SCU, the ECU estimation showed a relatively lower performance, with a relatively lower R^2 (60%–84%) and a higher RMSE (5.3–8.5 days) (Figure 5 A, C, E, G).

The CFP estimation performance varied among different biomes. The estimation RMSE for both SCU and ECU was less than 9 days for deciduous broadleaf forest (Figure 5 A, B). Evergreen needleleaf forest had a lower RMSE for SCU but a higher RMSE for ECU (Figure 5 C, D). Grasslands had the highest R^2 but also the highest RMSE for ECU (Figure 5 E, F), while croplands had the lowest RMSE for both SCU and ECU (Figure 5 G, H).

Discussion

The optimal explanatory variables in estimating CFP dates

The temporal and spatial variation in CFP metrics (i.e., SCU and ECU) is controlled by many factors, including the biome type, canopy structures, species compositions, soil type, forest age and meteorological factors (e.g., temperature, precipitation, etc.) [3,18,19,41]. Wu et al. [3] demonstrated that the interannual variation in NEE and phenological indicators at a study site could be mainly resulted from the meteorological factors, while differences of canopy structures and species compositions had no significant impacts. On the contrary, when the spatial variation was considered, the primary controlling factors may be site-specific differences in canopy structures, soil prosperities and biome types. Therefore, they suggested a separated analysis for spatial and temporal variation in the response of annual NEE to CUP and its transitions. In fact, the most challenging aspect in estimating CFP dates is to identify the optimal explanatory variables whether or not the temporal and spatial variation in CFP metrics is separately considered.

The satellite-derived phenological metrics reflect an integrated signal of a group of species (individuals) in a pixel because remote sensing can capture the spectral characteristics of green leaf and vegetation canopy structures at moderate to coarse spatial resolutions [19,42,43]. Our results showed high explained SCU/ECU variances with SOS/EOS (Table 1), which have been also found by Gonsamo et al. [19]. This suggested that satellite-derived LSP dates can effectively reflect the spatial and temporal variations in CFP dates and should be selected as one of the primary explanatory variables in estimating CFP dates, especially for the large-scale (e.g., regional or continental) studies. However, we should also note that large discrepancies exist in different LSP retrieval methods in terms of the CFP estimating performances (Text S2). Because of the different SOS/EOS definitions (Text S1), the satellite-derived LSP metrics with different methods do

Table 1. The coefficient of determination (R^2), Root Mean Square Error (RMSE) and Bias between Net Ecosystem Exchange (NEE)-derived carbon flux phenology dates and Normalized Difference Vegetation Index (NDVI)-derived land surface phenology dates based on the best performing retrieval method (i.e., the local mean midpoint threshold method) for different biomes.

Biome type	SOS vs. SCU[†]				EOS vs. ECU[†]			
	Samples	R^2 (%)	RMSE	Bias	Samples	R^2 (%)	RMSE	Bias
Deciduous broadleaf forest	24	74.3[*]	7.5	−10.5	20	51.4[*]	6.3	4.6
Evergreen needleleaf forest	16	78.4[*]	7.6	15.8	30	43.5[*]	13.1	35.6
Grasslands	16	43.1[*]	6.5	3.3	14	67.1[*]	10.3	9.4
Croplands	17	68.8[*]	2.7	−0.2	22	65.0[*]	5.6	14.7
All biomes	73	49.6[*]	17.1	0.7	86	43.5[*]	14.6	18.8

[†]SOS = Start of Season derived from satellite data, SCU = Start of Carbon Uptake derived from carbon flux data, EOS = End of Season derived from satellite data, ECU = End of Carbon Uptake derived from carbon flux data.
[*]Statistically significant at the 0.05 level.

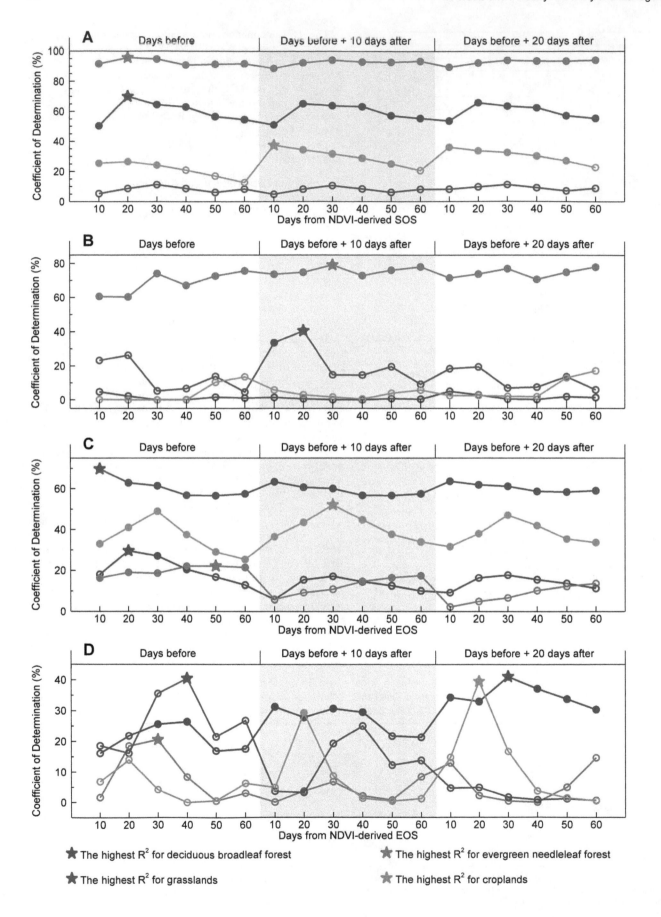

★ The highest R² for deciduous broadleaf forest   ★ The highest R² for evergreen needleleaf forest

★ The highest R² for grasslands   ★ The highest R² for croplands

Figure 4. The relationships between carbon flux phenology dates and climate drivers in different impact periods. A) The coefficient of determination (R^2) between Net Ecosystem Exchange (NEE)-derived Start of Carbon Uptake (SCU) and the cumulative daily air temperature (above 0°C) for different periods around Normalized Difference Vegetation Index (NDVI)-derived Start of Season (SOS). B) The R^2 between NEE-derived SCU and the total precipitation for different periods around NDVI-derived SOS. C) The R^2 between NEE-derived End of Carbon Uptake (ECU) and the cumulative daily air temperature (above 0°C) for different periods around NDVI-derived End of Season (EOS). D) The R^2 between NEE-derived ECU and the total precipitation for different periods around NDVI-derived EOS. Red colored line: Deciduous broadleaf forest; green: evergreen needleleaf forest; blue: grassland; orange: cropland. Stars indicate the locations with the highest R^2 for each biome and with a statistical significance at the 0.05 level. Solid circles indicate statistically significant R^2 at the 0.05 level, and hollow circles indicate statistically non-significant R^2.

not actually measure the same phenological traits [44]. Our results demonstrated that the SCU/ECU is much closer to and more consistent with the time when satellite-derived vegetation index reaches its midpoint in a growing season for the 4 tested biomes.

Meteorological factors are important candidate explanatory variables in estimating CFP dates. Previous results [18,19] and our results (Table 1, Figure 5) all indicated that combining optimal climate drivers with LSP dates can obviously improve the estimation robustness when the established regression models are applied in a wide spatial and temporal range. Our results also demonstrated the differences in the optimal impact periods of climate drivers on SCU/ECU for different biomes (Figure 4). Therefore, the key question is how to select the optimal explanatory climate drivers for different biomes. Gonsamo et al. [19] conducted a sensitivity analysis to search the optimal impact period based on human calendar month (e.g., the mean air temperature in April and May for SCU). Rather than applying the human calendar month, we used the distance (in days) from the LSP dates to identify the optimal impact periods. This is because the LSP dates retrieved with the best performing method are close to the CFP dates and vary simultaneously with CFP dates when across regions and/or biomes, while the unified human calendar month cannot reflect the variations in vegetation phenology across large heterogeneous areas and thus may fail to describe the actual impact period of climate drivers on SCU/ECU.

The difference in CFP estimation performance among different biomes

Large discrepancies existed in the CFP estimation performance for different biomes. The CFP dates are relatively easy to be estimated for deciduous broadleaf forest because of its distinct seasonal variation in canopy structure and carbon flux which can be effectively captured by remote sensing and eddy covariance system. Our estimated CFP dates based on both LSP dates and climate drivers for deciduous broadleaf forest was comparable with that based on observed carbon flux and meteorology data (5.20 vs. 5.12 for SCU, 5.30 vs. 6.65 for ECU in RMSE) and slightly better than that solely based on satellite data (5.20 vs. 6.98 for SCU, 5.30 vs. 8.88 for ECU in RMSE) from Gonsamo et al. [19].

Theoretically, it is difficult to define CFP metrics for conifers because NEE may be negative throughout the year. The transition from net positive to net negative NEE is more flexible in coniferous than in deciduous forest because the seasonality of coniferous forest is not related to changes in canopy structure [45]. However, in high latitude snow-dominated coniferous forests, the annual cycle from near-total snow cover to a mature canopy provides a distinct and detectable VI and NEE cycle [27]. The flux sites for evergreen needleleaf forest in this study are distributed above 45°N and are fully covered by snow in winter. The satellite-derived LSP dates are closely related to the timing of spring snowmelt and winter snow, which substantially reflects the change in air temperature. Therefore, the CFP dates estimation perfor-

mance with LSP dates and climate drivers for evergreen needleleaf forest in this study was relatively high, especially in estimating SCU. It should be noted that the high estimation performance for evergreen needleleaf forest may be only suitable for snow-dominated ecosystems and cannot be extrapolated to other coniferous forests, such as temperate and subtropical evergreen needleleaf forest.

Out results also demonstrated a high estimation performance for CFP dates of herbaceous biomes (i.e., grassland, crop). The high estimation performance may result from the primary control of satellite-derived LSP dates on estimating CFP dates (Table 1). For example, the phenological development for the 6 selected grassland sites was mainly precipitation-driven (Figure 4 A, B) and their leaf-out and NEE transition was usually occurred in a short period. Moreover, herbaceous biomes do not have understory plants that could confound the spectral signal. Therefore, LSP dates showed a significant variance explanatory rate in estimating CFP dates for herbaceous biomes.

The difference in estimation performance between SCU and ECU

The SCU estimation performance was generally better than ECU (Figure 5). The explained variance by LSP dates (Table 1) and climate drivers (Figure 4) for ECU was generally lower than that for SCU, implying that the satellite-derived EOS and climate drivers had relatively weak relationships with ECU. During the greenup phase, increasing greenness is closely related to chlorophyll, leaf area and changes in canopy structure, which scale rather well with photosynthesis and respiration [46,47]. However, during the leaf senescence phase, changes in leaf color, environmental stress (e.g., drought stress), and meteorological conditions (e.g., cooler air temperature) may complicate the relationships between canopy structure-based phenology metrics and carbon fluxes, and in general make the detection of senescence events more difficult [41,48]. In fact, the factors controlling senescence and dormancy are not well-documented in all biomes [41]. A mechanistic understanding of the drivers controlling senescence and dormancy is urgently needed.

Potentials of the optimization method in estimating CFP metrics

The optimization method proposed in this study can be used to extrapolate regional CFP metrics through extending the footprints of flux towers. As for a given biome over large heterogeneous areas, a synoptic train of thought is first to classify the biome to smaller ecoregions, or use existing ecoregion maps (e.g., the terrestrial ecoregions compiled by the World Wildlife Fund (WWF) (available at: http://worldwildlife.org/biome-categories/terrestrial-ecoregions)), since vegetation phenology may differ significantly even within the same biome. Then for each ecoregion, time-series satellite data, climate data and NEE data for the involved eddy flux sites can be used to build an optimized empirical model to predict CFP metrics beyond the footprints of flux towers.

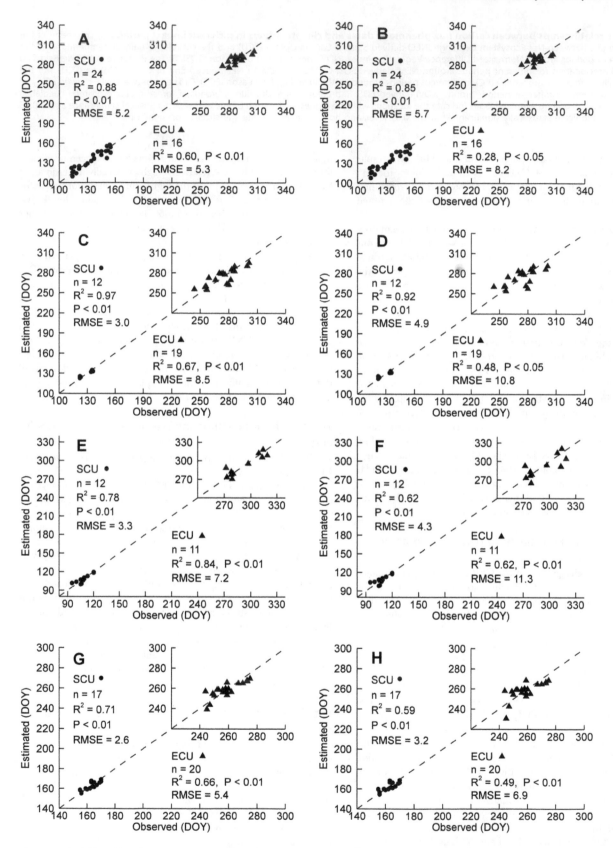

Figure 5. Relationships between observed and estimated carbon flux phenology dates for different biomes. A) and B) Deciduous broadleaf forest, C) and D) Evergreen needleleaf forest, E) and F) Grasslands, and G) and H) Croplands. The left panel (i.e., A, C, E and G) indicates the relationships between observed Start/End of Carbon Uptake (SCU/ECU) in Julian Day of Year (DOY) and estimated with the best performing explanatory variables given in Figure 4, and the right panel (i.e., B, D, F and H) indicates the relationships between observed SCU/ECU and estimated with the best performing explanatory variables based on the leave-one-out cross-validation approach.

Conclusions

This study provided a methodology to estimate CFP metrics with satellite-derived LSP metrics and climate drivers for different biomes through a synthesis of AmeriFlux observations. LSP metrics combined with optimal climate drivers can explain SCU variability by more than 70% (ranging from 71% to 97% for different biomes) and ECU variability by more than 60% (ranging from 60% to 84% for different biomes). The RMSE of the estimations ranged from 2.6 to 5.2 days for SCU and from 5.3 to 8.5 days for ECU. The results of our study highlighted the relative strengths and weaknesses of LSP metrics derived from different methods and climate drivers in different impact periods in estimating a specific CFP metric for different biomes. The estimation performance for the methodology was primarily dependent on the optimal combination of the LSP retrieval methods, the explanatory climate drivers, the biome types, and the specific CFP metric. Although the proposed methodology showed high performance in estimating CFP metrics for biomes with a distinct and detectable VI and NEE cycle, a better mechanistic understanding of the drivers controlling vegetation phenology is urgently needed in order to improve the estimation performance, especially for senescence and dormancy phenology.

Acknowledgments

Special thanks to the data archive at the Carbon Dioxide Information Analysis Center (CDIAC) for providing flux NEE data, and the Oak Ridge National Laboratory for the National Aeronautics and Space Administration for providing MODIS data. We also appreciate the three anonymous reviewers for providing very useful comments and suggestions, which led to a major improvement of this paper.

Author Contributions

Conceived and designed the experiments: WQZ. Performed the experiments: WQZ MJM NJ JHL. Analyzed the data: WQZ GSC. Wrote the paper: WQZ GSC. Edited manuscript: WQZ GSC..

References

1. Richardson AD, Black TA, Ciais P, Delbart N, Friedl MA, et al. (2010) Influence of spring and autumn phenological transitions on forest ecosystem productivity. Philosophical Transactions of the Royal Society B-Biological Sciences 365: 3227–3246.
2. Dragoni D, Schmid HP, Wayson CA, Potter H, Grimmond C, et al. (2011) Evidence of increased net ecosystem productivity associated with a longer vegetated season in a deciduous forest in south-central Indiana, USA. Global Change Biology 17: 886–897.
3. Wu CY, Gonsamo A, Chen JM, Kurz WA, Price DT, et al. (2012) Interannual and spatial impacts of phenological transitions, growing season length, and spring and autumn temperatures on carbon sequestration: A North America flux data synthesis. Global and Planetary Change 92–93: 179–190.
4. Wu CY, Chen JM, Gonsamo A, Price DT, Black TA, et al. (2012) Interannual variability of net carbon exchange is related to the lag between the end-dates of net carbon uptake and photosynthesis: Evidence from long records at two contrasting forest stands. Agricultural and Forest Meteorology 164: 29–38.
5. Wu C, Chen JM, Black TA, Price DT, Kurz WA, et al. (2013) Interannual variability of net ecosystem productivity in forests is explained by carbon flux phenology in autumn. Global Ecology and Biogeography 22: 994–1006.
6. Chen WJ, Black TA, Yang PC, Barr AG, Neumann HH, et al. (1999) Effects of climatic variability on the annual carbon sequestration by a boreal aspen forest. Global Change Biology 5: 41–53.
7. Black TA, Chen WJ, Barr AG, Arain MA, Chen Z, et al. (2000) Increased carbon sequestration by a boreal deciduous forest in years with a warm spring. Geophysical Research Letters 27: 1271–1274.
8. Bergeron O, Margolis HA, Black TA, Coursolle C, Dunn AL, et al. (2007) Comparison of carbon dioxide fluxes over three boreal black spruce forests in Canada. Global Change Biology 13: 89–107.
9. Richardson AD, Hollinger DY, Dail DB, Lee JT, Munger JW, et al. (2009) Influence of spring phenology on seasonal and annual carbon balance in two contrasting New England forests. Tree Physiology 29: 321–331.
10. Kramer K (1995) Modeling comparison to evaluate the importance of phenology for the effects of climate-change on growth of temperate-zone deciduous trees. Climate Research 5: 119–130.
11. White MA, Running SW, Thornton PE (1999) The impact of growing-season length variability on carbon assimilation and evapotranspiration over 88 years in the eastern US deciduous forest. International Journal of Biometeorology 42: 139–145.
12. Cramer W, Bondeau A, Woodward FI, Prentice IC, Betts RA, et al. (2001) Global response of terrestrial ecosystem structure and function to CO$_2$ and climate change: results from six dynamic global vegetation models. Global Change Biology 7: 357–373.
13. White MA, Nemani AR (2003) Canopy duration has little influence on annual carbon storage in the deciduous broad leaf forest. Global Change Biology 9: 967–972.
14. Piao SL, Ciais P, Friedlingstein P, Peylin P, Reichstein M, et al. (2008) Net carbon dioxide losses of northern ecosystems in response to autumn warming. Nature 451: 49–52.
15. Baldocchi DD, Wilson KB (2001) Modeling CO$_2$ and water vapor exchange of a temperate broadleaved forest across hourly to decadal time scales. Ecological Modelling 142: 155–184.
16. Zhang WJ, Wang HM, Yang FT, Yi YH, Wen XF, et al. (2011) Underestimated effects of low temperature during early growing season on carbon sequestration of a subtropical coniferous plantation. Biogeosciences 8: 1667–1678.
17. Baldocchi D, Falge E, Gu LH, Olson R, Hollinger D, et al. (2001) FLUXNET: A new tool to study the temporal and spatial variability of ecosystem-scale carbon dioxide, water vapor, and energy flux densities. Bulletin of the American Meteorological Society 82: 2415–2434.
18. Garrity SR, Bohrer G, Maurer KD, Mueller KL, Vogel CS, et al. (2011) A comparison of multiple phenology data sources for estimating seasonal transitions in deciduous forest carbon exchange. Agricultural and Forest Meteorology 151: 1741–1752.
19. Gonsamo A, Chen JM, Wu CY, Dragoni D (2012) Predicting deciduous forest carbon uptake phenology by upscaling FLUXNET measurements using remote sensing data. Agricultural and Forest Meteorology 165: 127–135.
20. Gonsamo A, Chen JM, Price DT, Kurz WA, Wu CY (2012) Land surface phenology from optical satellite measurement and CO$_2$ eddy covariance technique. Journal of Geophysical Research-Biogeosciences 117, G03032.
21. Churkina G, Schimel D, Braswell BH, Xiao XM (2005) Spatial analysis of growing season length control over net ecosystem exchange. Global Change Biology 11: 1777–1787.
22. Baldocchi DD, Black TA, Curtis PS, Falge E, Fuentes JD, et al. (2005) Predicting the onset of net carbon uptake by deciduous forests with soil temperature and climate data: A synthesis of FLUXNET data. International Journal of Biometeorology 49: 377–387.
23. Holben B, Fraser RS (1984) Red and near-infrared sensor response to off-nadir viewing. International Journal of Remote Sensing 5: 145–160.
24. Kobayashi H, Dye DG (2005) Atmospheric conditions for monitoring the long-term vegetation dynamics in the Amazon using normalized difference vegetation index. Remote Sensing of Environment 97: 519–525.
25. Hird JN, McDermid GJ (2009) Noise reduction of NDVI time series: An empirical comparison of selected techniques. Remote Sensing of Environment 113: 248–258.
26. Chen J, Jönsson P, Tamura M, Gu Z, Matsushita B, et al. (2004) A simple method for reconstructing a high-quality NDVI time-series data set based on the Savitzky-Golay filter. Remote Sensing of Environment 91: 332–344.
27. White MA, Beurs KM, Didan K, Inouye DW, Richardson AD, et al. (2009) Intercomparison, interpretation, and assessment of spring phenology in North America estimated from remote sensing for 1982–2006. Global Change Biology 15: 2335–2359.
28. Zhang XY, Friedl MA, Schaaf CB, Strahler AH, Hodges J, et al. (2003) Monitoring vegetation phenology using MODIS. Remote Sensing of Environment 84: 471–475.
29. Cong N, Wang T, Nan HJ, Ma YC, Wang XH, et al. (2013) Changes in satellite-derived spring vegetation green-up date and its linkage to climate in China from 1982 to 2010: a multimethod analysis. Global Change Biology 19: 881–891.

30. Lloyd D (1990) A phenological classification of terrestrial vegetation cover using shortwave vegetation index imagery. International Journal of Remote Sensing 11: 2269–2279.

31. Fischer A (1994) A model for the seasonal variations of vegetation indices in coarse resolution data and its inversion to extract crop parameters. Remote Sensing of Environment 48: 220–230.

32. Markon CJ, Fleming MD, Binnian EF (1995) Characteristics of vegetation phenology over the Alaskan landscape using AVHRR time-series data. Polar Record 31: 179–190.

33. White MA, Thornton PE, Running SW (1997) A continental phenology model for monitoring vegetation responses to interannual climatic variability. Global Biogeochemical Cycles 11: 217–234.

34. Reed BC, Brown JF, VanderZee D, Loveland TR, Merchant JW, et al. (1994) Measuring phenological variability from satellite imagery. Journal of Vegetation Science 5: 703–714.

35. Beck PSA, Atzberger C, Hogda KA, Johansen B, Skidmore AK (2006) Improved monitoring of vegetation dynamics at very high latitudes: A new method using MODIS NDVI. Remote Sensing of Environment 100: 321–334.

36. Jönsson P, Eklundh L (2002) Seasonality extraction by function fitting to time-series of satellite sensor data. IEEE Transactions on Geoscience and Remote Sensing 40: 1824–1832.

37. Piao SL, Fang JY, Zhou LM, Ciais P, Zhu B (2006) Variations in satellite-derived phenology in China's temperate vegetation. Global Change Biology 12: 672–685.

38. Zhang XY, Friedl MA, Schaaf CB, Strahler AH (2004) Climate controls on vegetation phenological patterns in northern mid- and high latitudes inferred from MODIS data. Global Change Biology 10: 1133–1145.

39. Zhu W, Tian H, Xu X, Pan Y, Chen G, et al. (2012) Extension of the growing season due to delayed autumn over mid and high latitudes in North America during 1982–2006. Global Ecology and Biogeography 21: 260–271.

40. Shao J (1993) Linear-model selection by cross-validation. Journal of the American Statistical Association 88: 486–494.

41. Richardson AD, Keenan TF, Migliavacca M, Ryu Y, Sonnentag O, et al. (2013) Climate change, phenology, and phenological control of vegetation feedbacks to the climate system. Agricultural and Forest Meteorology 169: 156–173.

42. Campbell JB (2007) Introduction to Remote Sensing, 4th edn., The Guilford Press, New York.

43. Myneni RB, Maggion S, Iaquinto J, Privette JL, Gobron N, et al. (1995) Optical remote-sensing of vegetation: modeling, caveats, and algorithms. Remote Sensing of Environment 51: 169–188.

44. Badeck FW, Bondeau A, Bottcher K, Doktor D, Lucht W, et al. (2004) Responses of spring phenology to climate change. New Phytologist 162: 295–309.

45. Tanja S, Berninger F, Vesala T, Markkanen T, Hari P, et al. (2003) Air temperature triggers the recovery of evergreen boreal forest photosynthesis in spring. Global Change Biology 9: 1410–1426.

46. Lindroth A, Lagergren F, Aurela M, Bjarnadottir B, Christensen T, et al. (2008) Leaf area index is the principal scaling parameter for both gross photosynthesis and ecosystem respiration of Northern deciduous and coniferous forests. Tellus Series B-Chemical and Physical Meteorology 60: 129–142.

47. Peng Y, Gitelson AA, Keydan G, Rundquist DC, Moses W (2011) Remote estimation of gross primary production in maize and support for a new paradigm based on total crop chlorophyll content. Remote Sensing of Environment 115: 978–989.

48. Sparks TH, Menzel A (2002) Observed changes in seasons: An overview. International Journal of Climatology 22: 1715–1725.

8

Temperature Sensitivity and Basal Rate of Soil Respiration and Their Determinants in Temperate Forests of North China

Zhiyong Zhou[1,2]*, Chao Guo[1,2], He Meng[3]

1 Ministry of Education Key Laboratory for Silviculture and Conservation, Beijing Forestry University, Beijing, China, **2** The Institute of Forestry and Climate Change Research, Beijing Forestry University, Beijing, China, **3** College of Forestry, Inner Mongolia Agriculture University, Hohhot, China

Abstract

The basal respiration rate at 10°C (R_{10}) and the temperature sensitivity of soil respiration (Q_{10}) are two premier parameters in predicting the instantaneous rate of soil respiration at a given temperature. However, the mechanisms underlying the spatial variations in R_{10} and Q_{10} are not quite clear. R_{10} and Q_{10} were calculated using an exponential function with measured soil respiration and soil temperature for 11 mixed conifer-broadleaved forest stands and nine broadleaved forest stands at a catchment scale. The mean values of R_{10} were 1.83 µmol CO_2 m^{-2} s^{-1} and 2.01 µmol CO_2 m^{-2} s^{-1}, the mean values of Q_{10} were 3.40 and 3.79, respectively, for mixed and broadleaved forest types. Forest type did not influence the two model parameters, but determinants of R_{10} and Q_{10} varied between the two forest types. In mixed forest stands, R_{10} decreased greatly with the ratio of coniferous to broadleaved tree species; whereas it sharply increased with the soil temperature range and the variations in soil organic carbon (SOC), and soil total nitrogen (TN). Q_{10} was positively correlated with the spatial variances of herb-layer carbon stock and soil bulk density, and negatively with soil C/N ratio. In broadleaved forest stands, R_{10} was markedly affected by basal area and the variations in shrub carbon stock and soil phosphorus (P) content; the value of Q_{10} largely depended on soil pH and the variations of SOC and TN. 51% of variations in both R_{10} and Q_{10} can be accounted for jointly by five biophysical variables, of which the variation in soil bulk density played an overwhelming role in determining the amplitude of variations in soil basal respiration rates in temperate forests. Overall, it was concluded that soil respiration of temperate forests was largely dependent on soil physical properties when temperature kept quite low.

Editor: Dafeng Hui, Tennessee State University, United States of America

Funding: This study was jointly funded by the National Foundation of Natural Science of China (Grant No. 41003029) and by the Special Research Program for Public-Welfare Forestry of the State Forestry Administration of China (Grant No. 201104008). The funders had no role in study design, data collection and analysis, decision to publish, or preparation of the manuscript.

Competing Interests: The authors have declared that no competing interests exist.

* E-mail: zhiyong@bjfu.edu.cn

Introduction

CO_2 emission from soil and plants to the atmosphere determines the amplitude of feedbacks of forest ecosystems to global climate change. Accurate prediction of the amount of CO_2 respired by forest soil is of great importance in evaluating the carbon balance of forest ecosystems. In most cases, soil respiration rate at a given temperature can be estimated by the empirical functions using soil basal respiration rate (R_{10}, soil respiration rate at 10°C) and the temperature sensitivity of soil respiration (Q_{10}, a proportional change in soil respiration with a 10°C increase in temperature) [1,2,3]. Therefore, it seems vital to identify the biophysical variables driving these two parameters to advance the research on soil carbon turnover.

Soil respiration is mostly controlled by soil temperature [3,4], secondarily by soil moisture, nutrients [5], vegetation type [6], tree species composition [7], topography, and climate [8]. To increase the comparability of soil respiration rate under different environmental conditions, a standardized parameter (e.g. R_{10}) is proposed when emphasizing the effects of biophysical factors other than temperature. Although soil basal respiration may also be influenced by the similar variables mentioned above [9], it is still of importance to make clear the relationship of soil basal respiration with biophysical variables in improving the precision of simulation models. This is because, for a specific forest ecosystem, some biophysical factors can be considered as additional predictive variables when estimating soil respiration rate using empirical methods [3,10].

Great effort has been exerted to the response of soil respiration to a change in temperature in recent decades [2,11], which is denoted in most studies to be the temperature sensitivity of soil respiration, and is theoretically represented by an invariant coefficient (Q_{10}) of ~2, especially in coupled climate-carbon cycle models [12,13]. The extensive use of a fixed Q_{10} has brought large convenience in calculating the amount of CO_2 respired from soil, but it has also evoked a controversy between theoretical studies and incubation experiments or field measurements [14]. It is demonstrated that the temperature sensitivity of soil respiration (Q_{10}) can be influenced in ecosystems by many biophysical or physicochemical factors, including the forest floor conditions [15], soil physical properties [16], soil nutrients [17], and vegetation type [18]. Therefore, the Q_{10} originated from the temperature dependence equation shows distinct intersite difference or temporal variation [16,17,18]. Obviously, the application of a constant Q_{10} can not lead to an unbiased estimation of soil respiration rate for the studying ecosystem type any more.

Being illustrated by the calculation process of the common empirical function, an inherent correlation apparently exists between basal soil respiration and the temperature sensitivity [1,19]. Mathematically, Q_{10} is dependent on, and acts as a multiplier of R_{10} [19]. Any effort paid on the single parameter has limited use in improving the estimating precision of the extensively applied empirical functions.

Temperate forests in northern China mainly extend along the mountain ridge with heterogeneous growing conditions, which provide a natural experimental place for continuing similar research work on model parameters of soil respiration. In this study, we investigated the instantaneous rate of soil respiration and environmental variables at a representative catchment of the temperate forests in China, and calculated the two model parameters using the temperature dependent function. Herein hypotheses were proposed that the apparent temperature sensitivity of soil respiration could display detectable variations among forest types with different micro-environmental properties, and biophysical variables other than soil temperature could play an important role in determining soil basal respiration rate when the temperature decreased to a comparatively low level. Accordingly, our main objectives of this paper were to: 1) quantify the changing magnitude of model parameters of soil respiration within or between forest types; 2) identify the predominant variables controlling the spatial heterogeneity of the two parameters on the catchment scale in temperate forests.

Materials and Methods

Ethics statement

This research was conducted on field sites with the permission of the Taiyueshan Long-Term Forest Ecosystem Research Station. We declare that no privately owned land was used in this study, and that the field investigation did not involve any protected or endangered plant and animal species, and that no human or animal subjects were used in this study. The research has adhered to the legal requirements of China during the field study period.

Study site and experimental layout

This study was carried out at the catchment named after Xiaoshegou near the Taiyueshan Long-Term Forest Ecosystem Research Station (latitude 36°04′N, longitude 112°06′E; elevation 600 – 2600 m a.s.l), which is about 190 km southwest of Taiyuan in Shanxi province of China. Annual mean temperature varies between 10°C and 11°C, with 26°C in the warmest month of July and −23°C in coldest month of January; whilst mean annual precipitation ranges from 500 mm to 600 mm [20]. The hill in the study area is at an elevation of 1800 m with its bottom of 1200 m a.s.l. The soil type of the hill slope belongs to a Eutric Cambisols (FAO classification) or a Cinnamon soil (Chinese classification) with the mean soil depth of 30 cm to 50 cm, soil organic carbon content (SOC) of 0.77% to 5.47%, total nitrogen content (TN) of 0.036% to 0.232%, and soil pH from 6.9 to 7.6. The proportion of <0.01 mm and <0.001 mm soil fraction varies within the range of 46.54% to 63.10% and of 18.88% to 41.45%, respectively [21]. The dominant tree species in the forests are *Pinus tabuliformis*, *Quercus wutaishanica*, *Betula dahurica*, *Larix gmelinii var. principis-rupprechtii*, *Tilia mongolica*. The understory shrub community mainly consists of *Corylus mandshurica*, *Corylus heterophylla*, *Acer ginnala*, *Lespedeza bicolor*, *Philadelphus incanus*, *Rosa bella*, *Lonicera chrysantha*.

The herbaceous community is commonly composed of *Carex lanceolata*, *Spodiopogon sibiricus*, *Rubia chinensis*, *Thalictrum petaloideum*, *Melica pappiana*.

Twenty 20×20 m plots spread along four hill ridges with different topography at the small catchment, including 9 broad-leaved forest stands and 11 mixed conifer-broadleaved forest stands. The forest type was classified by the basal area ratio of coniferous to broadleaved tree species. The forest community was classified as the mixed forest type when its ratio fell within the range of 20% to 80%. Forest community structure was investigated in later Aug-2009. Each plant with diameter at breast height (DBH) >5 cm was measured for values of DBH and height respectively basing on tree species for these 20 plots. On each plot, five 5×5 m subplots were established for the investigation of shrub community, and five 1×1 m subplots for herbaceous community.

Measurements of soil respiration

Soil respiration rate was measured once per month for each forest stand during the growing season of May to November in 2008 and 2009, using a Li-Cor infrared gas analyzer (LI-8100, Li-Cor Inc., Lincoln, NE, U.S.A.) equipped with a portable chamber. The chamber was put on the top of installed collars for 2 minutes before measurements. In early April, nine polyvinyl chloride (PVC) collars were evenly placed on each plot with eight collars arranged in a circle at 5 m to the plot center and one right at the center. The PVC collar of 10 cm in diameter and 5 cm in height was permanently inserted 3 cm into the soil with 2 cm remaining above the surface of the forest floor. The live herbs or seedlings were carefully removed out the collars to avoid bias due to its respiratory activity just after plant growth occurred. Concurrently, soil temperature at 10 cm depth adjacent to each PVC collar was monitored using a thermocouple probe attached to LI-8100 system. The averaged data of soil respiration and soil temperature across the nine PVC collars per month were fitted to the following exponential model [1,10] to calculate basal parameters of soil respiration for each forest plot.

$$R_s = \alpha \times e^{\beta T} \tag{1}$$

where R_s is in situ soil respiration rate measured in the field, α and β are model parameters, T is the measured soil temperature. According to equation (1), the temperature sensitivity of soil respiration was calculated by:

$$Q_{10} = e^{10\beta} \tag{2}$$

Soil basal respiration was calculated by:

$$R_{10} = \alpha \times e^{10\beta} \tag{3}$$

Measurements of environmental variables

Shrub community was investigated by species for plant density and biomass of a representative sampling plant. The sampling plants were harvested and brought back to laboratory, and oven dried at 75°C to constant weight. The biomass of each shrub species within the community was estimated basing on plant density and its mean weight. The herbaceous plants in the 1×1 m subplot were all harvested for aboveground components. Additionally, litter on the forest floor was also collected in five 30×30 cm subplots on each plot. The herbaceous plant samples and litter were separately placed in envelope, transported to laboratory, and oven dried at 75°C for at least 48 h before weighing.

Soil cores of 4 cm in diameter and 20 cm in depth were sampled at five measurement points on each plot in later growing season of 2009. The air dried soil samples were mound to pass a 0.2 mm sieve for nutrient analysis after visible litter segments were picked out by hand. SOC and TN were determined separately following the modified Mebius method [22] and the Kjeldahl digestion procedure [23]. Soil phosphorus (P) was measured using the colorimetric determination method described by John [24]. Soil pH was measured in deionized H_2O using Sartorius AG (PB-10, Sartorius, Germany), after equilibration for 1 h in a water: soil ratio of 2.5:1. Soil cores were additionally excavated by a cylindrical sampler of 100 cm^3 at five sampling positions on each plot, and oven dried at 110°C for at least 48 h in laboratory to measure the soil bulk density.

Data analyses

Soil physicochemical properties were also averaged for each plot when their effects on R_{10} and Q_{10} were analyzed. A two-tailed t-test was applied to detect the differences of R_{10} and Q_{10} between these two forest types at $\alpha = 0.05$. The spatial variability was expressed using the coefficient of variation (CV) calculated as the following.

$$CV = \frac{SD}{M} \times 100\% \qquad (4)$$

where SD means standard deviation, and M represents mean value.

All these data analyses were carried out using the software of SPSS 15.0. Figures were made using the software of SigmaPlot in version 10.0.

In order to test the combined contribution of biophysical variables to the variability of R_{10} and Q_{10}, redundancy analysis (RDA) [25] was conducted with R_{10} and Q_{10} as dependent variables and with selected biophysical variables, i.e. DBH, soil pH, variances in soil bulk density, soil TN and soil pH, as explanatory variables. RDA was performed using the software of Canoco for Windows 4.5.

Results

Inter- and intra-forest-type variations in basal parameters of soil respiration

R_{10} and Q_{10} were on average 10% and 11% higher in the broadleaved forest stands than in the mixed forest stands, although no statistically significant difference was detected between these two forest types ($P = 0.25$ for R_{10} and 0.91 for Q_{10}). There existed large spatial heterogeneity in temperature sensitivity and basal rate of soil respiration among forest stands. The CV of R_{10} ranged from 11% in the broadleaved forest to 19% in the mixed forest, and the CV of Q_{10} varied from 24% in the broadleaved forest to 29% in the mixed forests (Table 1).

Particularly, in the mixed forest stands, R_{10} was significantly affected by the basal area ratio between coniferous and broadleaved tree species, and greatly declined with the percentage of coniferous tree species. No significant correlation was found between Q_{10} and the basal area ratio in mixed forest stands (Fig. 1).

Determinant variables of soil basal respiration rate

R_{10} was mainly influenced by soil nutrient content and rose linearly with CV of SOC and CV of TN; soil temperature range during which soil respiration was monitored was significantly correlated with R_{10} in the mixed forest (Fig. 2A, 2B, and 3). Contrarily, in the broadleaved forest stands, R_{10} was largely determined by the basal area and the spatial variations of shrub carbon stock and soil phosphorus content (Fig. 2C and 4). There was a linearly inverse relationship between R_{10} and the basal area, whereas R_{10} increased differentially with increasing variations in shrub carbon stock and soil P in the broadleaved forest stands.

Determinant variables of the temperature sensitivity of soil respiration

In the mixed forest, Q_{10} was positively correlated with CV of herbaceous carbon stock and CV of soil bulk density (Fig. 5 and 6B), and negatively with soil C/N ratio (Fig. 7C). In the broadleaved forest, Q_{10} notably decreased with soil pH (Fig. 6A), but significant positive correlations were found between Q_{10} and CV of SOC and TN (Fig. 7A and B).

Combined relationships among R_{10}, Q_{10} and biophysical factors

Although many environmental factors were found in this study to independently exert significant effects on individual parameter of R_{10} or Q_{10}, 51% of the variations in both R_{10} and Q_{10} on the spatial scale were explained jointly by five biophysical variables, i.e., CV of soil bulk density, DBH, CV of soil TN, soil pH, and CV of soil pH, after forward selection of environmental variables. Particularly, most of the variations in R_{10} and Q_{10} were mainly ascribed to the variance of soil bulk density (Table 2). In addition, the importance of these selected factors was also highlighted by the result of Redundancy analysis, which showed that Axis 1 and Axis 2 accounted for 86.3% and 13.7% of the total variance in basal parameters of soil respiration, respectively (Fig. 8).

Discussion

Variation in R_{10} and its determining variables

Soil respiration rate at 10°C has received little attention in contrast to the instantaneous rate of soil respiration in the study of soil carbon cycle. Moreover, the comparability of R_{10} under changing circumstances is more reasonable than that of normally measured soil respiration rate. Even at the same temperature of 10°C, soil basal respiration still exhibits a large variation within or across forest types. On the scale of the catchment, R_{10} varies in a range of 1.25 $\mu mol\ CO_2\ m^{-2}\ s^{-1}$ to 2.30 $\mu mol\ CO_2\ m^{-2}\ s^{-1}$ in the mixed forest with a mean value of 1.83 $\mu mol\ CO_2\ m^{-2}\ s^{-1}$. R_{10} changes from 1.59 $\mu mol\ CO_2\ s^{-1}$ to 2.46 $\mu mol\ CO_2\ m^{-2}\ s^{-1}$ in the broadleaved forest with a mean rate of 2.01 $\mu mol\ CO_2\ m^{-2}\ s^{-1}$. These values of R_{10} just fall well in the range of R_{10} of different forests in the same region [3], but they are slightly higher than those for pine and oak forests in Brasschaat [26]. Given that the similar empirical function has been applied in calculating the basal rate of soil respiration, the variation of R_{10} is greatly induced by environmental factors other than soil temperature.

Stand structure has been indicated to be a dominant factor accounting for the spatial variation in soil respiration in beech and mixed-dipterocarp forests. Basal area and DBH exert a significant positive effect on soil respiration [4,27]. But, as to the specific results of this study, significant negative correlations are found between stand structure parameters and R_{10} for both forest types in the temperate region of North China. R_{10} declines differentially with the percentage of coniferous tree species in mixed forest community, and with the basal area across broadleaved forest stands. This intriguing scenario may be ascribed to the complexity

Table 1. Variation in basal parameters of soil respiration within or between forest types.

	R_{10} (μmol CO_2 m^{-2} s^{-1})	CV of R_{10} (%)	Q_{10}	CV of Q_{10} (%)
Broadleaved forest	2 a	11	4 a	24
Mixed forest	1 a	19	3 a	29

The significance of differences of basal parameters between forest types were separately tested by independent t - test (two - tailed) at $\alpha = 0.05$ (n = 9 in the broadleaved forest, and 11 in the mixed forest). Same lowercase letter means no significant difference is detected at $\alpha = 0.05$ within 95% confidence interval between the two forest types.

of CO_2 production in forest soils. Soil respiration consists of autotrophic respiration from roots and rhizosphere and heterotrophic respiration from microbial decomposition. Total rate and basal rate of soil respiration have been found to be slightly higher in the pure broadleaved forest stands than in the pure coniferous forest stands [3]. R_{10} is apparently depressed by the increasing admixed proportion of needle leaf tree species in the mixed forests. Perhaps, it is ascribed to the physiological differences between coniferous and broadleaved tree species. R_{10} is indicated to be modulated by plant photosynthesis (i.e. gross primary productivity) [9] via determining the activity of rhizosphere respiration [28]. Autotrophic respiration accounts for ~50% of total soil respiration, which may even be higher in growing season for temperate forests [29]. In cold weather with temperature at ~10°C, photosynthetic activity of the mixed forest stands with higher basal area can be heavily impeded, subsequently resulting in a lower R_{10}.

Conversely, R_{10} significantly increases with the heterogeneous properties of shrub carbon stock and soil P content in the broadleaved forest stands and by the spatial variations in soil organic carbon content and TN content in the mixed forest stands. This may be due to the dominance of microbial respiratory fraction in total soil respiration at lower temperature. It is the microbial community composition and climatic factors that control forest soil respiration in cold seasons [30]. Additionally, soil microbial biomass and respiration have been eventually influenced by soil biophysical properties [31] and by environmental biochemical processes [32] through substrate availability, which

indicates that soil respiration is essentially an enzymatic controlled process [9,19].

Variation in Q_{10} and its determinants

The temperature sensitivity of soil respiration, Q_{10}, shows evident intra- and inter-forest-type variations on the catchment scale in temperate forests of northern China. The average Q_{10} values across forest stands for each forest type are larger than those of young plantations and a secondary *Populus davidiana* stand in the semiarid Loess Plateau [3]. However, the average Q_{10} is comparable to the reported values by Peng et al. [18] and by Zheng et al. [17] through synthesizing a great number of studies about temperature sensitivity of soil respiration on the spatial scales from region to country. In essence, variant Q_{10} values demonstrate the deficiency of the temperature dependent functions in describing the sensitivity of soil respiration to temperature.

The apparent Q_{10} derived from field experiment is actually a combined temperature sensitivity of different fractions of soil CO_2 flux [10]. Particularly, the enzymatic reactivity of substrate decomposition to temperature is considered as the intrinsic Q_{10} [19]. Although the Q_{10} value of experimental study is suggested to be influenced by a wide range of ecological variables from molecular structure to climatic factors [18,19], the direct determinant of the temperature sensitivity of soil respiration is still dependent on the substrate availability [19]. In this study, we find that Q_{10} could be markedly influenced by soil C/N ratio, soil pH, and the spatial heterogeneous properties of herbaceous carbon stock, SOC, soil TN, and soil bulk density. It is also worth mentioning that the contributors to the variations of Q_{10} differ

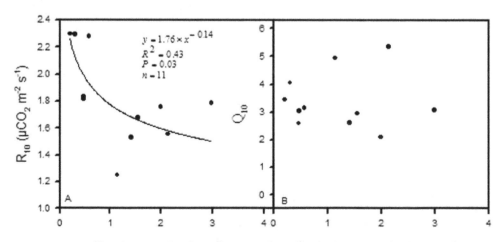

Figure 1. Trends of R_{10} and Q_{10} with basal area ratio of coniferous to broadleaved tree species.

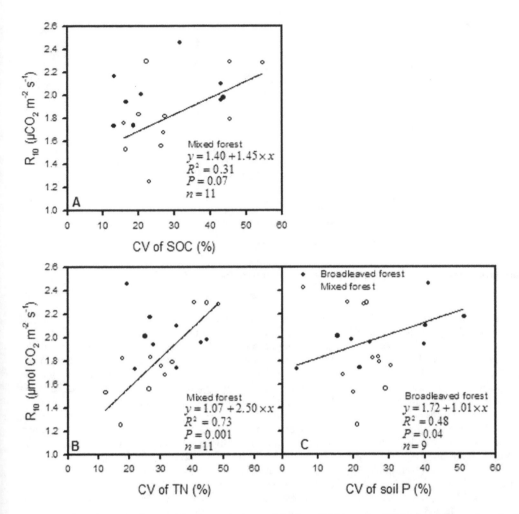

Figure 2. Correlations of R_{10} with the variations of SOC, soil TN, and soil P in broadleaved and mixed forests.

considerably between these two forest types. Similar results have been reported that the Q_{10} values changed with the alteration of ecosystems and vegetation types [17,18]. Indeed, the extrinsic

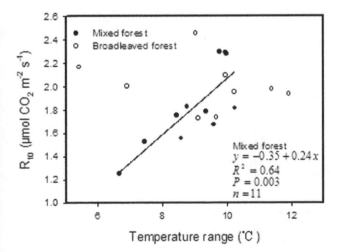

Figure 3. Correlations of R_{10} with soil temperature range for both forest types.

factors pose the effect on temperature sensitivity mainly via the primary control of substrate availability.

It has been recognized that the assumption of constant Q_{10} of soil respiration is incorrect [9,16], because the sensitivity of soil respiration to temperature is a complex reactivity; also it is more than being described by a simple parameter of temperature-dependent models. To date, a consensus has not been reached to clarify the mechanism underpinning the temperature sensitivity of soil respiration, but the study of easily monitored variables, such as soil physicochemical properties, SOC, soil TN, and forest type, etc., will help add extra predictive factors other than temperature in interpreting the variability of the apparent Q_{10}.

The effects of forest types on the correlations between biophysical variables and R_{10} and Q_{10}

R_{10} and Q_{10} have been demonstrated by our results to be influenced by biophysical factors and their spatial variation in forest stands. Although similar intrinsic mechanisms account for the variations of R_{10} and Q_{10} with forest microenvironments, the specific determining factors of soil basal respiration still vary with forest type. This is because forest type consisting of different tree species displays great distinctions in biotic and abiotic variables, which ultimately manipulate the changing gradient and direction of R_{10} and Q_{10}.

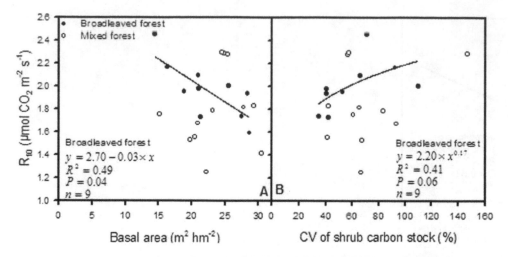

Figure 4. Correlations of R_{10} with basal area and CV of shrub carbon stock separately for both forest types.

At our study site, the mixed forest type is mainly composed of the coniferous tree species *P. tabuliformis* and the broadleaved tree species *Q. wutaishanica*. *P. tabuliformis* in forest ecosystem perhaps takes responsibility for the distinct correlations of R_{10} and Q_{10} with biophysical variables between broadleaved and mixed forest types, because tree species determines not only the microbial community structure but also the decomposition dynamics of forest litter [33,34]. Differences in the mycorrhizospheres and hyphospheres are the substantial way through which tree species affect soil microbial community including bacteria, archaea, fungi, and both free-living and symbiotic organisms [34]. Greater catabolic diversity and different bacterial and fungal communities have been found in the surface soil layer beneath mixed species plantations [35], and ectomycorrhizal fungi has been indicated to correlate with the presence of pine trees [36]. In addition, labile or soluble organic matter could also be affected by forest type with

quantity and quality differences in litter and root exudates. This may induce the variations of soil microbial and enzyme activities between broadleaved and mixed forest types [37,38]. Obviously, the anisotropic response of heterotrophic respiration derived from microbial activity to biophysical factors may account for the variant correlations of Q_{10} and R_{10} with measured variables between the two forest types.

In general, Q_{10} has a mathematical interrelationship with R_{10} and they also can be expressed by each other [10]. Furthermore, both Q_{10} and R_{10} display the confounding reactions of the complex process of soil respiration to the changes in exterior environmental factors. Therefore it can improve the overall understanding of the underlying mechanism driving soil respiration to concurrently analyze the variances of R_{10} and Q_{10} and their determinants. Although a single variable can explain larger variance of R_{10} or Q_{10}, the comparatively lower attribution of the

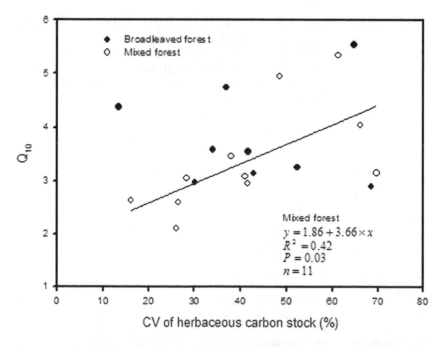

Figure 5. Relationships of Q_{10} with variation of herbaceous carbon stock in both forest types.

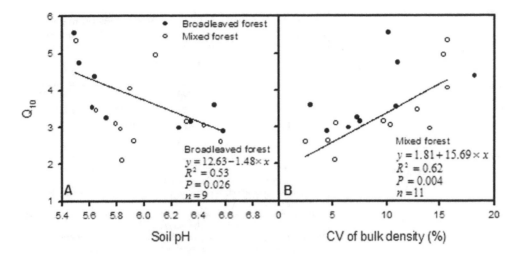

Figure 6. Relationships of Q_{10} with soil physical factors separately in broadleaved and mixed forest types.

variation of Q_{10} and R_{10} to the five selected variables demonstrates the inherent interactions existing among biotic and abiotic variables. It is the internal interaction that determines the amplitudes of soil basal respiration rates with varying environmental conditions across temporal or spatial scales [10]. This

viewpoint is also supported by the study of SØE and Buchmann [4]. Therefore, a more accurate estimation of soil CO_2 efflux cannot be achieved for a specific forest ecosystem until the changes in Q_{10} and R_{10} are concurrently taken into account with alterations of microenvironment.

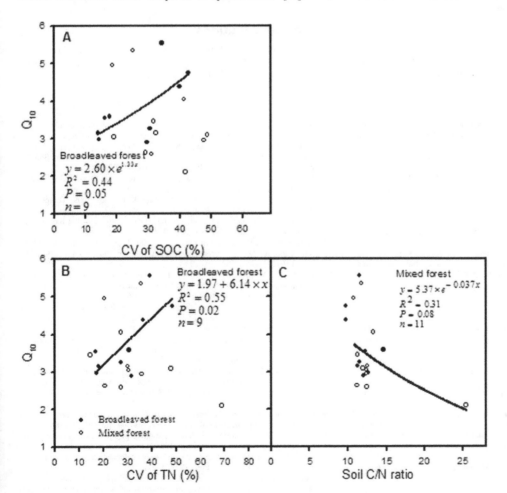

Figure 7. Relationships between soil chemical properties and Q_{10} respectively in broadleaved and mixed forest types.

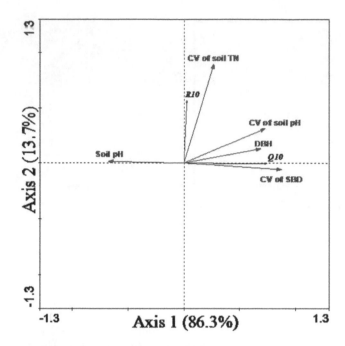

Figure 8. Redundancy analysis (RDA) among Q_{10}, R_{10} and the biophysical variables. DBH means diameter at breast height; SBD means soil bulk density.
doi:10.1371/journal.pone.0081793.g008

Table 2. The effects of biophysical variables on R_{10} and Q_{10} analyzed by the method of Redundancy Analysis (RDA).

Variables	Lambda-1[a]	Lambda-A[b]	P[c]	F[d]
CV of SBD	0.340	0.34	0.004	9.28
DBH	0.244	0.07	0.152	2.01
CV of soil TN	0.213	0.06	0.178	1.93
Soil pH	0.204	0.03	0.454	0.81
CV of pH	0.088	0.01	0.788	0.25

SBD: soil bulk density; DBH: diameter at breast height.
[a]Describe marginal effects, which shows the variance when the variable is used as the only factor.
[b]Describe conditional effects, which shows the additional variance each variable explains when it is included in the model.
[c]The level of significance corresponding to Lambda-A when performing Monte Carlo test (with 499 random permutations) at the 0.05 significance level.
[d]The Monte Carlo test statistics corresponding to Lambda-A at the 0.05 significance level.
doi:10.1371/journal.pone.0081793.t002

figure edition. The Taiyueshan Long-Term Forest Ecosystem Research Station is also appreciated for the access to conduct field work on its experimental sites. Fengfeng Kang is especially acknowledged by us for editing an earlier version of the manuscript. We also thank two anonymous reviewers for their valuable comments and suggestions in improving the manuscript.

Acknowledgments

We are grateful to Zhongkui Luo for his assistance in field investigation and to Yeming You and Hua Su for their advices on Redundancy analysis and

Author Contributions

Conceived and designed the experiments: ZZ HM. Performed the experiments: ZZ CG. Analyzed the data: ZZ. Contributed reagents/ materials/analysis tools: ZZ CG. Wrote the paper: ZZ.

References

1. van't Hoff, JH (1898) Lectures on theoretical and physical chemistry. In: Lehfeldt, RA. (Ed.), Chemical Dynamics: Part 1. Edwart Arnold, London, UK, pp. 224–229 (translated from German).
2. Boone RD, Nadelhoffer KJ, Canary JD, Kaye JP (1998) Roots exert a strong influence on the temperature sensitivity of soil respiration. Nature 396: 570–572.
3. Zhou Z, Zhang Z, Zha T, Luo Z, Zheng J, et al. (2013) Predicting soil respiration using carbon stock in roots, litter and organic matter in forests of Loess Plateau in China. Soil Biol Biochem 57: 135–143. doi:10.1016/j.soilbio.2012.08.010
4. SØE ARB, Buchmann N (2005) Spatial and temporal variations in soil respiration in relation to stand structure and soil parameters in an unmanaged beech forest. Tree Physiology 25: 1427–1436.
5. Rodeghiero M, Cescatti A (2005) Main determinants of forest soil respiration along an elevation/temperate gradient in the Italian Alps. Global Change Biol 11: 1024–1041. doi:10.1111/j.1365-2486.2005.00963.x
6. Jenkins M, Adams MA (2010) Vegetation type determines heterotrophic respiration in subalpine Australian ecosystems. Global Change Biol 16: 209–219. doi:10.1111/j.1365-2486.2009.01954.x
7. Bréchet L, Ponton S, Roy J, Freycon V, Coûteaux M-M, et al. (2009) Do tree species characteristics influence soil respiration in tropical forests? A test based on 16 tree species planted in monospecific plots. Plant Soil 319: 235–246. doi: 10.1007/s11104-008-9866-z
8. Kang SY, Doh S, Lee D, Lee D, Jin VL, et al. (2003) Topographic and climatic controls on soil respiration in six temperate mixed-hardwood forest slopes, Korea. Global Change Biol 9: 1427–1437.
9. Sampson DA, Janssens IA, Curiel Yuste J, Ceulemans R (2007) Basal rates of soil respiration are correlated with photosynthesis in a mixed temperate forest. Global Change Biol 13: 2008–2017. doi: 10.1111/j.1365-2486.2007.01414.x
10. Davidson EA, Janssens IA, Luo YQ (2006) On the variability of respiration in terrestrial ecosystems: moving beyond Q_{10}. Global Change Biol 12: 154–164.
11. Lloyd J, Taylor JA (1994) On the temperature dependence of soil respiration. Funct Ecol 8: 315–323.
12. Tjoelker MG, Oleksyn J, Reich PB (2001) Modelling respiration of vegetation: evidence for a general temperature-dependent Q_{10}. Global Change Biol 7: 223–230.
13. Friedlingstein P, Cox P, Betts R, Bopp L, von Bloh W, et al. (2006) Climate-carbon cycle feedback analysis: Results from the C^4MIP model intercomparison. J Climate 16: 3337–3353.

14. Knorr W, Prentice IC, House JI, Holland EA (2005) Long-term sensitivity of soil carbon turnover to warming. Nature 433: 298–301. doi:10.1038/nature03226
15. Malcolm GM, López-Gutiérrez JC, Koide RT (2009) Temperature sensitivity of respiration differs among forest floor layers in a *Pinus resinosa* plantation. Soil Biol Biochem 41: 1075–1079. doi:10.1016/j.soilbio.2009.02.011
16. Janssens IA, Pilegaard K (2003) Large seasonal changes in Q_{10} of soil respiration in a beech forest. Global Change Biol 9: 911–917.
17. Zheng ZM, Yu GR, Fu YL, Wang YS, Sun XM, et al. (2009) Temperature sensitivity of soil respiration is affected by prevailing climatic conditions and soil organic carbon content: A trans-China based case study. Soil Biol Biochem 41: 1531–1540. doi:10.1016/j.soilbio.2009.04.013
18. Peng S, Piao S, Wang T, Sun J, Shen Z (2009) Temperature sensitivity of soil respiration in different ecosystems in China. Soil Biol Biochem 41: 1008–1014. doi:10.1016/j.soilbio.2008.10.023
19. Davidson EA, Janssens IA (2006) Temperature sensitivity of soil carbon decomposition and feedbacks to climate change. Nature 440: 165–173. doi:10.1038/nature04514
20. Ma QY (1983) Determination of the biomass of individual trees in stands of *Pinus Tabulaeformis* plantation in north China. J Beijing Forestry College 4: 1–18. (in Chinese with English abstract)
21. Guo JT, Wang WX, Cao XF, Wang ZQ, Lu BL (1992) A study on forest soil groups of Taiyue mountain. Journal of Beijing Forestry University 14: 134–142. (in Chinese with English abstract)
22. Nelson DW, Sommers LE (1982) Total carbon, organic carbon, and organic matter. In: Page AL, Miller RH, Keeney DR (eds) Methods of soil analysis. American Society of Agronomy and Soil Science Society of American, Madison, pp 101–129
23. Gallaher RN, Weldon CO, Boswell FC (1976) A semiautomated procedure for total nitrogen in plant and soil samples. Soil Sci Soc Am J 40: 887–889.
24. John MK (1970) Colorimetric determination of phosphorus in soil and plant materials with ascorbic acid. Soil Sci 109: 214–220.
25. Lep J, milauer P (2003) Multivariate analysis of ecological data using Canoco, 1st edn. Cambridge University Press, New York.
26. Curiel Yuste J, Janssens IA, Carrara A, Ceulemans R (2004) Annual Q_{10} of soil respiration reflects plant phenological patterns as well as temperature sensitivity. Global Change Biol 10: 161–169. doi:10.1111/j.1529-8817.2003.00727.x
27. Katayama A, Kume T, Komatsu H, Ohashi M, Nakagawa M, et al. (2009) Effect of forest structure on the spatial variation in soil respiration in a Bornean

tropical rainforest. Agr Forest Meteorol 149: 1666–1673. doi: 10.1016/j.agroformet.2009.05.007

28. Levy-Varon JH, Schuster WSF, Griffin KL (2012) The autotrophic contribution to soil respiration in a northern temperate deciduous forest and its response to stand disturbance. Oecologia 169: 211–220. doi:10.1007/s00442-011-2182-y

29. Hanson PJ, Edwards NT, Garten CT, Andrews JA (2000) Separating root and soil microbial contribution to soil respiration: A review of methods and observations. Biogeochemistry 48: 115–146.

30. Monson RK, Lipson DL, Burns SP, Turnipseed AA, Delany AC, et al. (2006) Winter forest soil respiration controlled by climate and microbial community composition. Nature 439: 711–714. doi:10.1038/nature04555

31. Dupuis EM, Whalen JK (2007) Soil properties related to the spatial pattern of microbial biomass and respiration in agroecosystems. Can J Soil Sci 87: 479–484.

32. Resat H, Bailey V, McCue LA, Konopka A (2012) Modeling microbial dynamics in heterogeneous environments: growth on soil carbon sources. Microb Ecol 63: 883–897. doi:10.1007/s00248-011-9965-x

33. Vivanco L, Austin AT (2008) Tree species identity alters forest litter decomposition through long-term plant and soil interactions in Patagonia, Argentina. J Ecol 96: 727–736. doi: 10.1111/j.1365-2745.2008.01393.x

34. Prescott CE, Grayston SJ (2013) Tree species influence on microbial communities in litter and soil: Current knowledge and research needs. Forest Ecol Manag 309: 19–27.

35. Jiang Y, Chen C, Xu Z, Liu Y (2012) Effects of single and mixed species forest ecosystems on diversity and function of soil microbial community in subtropical China. J Soil Sediment 12: 228–240.

36. Hackl E, Pfeffer M, Donat C, Bachmann G, Zechmeister-Boltenstern S (2005) Compostion of the microbial communities in the mineral soil under different types of natural forest. Soil Biol Biochem 37: 661–671.

37. Xing SH, Chen CR, Zhou BQ, Zhang H, Nang ZM, et al. (2010) Soil soluble organic nitrogen and active microbial characteristics under adjacent coniferous and broadleaf plantation forests. J Soil Sediment 10: 748–757.

38. Cheng F, Peng XB, Zhao P, Yuan J, Zhong CG, et al. (2013) Soil microbial biomass, basal respiration and enzyme activity of main forest types in the Qinling Mountains. Plos One 8: e67353. doi:10.1371/journal.pone.0067353.

9

Phylobetadiversity among Forest Types in the Brazilian Atlantic Forest Complex

Leandro Da Silva Duarte[1]*, **Rodrigo Scarton Bergamin**[1], **Vinícius Marcilio-Silva**[2], **Guilherme Dubal Dos Santos Seger**[1], **Márcia Cristina Mendes Marques**[2]

1 Departamento de Ecologia, Universidade Federal do Rio Grande do Sul, Porto Alegre, Brazil, 2 Departamento de Botânica, Universidade Federal do Paraná, Curitiba, Brazil

Abstract

Phylobetadiversity is defined as the phylogenetic resemblance between communities or biomes. Analyzing phylobeta-diversity patterns among different vegetation physiognomies within a single biome is crucial to understand the historical affinities between them. Based on the widely accepted idea that different forest physiognomies within the Southern Brazilian Atlantic Forest constitute different facies of a single biome, we hypothesize that more recent phylogenetic nodes should drive phylobetadiversity gradients between the different forest types within the Atlantic Forest, as the phylogenetic divergence among those forest types is biogeographically recent. We compiled information from 206 checklists describing the occurrence of shrub/tree species across three different forest physiognomies within the Southern Brazilian Atlantic Forest (Dense, Mixed and Seasonal forests). We analyzed intra-site phylogenetic structure (phylogenetic diversity, net relatedness index and nearest taxon index) and phylobetadiversity between plots located at different forest types, using five different methods differing in sensitivity to either basal or terminal nodes (phylogenetic fuzzy weighting, COMDIST, COMDISTNT, UniFrac and Rao's H). Mixed forests showed higher phylogenetic diversity and overdispersion than the other forest types. Furthermore, all forest types differed from each other in relation phylobetadiversity patterns, particularly when phylobetadiversity methods more sensitive to terminal nodes were employed. Mixed forests tended to show higher phylogenetic differentiation to Dense and Seasonal forests than these latter from each other. The higher phylogenetic diversity and phylobetadiversity levels found in Mixed forests when compared to the others likely result from the biogeographical origin of several taxa occurring in these forests. On one hand, Mixed forests shelter several temperate taxa, like the conifers *Araucaria* and *Podocarpus*. On the other hand, tropical groups, like Myrtaceae, are also very representative of this forest type. We point out to the need of more attention to Mixed forests as a conservation target within the Brazilian Atlantic Forest given their high phylogenetic uniqueness.

Editor: Keping Ma, Institute of Botany, Chinese Academy of Sciences, China

Funding: The study was funded by the Brazilian National Research Foundation (CNPq, http://www.cnpq.br) via researcher fellowships granted to Leandro Duarte and Márcia Marques. The funders had no role in study design, data collection and analysis, decision to publish, or preparation of the manuscript.

Competing Interests: The authors have declared that no competing interests exist.

* Email: duarte.ldas@gmail.com

Introduction

Phylobetadiversity can be defined as the phylogenetic resemblance between communities or biomes [1]. The distribution of species belonging to different phylogenetic clades across the biogeographic space is often explained by major climatic conditions [2,3]. On the other hand, the geographical distribution of different taxa also depends on historical processes promoting speciation and dispersal [4,5]. Indeed, in some cases past ecological conditions can explain the diversity patterns of different lineages better than current environmental gradients [6]. Thus, environmental, evolutionary, historical and neutral factors likely interact to determine species composition patterns [5]. In this context, any biome might be thought as a snapshot of multiple interactions among those factors molding the distribution of taxa.

The resemblance between different phylogenetic lineages along a phylogenetic tree expresses a temporal accumulation of evolutionary divergence among clades. Evolutionary divergence near the root node of the phylogenetic tree reflects events occurred in remote past, while divergence near the terminal tips indicates recent evolutionary events. Further, we are likely to find contrasting phylobetadiversity patterns, depending on the analytical approach used to assess them [3]. Therefore, using different phylobetadiversity measures might help us to investigate whether the phylogenetic divergence between an array of sites has occurred more recently or deeper in the past. Nothing else being different, two sites located in the same biome are expected to vary more in relation to the occurrence of more recent nodes (e.g. families, genera), than in relation to more basal nodes (*e.g.* superorders, classes). On the other hand, sites located in different biomes might be expected to differ more in relation to more basal phylogenetic nodes than local sites within the same biome, as the respective biomes diverged earlier in terms of historical development than local sites within the same biome.

The Atlantic Forest is one of the most widely distributed tropical forests in Southern America, occupying almost all Brazilian Eastern coast besides inland areas. It is considered a hotspot for biodiversity conservation due to its high endemism and threatened areas [7,8]. It shelters about 15,000 vascular plants, from which

48% of species are endemic [9]. Actually, endemism levels in Atlantic Forest are among the highest observed in the world [10,11]. The Atlantic Forest biota is composed by taxa from different biogeographic origins, notoriously from the Amazonian Forest, the gallery forests of Cerrado, and the Andean areas in the austral portion of the biome [12,13]. Based on species distribution, the vegetation of the Atlantic Forest is recognized as composed by three forest types resulting from the differential influence of bordering floras: dense, mixed and seasonal forests [14–16]. In Material and Methods we provide a more detailed description of these different forest types. Floristic variation within and among different forest types within the Brazilian Atlantic Forest is strongly determined by environmental gradients [15,17,18]. On the other hand, it is widely recognized the biogeographically common origin of the different vegetation types within the Atlantic Forest [15,19]. Climate in South-America had been wetter and hotter by the beginning of the Eocene, and the Atlantic and the Amazonian Forest formed a unique large forest from Pacific to Atlantic oceans [20,21]. However, from the Pliocene, with the global climatic cooling and drying, an expansion of open vegetation types of Cerrado (Brazilian savanna), Caatinga and Chaco had occurred, which have disrupted the connection between the Atlantic Forest from other South-American forests. Since then, the Atlantic Forest is likely to have evolved as a single biogeographic unit [20].

To our knowledge, no attempts of analyzing a possible phylogenetic differentiation among these floras have yet been done. In this study we aim at carrying out such analysis, focusing mainly on phylobetadiversity patterns. Analyzing phylogenetic gradients among different forest physiognomies within the Atlantic Forest is crucial to understand the historical affinities between them. Based on the widely accepted idea that different forest physiognomies within the Atlantic Forest constitute different facies of a single eco-evolutionary entity, we hypothesize that recent nodes should drive phylobetadiversity gradients between the different forest types within the Southern Brazilian Atlantic Forest, as the phylogenetic divergence among them is biogeographically recent. To test this hypothesis, we compiled information from 206 floristic checklists describing the occurrence of shrub/tree species across the Southern Brazilian Atlantic Forest. Based on that da we evaluated the phylogenetic structure of different Atlantic Forest types and compared those forest types in relation to phylobetadiversity using five distinct analytical methods, which captured phylobetadiversity patterns more related to either basal or recent phylogenetic nodes. A second goal of this study is methodological. Although we have previously employed phylogenetic fuzzy weighting [22] to evaluate phylogenetic gradients across sets of communities or ecoregions [18,23,24], we have never compared the patterns we found with those generated by other methods for phylobetadiversity analysis. Given the first goal of the study, we think we have an excellent opportunity of providing such comparison, which can improve the general understanding on the method.

Materials and Methods

The Southern Brazilian Atlantic Forest

The Atlantic Forest extends along the Brazilian coast and inwards to eastern Paraguay and Northeastern Argentina, across variable climatic conditions with elevations ranging from sea level to 2,900 m [14]. This includes, approximately, latitudes ranging from 5° N to 33° S, longitudes from 35° W to 52° W and altitudes from 0 to 2,200 m [14]. Such broad geographical variation determines a climatic gradient related to annual rainfall (approximately from 800 to 4,000 mm) and mean annual temperatures

(averages from 15° to 25°C), which influence species distributions [25–27]. In the south and southeast Brazil the Atlantic Forest is marked by the occurrence of three forest types [15], the Dense Rain Forest (hereafter Dense forests), the Mixed Rain Forest (hereafter Mixed forests) and the Seasonal Deciduous and Semideciduous Forest (hereafter Seasonal forests).

The Dense forests

Dense forests are associated with the Atlantic coast and include a large area of lowland (until ~50 m a.s.l.) and slope (~50 to 2,200 m a.s.l.) forests from the Northeastern to the Southern regions of Brazil. The climate is variable, but generally hot and wet in lowlands and cold and wetter in slopes [14,15]. This biome shows floristic affinities with the Amazon Forest and Caatinga in the North [26,28,29] and it is influenced by the flora of other regions, such as the Andes and elements of the ancient southern Gondwana in the South [30]. The vegetation in lowlands comprises forests and scrubs that occur in drier climates (restingas) and in wetter climates (rain forests), determined by rainfall and soil sandiness [27]. Among species that determine vegetation in the coastal plain are *Maytenus obtusifolia*, *Byrsonima sericea*, *Ilex theazans*, *Calophyllum brasiliense*, *Ocotea pulchella* and *Myrcia multiflora* [27]. In the slopes, forests are highly differentiated by altitude, and species such as *Drimys brasiliensis*, *Ilex microdonta*, *Weinmannia paulliniifolia* characterize the vegetation [31].

Mixed forests

Mixed forests, also known as *Araucaria* forests, constitute the main forest type on the highland plateau in southern Brazil at elevations above 500 m a.s.l. [32]. Its northern distribution limit is in the Serra da Mantiqueira in south-eastern Brazil (latitude 20°S), where it occurs as vegetation patches or as isolated individuals in high-altitude grasslands, above 1,000 m. Southwards, mixed forests extend to latitude 29° S [32]. These forests are subjected to tropical and sub-tropical humid climates without pronounced dry periods. The annual rainfall ranges from 1400 to 2200 mm, and the annual mean temperature ranges mainly from 12°C to 18°C [33]. The presence of species phytogeographically related to temperate Austral-Antarctic and Andean floras distinguishes communities within the Mixed Forest from more tropical facies of Brazilian Atlantic forests [34]. Besides *Araucaria angustifolia*, some other typical species found in those forests are *Podocarpus lamberti* (conifer), *Dicksonia sellowiana* (tree fern), *Drimys* spp. (Winteraceae), and several species of Myrtaceae, Melastomataceae and Lauraceae.

Seasonal forests

Seasonal forests are related to the hinterland Parana River basin in the south and southeast Brazil. These forests are characterized by two distinct seasons with marked alternation from tropical with intense summer rainfalls to subtropical with low winter temperatures and scarce precipitation. During the cold and dry period, 20% to 50% of the canopy trees fall their leaves (deciduous) [14]. The mean temperature in the winter is lower than 15°C. The flora of Seasonal forests is often influenced by taxa typical from Cerrado (Brazilian savannah) and the alternation between wet/hot summers and dry/cold winters influences the leaf longevity causing leaf fall on winter [15]. This forest type has a dominance of species of *Parapiptadenia*, *Peltophrum*, *Cariniana*, *Lecythis*, *Tabebuia*, *Astronium* among others [14].

Species occurrences in floristic plots

We compiled information from 206 floristic checklists (Appendix S1) describing the occurrence of shrub/tree species across the geographic range of the Southern Brazilian Atlantic Forest biome (63 Dense forests, 50 Mixed forests, and 96 Seasonal forests). Floristic data were obtained by employing several distinct methodologies (Appendix S1). For instance, some authors used quadrats while others had no pre-defined surveying area; some used diameter at breast height as inclusion criteria while others used plant height. For this reason we only considered species presence/absence in sites. We checked for recent synonyms in the Missouri Botanical Garden (http://www.tropicos.org), The Plant List (http://www.theplantlist.org/), and Flora do Brasil databases (http://floradobrasil.jbrj.gov.br). Undetermined species, which represented in average less than 4% of the number of species in each checklist, were not included in the floristic dataset. Clade names followed Smith et al. [35] and Chase & Reveal [36]. Thus, the complete floristic data set was arranged in sites-by-species matrix of 206 sites described by 1,916 species, which was used for the analyses.

We compared the forest types in relation to the logarithmic number of species recorded in each plot by using one-way ANOVA. P-values were obtained by a permutation test with 999 iterations [37]. P-values were calculated based on the number of times the observed F-value was lower than the random F-values computed at each permutation procedure. We also compared forest types in relation to the occurrence of species in the plots. For this, we performed a PERMANOVA with permutation test (999 iterations), using Jaccard index as resemblance measure [37,38]. For both analyses, whenever a significant P-value was obtained, we performed pairwise contrast analysis to test which group differed from others. The significance of contrasts was also evaluated by permutation, in a similar way as in ANOVA and PERMANOVA [37]. Analyses were performed in the R environment (available at http://www.r-project.org), using package vegan 2.0–10 ([39], available at http://cran.r-project.org/web/packages/vegan/).

Building a phylogenetic tree for Atlantic Forest plants

To define phylogenetic affinities among plant species we used the phylogenetic hypothesis of APG III [40] for angiosperms and the hypothesis of Burleigh et al. [41] for gymnosperms, which solve phylogenetic relationships to the family level. For this, we used the megatree R20120829 (available at https://github.com/camwebb/tree-of-trees/blob/master/megatrees/R20120829.new), removed outdated intrafamilial resolution and included the gymnosperms tree into the megatree. Since phylogenetic uncertainties influence different phylogenetic metrics, to reach intrafamilial node resolution we also included 51 constructed angiosperms families' trees based on recent studies (families with more than one species and for which reliable phylogenetic hypotheses are available) (references in Appendix S2). This procedure solved genera relationships for 84% of the species in the database. We defined branch lengths using node age estimates proposed by Bell et al. [42] and the age estimates of Magallón et al. [43] for clades older than angiosperms, using only clade age estimates that were consistent with the APG III tree topology. We also included clade age estimates within some of the 51 families added to the megatree (references in Appendix S2). Undated nodes were adjusted using the BLADJ algorithm of Phylocom 4.2 software [44] and the phylogenetic tree was obtained using the Phylomatic 2 module of Phylocom 4.2 software [45]. Then we computed the phylogenetic pairwise patristic distances between species.

Analyzing phylogenetic structure within Atlantic Forest types

We analyzed the phylogenetic structure of forest plots using different methods, in order to capture distinct properties of the phylogenetic structure of the plots. Since our species-by-sites matrix had only occurrences, no methods employed took into account species abundances. Phylogenetic diversity (PD) was computed as the total sum of branch lengths for species occurring in each plot [46]. Phylogenetic clustering/overdispersion was measured using the two metrics proposed by Webb et al. [47]: mean phylogenetic distances (MPD) between the species present in each plot, and mean phylogenetic distance between each species and its phylogenetically nearest species (MNTD). For PD, MPD and MNTD values we computed standardized effect sizes (SES) based on 999 null values obtained from a null model that keeps the species composition of the plot while the position of each species in the phylogenetic tree for the regional species pool (defined by all species present in the dataset) is freely shuffled ("taxa.label" model), as follows:

$$\text{SES} = \frac{\text{Observed value } (x) - \text{Mean null value } (\bar{x}_0)}{\text{Standard deviation of null values } (s_0)}$$

Hereafter, we refer only to the standardized values of theses methods, respectively SES.PD, NRI (net relatedness index) and NTI (nearest taxon index). Positive or negative SES.PD values indicate, respectively, phylogenetic diversity higher or lower than expected by the null model. Positive or negative NRI/NTI values indicate, respectively, phylogenetic clustering or overdispersion of species in the plot. While NRI captures the influence of deeper phylogenetic nodes to the phylogenetic structure of the plot, NTI characterizes the effect of shallower phylogenetic nodes [47]. Phylogenetic structure measures were computed in the R environment (available at http://www.r-project.org), using the package picante 1.6–2 [48], available at http://cran.at.r-project.org/web/packages/picante/).

We compared the forest types in relation to phylogenetic structure methods (SES.PD, NRI and NTI) by using one-way ANOVA. P-values were obtained by a permutation test with 999 iterations [37]. For both analyses, whenever a significant P-value was obtained, we performed pairwise contrast analysis to test which group differed from others [37]. The significance of contrasts was also evaluated by permutation, in a similar way as in ANOVA [37]. Analyses were performed in the R environment (available at http://www.r-project.org), using package vegan 2.0–10 ([39], available at http://cran.r-project.org/web/packages/vegan/).

Analyzing phylobetadiversity among Atlantic Forest types

We compared the different forest types in relation to phylobetadiversity patterns using five methods: phylogenetic fuzzy weighting [22], COMDIST [44], COMDISTNT [44], UniFrac [49] and Rao's H [50]. As our species-by-sites matrix contained only species occurrences, all phylobetadiversity metrics were defined to do not consider species abundances. As some methods are more sensitive to variation in deeper phylogenetic nodes (COMDIST) while others capture variation mostly associated with shallower nodes (COMDISTNT, UniFrac and Rao's H), using several indices to analyze phylobetadiversity patterns might help us to understand to what extent phylobetadiversity levels are explained by more basal or recent nodes [3]. On the other hand, phylogenetic fuzzy weighting is likely to capture phylobetadiversity patterns associated with both basal and more terminal nodes [18]. Therefore, using these five different methods enabled us to test our hypothesis on the phylogenetic relationships of different forest types within the Southern Brazilian Atlantic Forest.

Phylogenetic fuzzy weighting is a method developed to analyze phylobetadiversity patterns across metacommunities, based on fuzzy set theory [22]. The method is based on the computation of matrix **P** from the species-by-sites incidence matrix [22,24]. The procedure consists of using pairwise phylogenetic similarities between species to weight their occurrence in the plots. The first step involves transforming pairwise phylogenetic distances into similarities ranging from 0 to 1. For this, each distance value d_{ij} is converted into a similarity s_{ij} using.

$$s_{ij} = 1 - \left(\frac{d_{ij}}{\max\left(d_{ij}\right)} \right)$$

where max (d_{ij}) is the maximum observed distance between two species in the tree.

Each phylogenetic similarity between a pair of species (s_{ij}) is then divided by the sum of similarities between the species i and all other k species. This procedure generates phylogenetic weights for each species in relation to all others, expressed as.

$$q_{ij} = \frac{s_{ij}}{\sum_{k=1}^{n} s_{kj}}$$

Such phylogenetic weights (q_{ij}) expresses the degree of phylogenetic belonging of each taxon i in relation to all others [22]. The degree of phylogenetic belonging reflects the amount of evolutionary history shared between a given species and all others in the dataset. The second analytical step consists of incorporating those standardized phylogenetic weights into the species-by-sites matrix. The occurrence of each species i in a plot k (w_{ik}) is distributed among all other j species occurring in that plot, proportionally to the degree of phylogenetic belonging between each pair of species as follows:

$$p_{ik} = (q_{ii}w_{ik}) + \sum_{j=1}^{n} q_{ij}w_{jk}$$

This procedure generates a matrix describing phylogeny-weighted species composition for each plot (matrix **P**), which expresses the representativeness of different lineages across the sites (see Duarte *et al.* [24] for a detailed description). Phylogenetic fuzzy weighting was performed in the R environment (available at http://www.r-project.org), using the package *SYNCSA* 1.3.2 ([51], available at http://cran.r-project.org/web/packages/SYNCSA/). Pairwise phylobetadiversity between plots was obtained by computing squared-rooted Bray-Curtis dissimilarities (or other appropriate resemblance measure, see Legendre & Anderson [52]) for every pair of plots in matrix **P** (Table 1).

We adopted this method to analyze phylobetadiversity because it allows to decompose phylogenetic gradients across an array of plots into orthogonal eigenvectors and, more importantly, to evaluate which clades are related to each phylogenetic eigenvector [24]. We achieved this by performing a PCoA [53] based on the square-rooted Bray-Curtis dissimilarities between pairs of plots previously computed on matrix **P**. Such procedure generated

principal coordinates of phylogenetic structure (PCPS) for each floristic plot. Each PCPS is a vector describing an orthogonal phylogenetic gradient in the dataset [18,23]. The PCPS with the highest eigenvalue describes broader phylogenetic gradients related to the split of the deepest tree nodes across the dataset, such as that connecting conifers and angiosperms. As the eigenvalues of the other PCPS decrease, finer phylogenetic gradients related to splits of shallower nodes (*e.g.* families, genera) are described [18]. By relating the correlation between species from major clades and the PCPS eigenvectors, we can draw a scatterplot relating directly sites and species grouped in clades. PCPS analysis was performed using the package *PCPS* (available at http://cran.r-project.org/web/packages/PCPS/) of the R environment (available at http://www.r-project.org). Further, we compared the forest types in relation to the PCPS eigenvectors containing more than 5% of total variation in matrix **P** using one-way ANOVA. *P*-values were obtained by a permutation test with 999 iterations [37]. Such analysis allowed us to define which phylogenetic gradients were mostly related to different Atlantic forest types. ANOVA was performed in the R environment (available at http://www.r-project.org), using package *vegan* 2.0–10 ([39], available at http://cran.r-project.org/web/packages/vegan/).

Furthermore, we employed other four well-known phylobetadiversity measures to compare the forest types within the Southern Brazilian Atlantic Forest (see Table 1). COMDIST is a phylobetadiversity measure that computes the mean phylogenetic distance among species occurring in two different sites [44]. For this reason, this phylobetadiversity measure captures variation associated with the more basal nodes linking species [3]. Computing COMDIST values without considering the variation in species abundances is equivalent to compute the phylogenetic distinctness (Rao's D) proposed by Hardy & Senterre [50]. Thus, we opted for using only the former in this study. On the other hand, by standardizing Rao's D values by the mean within-site phylogenetic diversity it is possible to obtain another phylobetadiversity measure (Rao's H, [50]), which captures phylobetadiversity patterns related to more terminal nodes in the tree [3]. COMDISTNT [44] measures the mean phylogenetic distance between every species in a plot and the nearest phylogenetic neighbor in another site (Table 1). It is, therefore, a "terminal node" metric [3]. The last phylobetadiversity method used in this study was UniFrac [49], which measures, for each pair of sites, the fraction of the total branch length of phylogenetic tree that is exclusive to each site (Table 1). Since more basal nodes are likely to be shared by most species, UniFrac captures phylobetadiversity patterns related to more terminal nodes [3]. This method is mathematically equivalent to the Jaccard index when a star phylogeny is considered [49]. UniFrac gives very similar (but not exactly similar) results when compared to PhyloSor [3], which is another well-known phylobetadiversity measure [54]. For this reason, we opted for using only the former. COMDIST, COMDISTNT and Rao's H were computed in the R environment (available at http://www.r-project.org), using the package *picante* 1.6–2 ([48], available at http://cran.at.r-project.org/web/packages/picante/). UniFrac was computed using the R package *GUniFrac* 1.0 (available at http://cran.r-project.org/web/packages/GUniFrac/index.html).

We carried out Mantel tests [53] based on Pearson correlations (999 permutations) to evaluate the association between pairwise phylobetadiversity values obtained from matrix **P** and all the other methods (COMDIST, COMDISTNT, UniFrac and Rao's H). Furthermore, we performed PERMANOVA with permutation test (999 iterations) [37,38] using each pairwise phylobetadiversity method as resemblance measures, to compare different forest types

Table 1. Phylobetadiversity methods used to compare different forest types within the Southern Brazilian Atlantic Forest.

Method	Formula	Description	Reference
Phylogenetic fuzzy weighting	$SqrtBC = \sqrt{\dfrac{\sum_{i=1}^{N}\lvert p_{ik1} - p_{ik2}\rvert}{\sum_{i=1}^{N}\lvert p_{ik1} + p_{ik2}\rvert}}$	Computes the square-rooted Bray-Curtis dissimilarity between plots k_1 and k_2 based on phylogenetically weighted incidence (p_{ik}) of N species i.	[22]
COMDIST	$MPD_{k1k2} = \frac{1}{2}\left[\left(\frac{1}{n}\sum_{i=1}^{n_{k1}} d_{ik2}\right) + \left(\frac{1}{n}\sum_{j=1}^{n_{k2}} d_{jk1}\right)\right]$	Computes the mean pairwise phylogenetic distance d between each species i of plot k_1 and all n species of plot k_2.	[44]
COMDISTNT	$MNTD_{k1k2} = \frac{1}{2}\left[\left(\frac{1}{n}\sum_{i=1}^{n_{k1}} \min d_{ik2}\right) + \left(\frac{1}{n}\sum_{j=1}^{n_{k2}} \min d_{jk1}\right)\right]$	Computes the mean pairwise phylogenetic distance d between each species i of plot k_1 and the phylogenetically nearest species of plot k_2 ($\min d_{ik2}$).	[44]
Rao's H	$Rao'sH = \dfrac{\frac{1}{2}\left[\left(\frac{1}{n}\sum_{i=1}^{n_{k1}} d_{ik2}\right) + \left(\frac{1}{n}\sum_{j=1}^{n_{k2}} d_{jk1}\right)\right]}{\frac{1}{2}\left[\left(\frac{1}{n}\sum_{i=1}^{n_{k1}} d_{ij}\right) + \left(\frac{1}{n}\sum_{i=1}^{n_{k2}} d_{ij}\right)\right]}$	Standardized measure of phylogenetic distinctness. The numerator is similar to COMDIST. The denominator is the mean phylogenetics distance within-plots.	[50]
UniFrac	$UniFrac = \sum_{i}^{n} BL_i\left(\dfrac{k_{1i}k_{2i}}{k_{1T}k_{2T}}\right)$	Computes the fraction of total branch length linking the species occurring in two plots, which is exclusive to each plot.	[49]

in relation to phylobetadiversity levels. Whenever a significant P-value was obtained for the general model, we performed pairwise contrast analysis to test which group differed from the others [34]. The significance of contrasts was also evaluated by permutation, in a similar way as in PERMANOVA [34]. Analyses were performed in the R environment (available at http://www.r-project.org), using package *vegan* 2.0–10 ([39], available at http://cran.r-project.org/web/packages/vegan/).

Results

From the 1,916 species occurring across the Southern Brazilian Atlantic Forest, eurosids (superorder Rosanae) comprised 58% of total number of species, asterids (superorder Asteranae) were represented by 25% of species in the dataset, and magnoliids (superorder Magnolinae) by 10%. Other phylogenetic clades occurring in the dataset were Caryophyllales and monocots (superorder Lilianae) (each comprising 2% of total richness), and, Proteanae, Santalanae, conifers (superorder Pinidae), Dillenianae, Chloranthanae and Ranunculanae, each with ≤1% of total number of species. The 10 more frequent species in the dataset were, in decreasing order, *Casearia sylvestris* (Salicaceae), *Myrsine umbellata* (Myrsinaceae), *Cupania vernalis* (Sapindaceae), *Allophylus edulis* (Sapindaceae), *Matayba elaeagnoides* (Sapindaceae), *Casearia decandra* (Salicaceae), *Zanthoxylum rhoifolium* (Rutaceae), *Campomanesia xanthocarpa* (Myrtaceae), *Guapira opposita* (Nyctaginaceae) and *Prunus myrtifolia* (Rosaceae).

We found 946 species in Mixed forests, 1,136 in Dense forests and 1,187 in Seasonal forests. ANOVA results showed that different forest types did not show significant variation in relation the number of species (Fig. 1a). This finding gives support to the significant variation found in relation to the three phylogenetic structure metrics analyzed. Mixed forests showed higher standardized phylogenetic diversity (Fig. 1b) and lower NRI values, indicating phylogenetic overdispersion, than the other forest types (Fig. 1c). By its turn, Seasonal forests showed lower standardized phylogenetic diversity and higher NRI values, indicating phylogenetic clustering. Dense forests presented intermediary values between Mixed and Seasonal forests. In relation to NTI, Seasonal

forests showed higher values than the other two forest types, indicating phylogenetic clustering (Fig. 1d), while Mixed and Dense forests did not vary in relation to each other.

Mantel tests showed that dissimilarities computed based on matrix **P** had significant Mantel correlations with all other phylobetadiversity methods. The highest correlation was between phylogenetic fuzzy weighting and COMDIST ($\rho = 0.59$; $P = 0.001$), followed by Rao's H ($\rho = 0.48$; $P = 0.001$), COMDISTNT ($\rho = 0.48$; $P = 0.001$) and UniFrac ($\rho = 0.39$; $P = 0.001$).

MANOVA indicated that species composition of floristic plots varied significantly ($P < 0.001$) between all forest types (Table 2). Nonetheless, the model fit for species composition was worse than for almost all phylobetadiversity methods (exception for COMDIST, see Table 2), indicating that phylobetadiversity patterns observed in this study were robust, and not merely an artifact of the variation in species composition between forest types. Among the phylobetadiversity methods, phylogenetic fuzzy weighting showed the best model fit ($R^2 = 0.42$; $F = 73.4$). Although PERMANOVA showed significant results for the other four methods, their model fit varied according to the properties of the method. COMDIST, a phylobetadiversity method that captures patterns related to more basal nodes, showed a very poor (although statistically significant) fit, while the other three metrics, which capture phylobetadiversity patterns related to terminal nodes showed better fit, especially Rao' H. Taking into account only the two methods with best model fit (phylogenetic fuzzy weighting and Rao's H), we found that most phylobetadiversity variation (higher F-value) was observed between Mixed and Seasonal forests. On the other hand, while phylogenetic fuzzy weighting showed a higher phylogenetic similarity between Dense and Seasonal forests (lower F-value), Rao's H showed a higher similarity between Mixed and Dense (Table 2).

The ordination of matrix **P** enabled us to explore the phylogenetic clades underlying phylobetadiversity patterns (Fig. 2). The four first PCPS axes contained more than 5% of total information in matrix **P** (explained together 59% of the total variation in matrix **P**). These four PCPS were then submitted to ANOVA. The test comparing the scores of PCPS 1 between forest

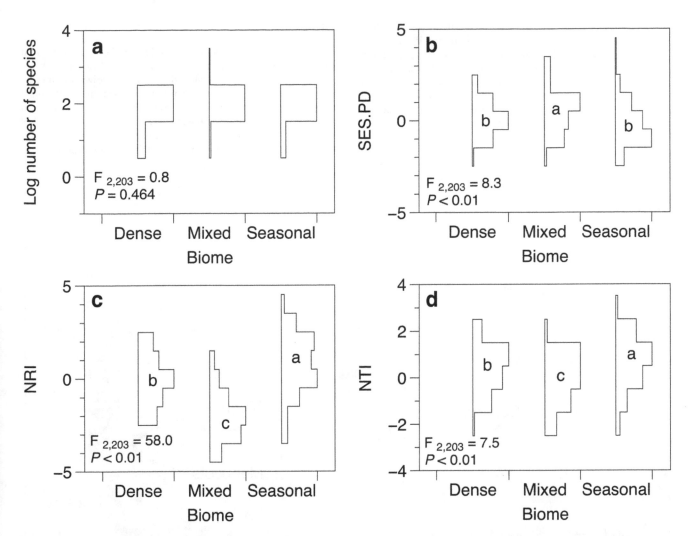

Figure 1. ANOVA with permutation tests for a) logarithmized species number, b) standardized phylogenetic diversity (SES.PD), c) net relatedness index (NRI) and d) nearest taxon index (NTI) for floristic plots occurring in different forest types within the Southern Brazilian Atlantic Forest. Probability plots drawn for each forest type define the relative frequency of values for each response variable. P-values obtained using 999 permutations. Different letters within the probability plots indicate significant difference between forest types (P≤0.05).

types showed the best fit ($F_{2,203} = 129.5$; $P<0.001$), followed by PCPS 3 ($F_{2,203} = 35.5$; $P<0.001$). The first PCPS (38% of total variation in matrix **P**) captured phylobetadiversity patterns related to the most basal node, *i.e.* the node separating conifers from angiosperms drove the variation between forest types, with Mixed forests (related to conifers) splitting from Dense and Seasonal

Table 2. PERMANOVA with permutation tests comparing species composition and five different phylobetadiversity methods between different forest types within the Southern Brazilian Atlantic Forest.

Response variable	Overall PERMANOVA model		F-values for pairwise contrasts		
	R^2	$F_{2,203}$	Mixed - Dense	Mixed - Seasonal	Dense - Seasonal
Species composition	0.081	9.0	9.5	8.3	9.3
Phylogenetic fuzzy weighting	0.420	73.4	65.8	128.8	16.2
COMDIST	0.019	2.0	1.4	2.2	2.1
COMDISTNT	0.230	30.3	32.5	37.9	23.1
Rao's H	0.340	52.2	18.2	68.6	58.1
UniFrac	0.135	15.9	19.1	18.6	11.6

All F-values showed P-values <0.001. P-values obtained by 999 permutations.

forests (related to angiosperms). The phylogenetic gradient along the third PCPS axis (8% of total variation in matrix **P**) was mostly driven by rosids (Fig. 2). While Dense forests were positively related to the occurrence of Myrtaceae and other Myrtales groups, Seasonal forests were positively associated with the occurrence of fabid rosids. PCPS 2 and 4 contained 9% and 5% of total variation in matrix **P**, respectively. ANOVA for these two PCPS showed poorer fit when compared to the former ones ($F_{2,203} = 8.1$ and $F_{2,203} = 22.6$, respectively).

Discussion

The classification of the Brazilian Atlantic Forest into different forest types was demonstrated here to have a phylogenetic basis. Except for COMDIST, all other phylobetadiversity metrics captured the variation between forest types within the Brazilian Atlantic Forest better than species composition alone (see also [24]). Actually, the most frequent species in the dataset are widely distributed across the Atlantic Forest, occurring in different forest types and under variable habitat conditions. Those species show

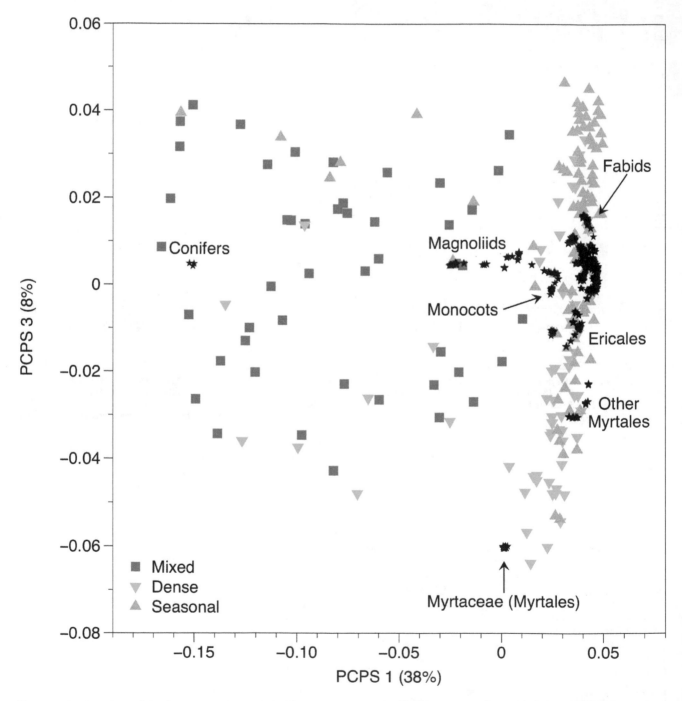

Figure 2. Scatter plots of the PCPS 1 and 3 generated from the ordination of matrix P describing phylogenetic weighted species composition of floristic plots located in different forest types (Mixed, Dense and Seasonal) within the Southern Brazilian Atlantic Forest.

high ecological plasticity, as they are capable to live under contrasting environmental conditions and soil types, are all dispersed by the fauna and show high tolerance to sunny environments. It is also noteworthy that none of these species are endemic from Brazil [55–57]. Considering the common biogeographic origin of different Atlantic Forest types [15,19], we hypothesized that more terminal phylogenetic nodes should drive phylobetadiversity patterns between different forest types within the Southern Brazilian Atlantic Forest. Indeed, the phylobetadiversity methods sensitive to phylogenetic gradients related to more terminal nodes (COMDISTNT, UniFrac and Rao's H, see [3]) captured phylobetadiversity variation between the forest types better than the "basal metric" (COMDIST). On the other hand, phylogenetically fuzzy weighting, which is likely to capture both the variation at basal and terminal nodes [18], showed the best model fit when we compared the different forest types. In general, all methods showed that Mixed forests differed more in relation to Dense and Seasonal forests than these latter from each other. The first PCPS captured phylogenetic gradient splitting conifers from other angiosperms (a basal node-driven gradient), which separated Mixed forests (related to conifers) from the other forest types (related to angiosperms), while the third PCPS captured a phylogenetic gradient related to more intermediary nodes (Myrtales related to Dense forests, fabids related to Seasonal forests). In general, the results from phylobetadiversity analysis showed that Mixed forests present a distinctive phylogenetic signature when compared to other Atlantic forests. To some extent, such patterns might be generated by the higher intra-site phylogenetic diversity found in Mixed forests when compared to other forest types. Nonetheless, the second phylobetadiversity method with higher fit in the comparison between forest types was Rao's H, which standardize phylobetadiversity by the mean intra-site phylogenetic diversity [3,50], reinforcing the patterns found here.

Mixed forests not only differed more in relation to phylobetadiversity from Dense and Seasonal forests than these latter from each other, but also showed higher intra-site phylogenetic diversity/overdispersion. This finding might be explained, on the one hand, by the importance of temperate conifers (*Araucaria*, *Podocarpus*) and magnoliids (e.g. *Drimys*, *Cinnamodendron*) for the flora of Mixed forests [34]. On the other hand, tropical Myrtaceae also constitutes an important eudicot group in Mixed forests, especially in those areas more directly connected with Dense forests [58]. Myrtaceae is the fourth largest plant family in Brazil [59], being the richest family in terms of tree species in several vegetation types, specially in Dense and Mixed Atlantic forests [60]. The floristic mixture found in Mixed forests is possibly influenced by the phylogenetic niche conservatism of the species occurring at more tropical sites of the Atlantic Forest, which precludes the advance of tropical species over the subtropical sites, allowing the permanence of several temperate taxa in Mixed forests [61]. As a consequence, Mixed forests is likely to show higher phylogenetic diversity and also higher degree of phylobetadiversity in relation to other Atlantic Forest types. The South American biota is formed by a northern tropical component and a southern temperate component, each with different biogeographic affinities [62–64]. The northern and southern portions of South America have always been connected, except during a brief period during the Cretaceous (100–80 Mya) when an epicontinental sea separated both halves of the American continent. The temperate taxa present in the Mixed forests had origin in the Southern Temperate Gondwana Province, namely Australia, New Zealand and New Caledonia [30]. The land connections between South America and West Antarctic continent allowed floristic exchanges

between Australia and South America until the late Eocene (~35 million years) or even the early Oligocene (30–28 million years). Such floristic connections provided a stock of subtropical taxa in South America [13]. On the other hand, the tropical taxa widely distributed across the Brazilian Atlantic Forest derived from Northern hemisphere ancestors through Laurasian migrations [13,65]. Thus, Mixed forests represent nowadays a unique mix of floras with distinct biogeographic and phylogenetic origins. The biogeographic features of Mixed forests increase the need of more effective conservation efforts to preserve that forest type, which has been suffered over the last century intensive human-made degradation due to logging, cattle-grazing and, more recently, silviculture [66]. Mixed forests need more attention from conservationists and decision-makers, as they have been often neglected as a conservation hotspot [67].

Our second goal in this study was to evaluate to what extent phylogenetic fuzzy weighting provides comparable values in relation to other phylobetadiversity methods. Swenson [3] showed that the association between different phylobetadiversity methods was due to the sensitivity of each method to more basal or terminal nodes of the phylogenetic tree. In this study, the highest correlation was observed between phylogenetic fuzzy weighting and COMDIST, which is a "basal" method [3]. Nonetheless, phylogenetic fuzzy weighting was also well correlated with the "terminal" methods (COMDISTNT, UniFrac and Rao's H), which reinforces the fact that phylogenetic fuzzy weighting captures phylobetadiversity from both basal and terminal nodes [18]. Such property of the method can be verified by means of the computation of principal coordinates of phylogenetic structure (PCPS), which provides independent phylogenetic gradients between a set of sites where each gradient captures node splits from basal to more terminal nodes [18,23,24]. Moreover, the possibility of exploring the identity of the different clades driving phylobetadiversity among the sites represents an advantage offered by phylogenetic fuzzy weighting in relation to other widely employed methods. Given the diversity of methods for assessing phylobetadiversity patterns, we think that employing different approaches simultaneously improves our capability to explore phylogenetic gradients across a set of communities, ecoregions or biomes.

Acknowledgments

The authors thank Dieter Liebsch for his help with the dataset organization. R. S. B. received PhD scholarship from CNPq; V. M. S. and G. D. S. S. received PhD scholarship from CAPES. L. D. S. D. and M. C. M. M. received fellowship from the Brazilian Research Council - CNPq (grants 303534/2012-5 and 304650/2012-9, respectively).

Author Contributions

Conceived and designed the experiments: LDSD MCMM. Performed the experiments: LDSD RSB VMS MCMM. Analyzed the data: LDSD RSB VMS GDSS. Contributed reagents/materials/analysis tools: LDSD. Wrote the paper: LDSD RSB VMS MCMM GDSS.

References

1. Graham CH, Fine PVA (2008) Phylogenetic beta diversity: linking ecological and evolutionary processes across space in time. Ecology Letters 11: 1265–1277.
2. Parmentier I, Hardy OJ (2009) The impact of ecological differentiation and dispersal limitation on species turnover and phylogenetic structure of inselberg's plant communities. Ecography 32: 613–622.
3. Swenson NG (2011) Phylogenetic Beta Diversity Metrics, Trait Evolution and Inferring the Functional Beta Diversity of Communities. PLoS ONE 6: e21264.
4. Ricklefs RE (1987) Community diversity: relative roles of local and regional processes. Science 235: 167–171.
5. Cavender-Bares J, Kozak KH, Fine PVA, Kembel SW (2009) The merging of community ecology and phylogenetic biology. Ecology Letters 12: 693–715.
6. Araújo MB, Nogués-Bravo D, Diniz-Filho JAF, Haywood AM, Valdes PJ, et al. (2008) Quaternary climate changes explain diversity among reptiles and amphibians. Ecography 31: 8–15.
7. Mittermeier RA, Gil PR, Hoffman M, Pilgrim J, Brooks T, et al. (2004) Hotspots revisited: earth's biologically richest and most endangered terrestrial ecoregions. Mexico City: CEMEX & Agrupacion Sierra Madre.
8. Ribeiro MC, Metzger JP, Martensen AC, Ponzoni FJ, Hirota MM (2009) The Brazilian Atlantic Forest: How much is left, and how is the remaining forest distributed? Implications for conservation. Biological Conservation 142: 1141–1153.
9. Stehmann JR, Forzza RC, Salino A, Sobral M, Costa DP, et al. (2009) Plantas da Floresta Atlântica. Rio de Janeiro: Instituto de Pesquisas Jardim Botânico do Rio de Janeiro.
10. Martini AMZ, Fiaschi P, Amorim AM, Paixão JLd (2007) A hot-point within a hot-spot: a high diversity site in Brazil's Atlantic Forest. Biodiversity and Conservation 16: 3111–3128.
11. Murray-Smith C, Brummitt NA, Oliveira-Filho AT, Bachman S, Moat J, et al. (2009) Plant diversity hotspots in the Atlantic coastal forests of Brazil. Conservation Biology 23: 151–163.
12. Costa LP (2003) The historical bridge between the Amazon and the Atlantic Forest of Brazil: a study of molecular phylogeography with small mammals. Journal of Biogeography 30: 71–86.
13. Fiaschi P, Pirani JR (2009) Review of plant biogeographic studies in Brazil. Journal of Systematics and Evolution 47: 477–496.
14. IBGE (1992) Manual técnico da vegetação brasileira. Série Manuais Técnicos em Geociências. Rio de Janeiro: IBGE.
15. Oliveira-Filho AT, Fontes MAL (2000) Patterns of floristic differentiation among Atlantic forests in southeastern Brazil and the influence of climate. Biotropica 32: 793–810.
16. Morrone JJ (2001) A proposal concerning formal definitions of the Neotropical and Andean regions. Biogeographica 77: 65–82.
17. Bergamin R, Müller S, Mello RS (2012) Indicator species and floristic patterns in different forest formations in southern Atlantic rainforests of Brazil. Community Ecology 13: 162–170.
18. Duarte LDS, Prieto PV, Pillar VDP (2012) Assessing spatial and environmental drivers of phylogenetic structure in Brazilian Araucaria forests. Ecography 35: 952–960.
19. Morrone JJ (2006) Biogeographic areas and transition zones of Latin America and the Caribbean islands based on panbiogeographic and cladistic analyses of the entomofauna. Annu Rev Entomol 51: 467–494.
20. Morley R (2000) Origin and evolution of tropical rainforests. Chichester: Wiley.
21. Burnham RJ, Johnson KR (2004) South American palaeobotany and the origins of neotropical rainforests. Philosophical Transactions of the Royal Society of London Series B: Biological Sciences 359: 1595–1610.
22. Pillar VD, Duarte LDS (2010) A framework for metacommunity analysis of phylogenetic structure. Ecology Letters 13: 587–596.
23. Duarte LDS (2011) Phylogenetic habitat filtering influences forest nucleation in grasslands. Oikos 120: 208–215.
24. Duarte LDS, Both C, Debastiani VJ, Carlucci MB, Gonçalves LO, et al. (2014) Climate effects on amphibian distributions depend on phylogenetic resolution and the biogeographical history of taxa. Global Ecology and Biogeography 23: 124–258.
25. Scudeller VV, Martins FR, Shepherd GJ (2001) Distribution and abundance of arboreal species in the atlantic ombrophilous dense forest in Southeastern Brazil. Plant Ecology 152: 185–199.
26. Oliveira Filho A, Tameirão Neto E, Carvalho W, Werneck M, Brina A, et al. (2005) Analise floristica do compartimento arboreo de areas de floresta atlantica sensu lato na regiao das Bacias do Leste (Bahia, Minas Gerais, Espirito Santo e Rio de Janeiro).(Floristic analysis of the tree component of atlantic forest areas in central eastern Brazil.). Rodriguésia 56: 185–235.
27. Marques MC, Swaine MD, Liebsch D (2011) Diversity distribution and floristic differentiation of the coastal lowland vegetation: implications for the conservation of the Brazilian Atlantic Forest. Biodiversity and conservation 20: 153–168.
28. Mori SA, Boom BM, Prance GT (1981) Distribution patterns and conservation of eastern Brazilian coastal forest tree species. Brittonia 33: 233–245.
29. Oliveira-Filho AD, Ratter J (1995) A study of the origin of central Brazilian forests by the analysis of plant species distribution patterns. Edinburgh Journal of Botany 52: 141–194.
30. SanMartín I, Ronquist F (2004) Southern Hemisphere biogeography inferred by event-based models: plant versus animal patterns. Systematic Biology 53: 216–243.
31. Bertoncello R, Yamamoto K, Meireles LD, Shepherd GJ (2011) A phytogeographic analysis of cloud forests and other forest subtypes amidst the Atlantic forests in south and southeast Brazil. Biodiversity and Conservation 20: 3413–3433.
32. Hueck K (1972) As Florestas da América do Sul. São Paulo: Ed. UnB/Ed. Polígono.
33. Behling H (2002) South and southeast Brazilian grasslands during Late Quaternary times: a synthesis. Palaeogeography, Palaeoclimatology, Palaeoecology 177: 19–27.
34. Rambo B (1951) O elemento andino no pinhal riograndense. Anais Botânicos do Herbário Barbosa Rodrigues 3: 7–39.
35. Smith AR, Pryer KM, Schuettpelz E, Korall P, Schneider H, et al. (2006) A classification for extant ferns. Taxon 55: 705–731.
36. Chase MW, Reveal JL (2009) A phylogenetic classification of the land plants to accompany APG III. Botanical Journal of the Linnean Society 161: 122–127.
37. Pillar VD, Orlóci L (1996) On randomization testing in vegetation science: multifactor comparisons of relevé groups. Journal of Vegetation Science 7: 585–592.
38. Anderson MJ (2001) A new method for non-parametric multivariate analysis of variance. Austral Ecology 26: 32–46.
39. Dixon P (2003) VEGAN, a package of R functions for community ecology. Journal of Vegetation Science 14: 927–930.
40. APG (2009) An update of the Angiosperm Phylogeny Group classification for the orders and families of flowering plants: APG III. Botanical Journal of the Linnean Society 161: 105–121.
41. Burleigh JG, Barbazuk WB, Davis JM, Morse AM, Soltis PS (2012) Exploring diversification and genome size evolution in extant gymnosperms through phylogenetic synthesis. Journal of Botany 2012: 6.
42. Bell CD, Soltis DE, Soltis PS (2010) The age and diversification of the angiosperms re-revisited. American Journal of Botany 97: 1296–1303.
43. Magallón S, Hilu KW, Quandt D (2013) Land plant evolutionary timeline: gene effects are secondary to fossil constraints in relaxed clock estimation of age and substitution rates. American Journal of Botany 100: 556–573.
44. Webb CO, Ackerly DD, Kembel SW (2008) Phylocom: software for the analysis of phylogenetic community structure and trait evolution. Bioinformatics 24: 2098–2100.
45. Webb CO, Donoghue MJ (2005) Phylomatic: tree assembly for applied phylogenetics. Molecular Ecology Notes 5: 181–183.
46. Faith DP (1992) Conservation evaluation and phylogenetic diversity. Biological Conservation 61: 1–10.
47. Webb CO, Ackerly DD, McPeek MA, Donoghue MJ (2002) Phylogenies and community ecology. Annual Review of Ecology and Systematics 33: 475–505.
48. Kembel SW, Cowan PD, Helmus MR, Cornwell WK, Morlon H, et al. (2010) Picante: R tools for integrating phylogenies and ecology. Bioinformatics 26: 1463–1464.
49. Lozupone C, Knight R (2005) UniFrac: a new phylogenetic method for comparing microbial communities. Applied and Environmental Microbiology 71: 8228–8235.
50. Hardy OJ, Senterre B (2007) Characterizing the phylogenetic structure of communities by an additive partitioning of phylogenetic diversity. Journal of Ecology 95: 493–506.
51. Debastiani VJ, Pillar VD (2012) SYNCSA–R tool for analysis of metacommunities based on functional traits and phylogeny of the community components. Bioinformatics 28: 2067–2068.
52. Legendre P, Anderson MJ (1999) Distance-based redundancy analysis: testing multispecies responses in multifactorial ecological experiments. Ecological Monographs 69: 1–24.
53. Legendre P, Legendre L (2012) Numerical Ecology. Amsterdam: Elsevier.
54. Bryant JA, Lamanna C, Morlon H, Kerkhoff AJ, Enquist BJ, et al. (2008) Microbes on mountainsides: Contrasting elevational patterns of bacterial and plant diversity. Proceedings of the National Academy of Sciences 105: 11505–11511.
55. Carvalho PER (2006) Espécies Arbóreas Brasileiras. Brasília: Embrapa.
56. Carvalho PER (2008) Espécies Arbóreas Brasileiras. Brasília: Embrapa.
57. Carvalho PER (2010) Espécies Arbóreas Brasileiras. Brasília: Embrapa.
58. Duarte LDS, Carlucci MB, Pillar VD (2009) Macroecological analyses reveal historical factors influencing seed dispersal strategies in Brazilian Araucaria forests. Global Ecology and Biogeography 18: 314–326.
59. Giulietti AM, Harley RM, Queiroz LP, Wanderley MGL, Van den Berg C (2005) Biodiversity and conservation of plants in Brazil. Conservation Biology 19: 632–639.
60. Scheer MB, Blum CT (2011) Arboreal Diversity of the Atlantic Forest of Southern Brazil: From the Beach Ridges to the Paraná River. In: Grillo O, Venora G, editors. The Dynamical Processes of Biodiversity - Case Studies of Evolution and Spatial Distribution. Rijeka: Intech. 109–134.
61. Giehl ELH, Jarenkow JA (2012) Niche conservatism and the differences in species richness at the transition of tropical and subtropical climates in South America. Ecography 35: 933–943.

62. Crisci JV, Cigliano MM, Morrone JJ, Roig-Junent S (1991) Historical biogeography of southern South America. Systematic Biology 40: 152–171.

63. Amorim D, Tozoni S (1994) Phylogenetic and biogeographic analysis of the Anisopodoidea (Diptera, Bibionomorpha), with an area cladogram for intercontinental relationships. Revista Brasileira de Entomologia 38: 517–543.

64. Lopretto E, Morrone J (1998) Anaspidacea, Bathynellacea (Crustacea, Syncarida), generalised tracks, and the biogeographical relationships of South America. Zoologica Scripta 27: 311–318.

65. Pennington RT, Dick CW (2004) The role of immigrants in the assembly of the South American rainforest tree flora. Philosophical Transactions of the Royal Society of London Series B: Biological Sciences 359: 1611–1622.

66. Carlucci MB, Jarenkow JA, Duarte LDS, Pillar VDP (2011) Conservation of Araucaria forests in the extreme South of Brazil. Natureza & Conservação 9: 111–113.

67. Myers N, Mittermeier RA, Mittermeier CG, Da Fonseca GA, Kent J (2000) Biodiversity hotspots for conservation priorities. Nature 403: 853–858.

10

Increased Drought Impacts on Temperate Rainforests from Southern South America: Results of a Process-Based, Dynamic Forest Model

Alvaro G. Gutiérrez[1,2*¤]**, Juan J. Armesto**[3,4]**, M. Francisca Díaz**[5]**, Andreas Huth**[1]

1 Department of Ecological Modeling, Helmholtz Centre for Environmental Research (UFZ), Leipzig, Germany, **2** Forest Ecology Group, Institute of Terrestrial Ecosystems, Department of Environmental Sciences, Swiss Federal Institute of Technology (ETH Zürich), Zürich, Switzerland, **3** Instituto de Ecología y Biodiversidad (IEB), Santiago, Chile, **4** Departamento de Ecología, Facultad de Ciencias Biológicas, Universidad Catolica de Chile, Santiago, Chile, **5** Departamento de Ciencias Biológicas, Facultad de Ciencias Biológicas, Universidad Andrés Bello, Santiago, Chile

Abstract

Increased droughts due to regional shifts in temperature and rainfall regimes are likely to affect forests in temperate regions in the coming decades. To assess their consequences for forest dynamics, we need predictive tools that couple hydrologic processes, soil moisture dynamics and plant productivity. Here, we developed and tested a dynamic forest model that predicts the hydrologic balance of North Patagonian rainforests on Chiloé Island, in temperate South America (42°S). The model incorporates the dynamic linkages between changing rainfall regimes, soil moisture and individual tree growth. Declining rainfall, as predicted for the study area, should mean up to 50% less summer rain by year 2100. We analysed forest responses to increased drought using the model proposed focusing on changes in evapotranspiration, soil moisture and forest structure (above-ground biomass and basal area). We compared the responses of a young stand (YS, ca. 60 years-old) and an old-growth forest (OG, >500 years-old) in the same area. Based on detailed field measurements of water fluxes, the model provides a reliable account of the hydrologic balance of these evergreen, broad-leaved rainforests. We found higher evapotranspiration in OG than YS under current climate. Increasing drought predicted for this century can reduce evapotranspiration by 15% in the OG compared to current values. Drier climate will alter forest structure, leading to decreases in above ground biomass by 27% of the current value in OG. The model presented here can be used to assess the potential impacts of climate change on forest hydrology and other threats of global change on future forests such as fragmentation, introduction of exotic tree species, and changes in fire regimes. Our study expands the applicability of forest dynamics models in remote and hitherto overlooked regions of the world, such as southern temperate rainforests.

Editor: Bruno Hérault, Cirad, France

Funding: This study was supported by grants from Millennium Scientific Initiative (P05-002) and CONICYT PFB-23 to the Institute of Ecology and Biodiversity. AGG was funded by a DAAD doctoral fellowship, by a Marie Curie Intra European Fellowship within the 7th European Community Framework Programme (PIEF-GA-2010-274798) and CONICYT-PAI grant (82130046). Additional funding was provided by the ERC advanced grant 233066. The funders had no role in study design, data collection and analysis, decision to publish, or preparation of the manuscript.

Competing Interests: The authors have declared that no competing interests exist.

* Email: bosqueciencia@gmail.com

¤ Current address: Instituto de Conservación Biodiversidad y Territorio, Facultad de Ciencias Forestales y Recursos Naturales, Universidad Austral de Chile, Valdivia, Chile

Introduction

Climate and forests are dynamically linked through the spatial and temporal variability of soil moisture [1], with climate system effects on ecological processes which are still poorly understood. Forest dynamics models, particularly those based on interactions among individual trees (i.e. gap models [2]), provide a simple, and general framework to assess the impacts of climate on forest dynamics. These models simulate the fate of single trees on the basis of species' life-history traits and limited resource availability (e.g. soil moisture), thereby facilitating the analysis of climate-forest interactions [3].

Forest gap models use a variety of approaches to model forest hydrology. While some gap models use a simple bucket water balance model [3,4], others include physiology-based representations of plant and soil controls on water uptake and evapotrans-

piration [5,6]. Regardless of the level of detail used to model forest hydrology, it seems necessary that forest gap models address water availability (i.e. soil moisture) as an integrating factor, with effects on canopy transpiration [7]. Changes in rainfall regimes, summarized by changes in the duration and frequency of periods of water stress during the year, should influence soil moisture dynamics limiting plant productivity [8]. Introducing dynamic linkages of ecological processes with soil moisture variation in gap models will contribute to predict drought-induced changes in forest dynamics. Such model improvements are increasingly relevant to understanding how forests can adapt to climate change [6,9].

Forest gap models have successfully simulated the dynamics of a variety of forest types including temperate rainforests of the southern hemisphere [10,11]. In southern South America (SSA, 37–43°S), the progressive loss, fragmentation and subsequent degradation of temperate rainforests due to unsustainable logging and fire is threatening the integrity of ecosystem functions [12,13] and modifying their hydrological balance [14,15]. Annual precipitation has decreased in the same region by about 40% in the last century (time period 1901–2005, [16]) and summer rainfall is expected to decrease up to 50% by the year 2100 [17,18]. SSA forests share similar structural characteristics with temperate rainforests of the Pacific Northwest of North America, Tasmania, and New Zealand [19]. In addition, SSA forests represent the largest area of temperate forest remaining in the southern hemisphere [20]. Floristic richness is the highest among evergreen temperate rainforests worldwide and the high concentration of endemism has given this region a unique global conservation value [20,21]. The global relevance of SSA forests and climate trends predicted for the coming decades make it urgent to expand model applications into this region, as a tool to predict temperate rainforest responses to impending declines in rainfall.

This study introduces a forest gap model specifically designed for assessing the responses of temperate rainforests in southern South America to increased drought. The model provides accurate estimates of forest water fluxes and incorporates dynamical linkages among rainfall regimes, soil moisture, and individual tree growth. We assessed model performance by comparing the results with detailed field measurements of water cycling in a stand located on northern Chiloé Island, Chile (41°50′S). We also conducted a sensitivity analysis of the response of current forests to drought, i.e. when rainfall is decreased. Model predictions of forest hydrology (evapotranspiration and soil moisture) and structure (above-ground biomass and basal area) under increased drought predicted for 2100 in the study area were compared for a young-secondary (YS) and an old-growth (OG) forest stand to analyze differences in responses to expected changes in rainfall.

Materials and Methods

Study area

The study was conducted on northern Chiloé Island, Chile (41°50′ S, Fig. 1) at the private protected area *Estación Biológica Senda Darwin* (EBSD), with permission granted by the owner. Fragments of secondary and primary forests occur over rolling hills of low altitude (50–100 m) dispersed in a matrix of bogs, shrublands and grazing pastures. The present landscape has been shaped by a history of widespread use of fire to clear land for pastures since the late 1800s, followed by selective logging of remaining forest patches [22]. Soils are generally thin (<0.5 m), originated from moraine fields and outwash plains from the last glaciation, often with poor drainage [23]. Soils have high organic matter content, soil texture loam to silty loam, and a 2–4 mm thick iron silicate layer or hardpan (found at ca. 52 cm depth), where roots cannot penetrate [24]. The prevailing climate is wet-temperate with strong oceanic influence [25]. Rainfall occurs throughout the year, with an annual average of 2158 mm (25% occurring in summer). Mean annual temperature is 9.1°C. Maximum and minimum monthly temperatures are 13.9°C (January) and 4.2°C (July) [26].

Floristically, forests of the study area belong to the North Patagonian temperate rainforest type [27]. The canopy is dominated by evergreen trees, mainly *Podocarpus nubigena* (Podocarpaceae), *Drimys winteri* (Winteraceae) and *Nothofagus nitida* (Nothofagaceae), with the common presence of *Tepualia stipularis* (Myrtaceae) and several Myrtaceae tree species in the understory. Ferns (e.g. *Hymenophyllum* spp., *Hymenoglossum cruentum, Polypodium feullei*) and angiosperms (e.g. Gesneriaceae and Bromeliaceae) growing epiphytically are frequent. Detailed

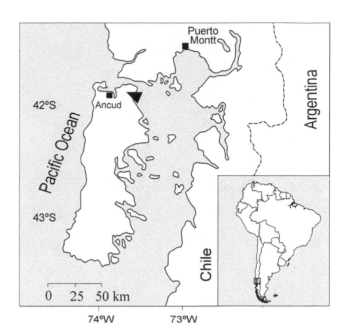

Figure 1. Location of study site (triangle) on northern Chiloé Island, Chile.

descriptions of structure and dynamics of this forest type have been previously published [26,28]. The study did not involve endangered or protected tree species.

The forest model

Here, we introduce an individual-oriented dynamic forest model (FORMIND-CL v.1.0) that includes calculations of hydrologic balance. The model is based on FORMIND, a forest model comprehensively tested to simulate the dynamics of temperate rainforests in SSA [11,13]. FORMIND is a generalized forest growth model that simulates the spatial and temporal dynamics of uneven-aged, mixed species forest stands [29–31]. The model simulates forest dynamics (in annual time steps, t) as a mosaic of interacting forest patches of 20×20 m, which is the approximate crown size of a large mature tree in the forest. Within these patches, stand dynamics is driven by competition for light and space following the gap model approach [2]. For the explicit modeling of the competition for light, each patch is vertically divided into height layers of 0.5 m, where leaf area is summed up and the light environment under the canopy is calculated via a light extinction law. The carbon balance of each individual tree is modeled explicitly, including the main physiological processes (photosynthesis and respiration [13]). Allometric functions and geometrical relations are used to calculate above-ground biomass, tree height, crown diameter and stem volume from the stem diameter at 1.3 m height of the tree (*dbh*). Tree mortality can occur either through self-thinning in densely populated stands, tree senescence, gap formation by large falling trees, slow tree grow, or external disturbances (e.g. windthrow). Gap formation links neighboring forest areas. Tree regeneration rates are formulated as maximum rates of recruitment of small trees at *dbh* threshold of 1 cm, with seed loss through predation and seedling mortality being incorporated implicitly [13]. Maximum recruitment rates are reduced by shading. Nutrient availability is considered to be homogeneous at the stand scale. A description of the core model and its equations is given elsewhere [11,13]. We focus below on the extensions added to incorporate forest hydrology.

Table 1. Structure and composition of the young-secondary North Patagonian forest stand (YS) in Senda Darwin Biological Station, compared to values reported for other secondary forest stands, elsewhere in Chiloé Island, Chile.

Species	Study site		Literature*			
	Density (trees ha^{-1})	Basal area (m^2 ha^{-1})	Density (trees ha^{-1})	Basal area (m^2 ha^{-1})	Density (trees ha^{-1})	Basal area (m^2 ha^{-1})
Drimys winteri	1550	32.3	1800	24.7	4532	61.4
Eucryphia cordifolia	175	0.9	112	1.4	480	14.7
Myrtaceae[†]	1014	3.5	252	2.3	827	4.4
Nothofagus nitida	713	17.7	4	0.1	260	4
Podocarpaceae[¥]	38	0.1	80	0.5		
Weinmannia trichosperma	88	0.3				
Others	287.5	1.9				
Total	3575	54.8	2431	46	7950	86.9

*Source: [26,77–80] only forest stands <100 years old. Only listed tree species present in YS.
†Tepualia stipularis, Myrceugenia spp. and Amomyrtus spp.;
¥Podocarpus nubigena, Saxegothaea conspicua.

The hydrologic submodel

Soil moisture dynamics is described at a daily timescale, treating soil as a reservoir with an effective storage capacity that is intermittently filled by rainfall events. Soil water losses occur via transpiration, interception by the forest canopy, and drainage below the root zone. We neglected lateral water flow, thus the model applies mainly to flat terrains. This is a reasonable assumption in forests of the study area because during the rainy season soils tend to be saturated and accumulated water cannot infiltrate the soil.

Soil moisture s (dimensionless, $0 \leq s \leq 1$), vertically averaged over the soil depth z (mm), was considered as central state variable [8]. Thus, the water balance equation for a given point in the forest can be expressed as [1]:

$$n \cdot z \cdot \frac{ds}{dd} = Pnet_d - Tr_d - Q(s,d) \qquad (1)$$

where d is the Julian day of the year, n is the porosity (volume of voids/total volume of soil, i.e. dimensionless, vertically averaged); $Pnet_d$ is the net precipitation falling to the soil surface (mm day^{-1}); Tr_d is the transpiration rate (mm day^{-1}); and $Q(s,d)$ is the soil drainage (mm day^{-1}). Both n and z are assumed to be time-invariant parameters [1]. The volumetric water content (θ, m^3 water/m^3 soil, i.e. dimensionless) can be calculated as follows [1]:

$$\theta = s \cdot n \qquad (2)$$

The normalized version of equation (1) is used through the text where all terms are divided by $n \cdot z$. Both the local vertical and horizontal variability of soil moisture are considered negligible at the daily timescale, assuming an equal propagation of the wetting front and equal soil moisture redistribution over the rooting zone [8,32].

Net precipitation. Daily net precipitation falling to the soil surface ($Pnet_d$) is described by,

$$Pnet_d = P_d - Ec_d \qquad (3)$$

where, Ec_d is the canopy interception (mm day^{-1}), defined here as the total daily rainfall (P_d, mm day^{-1}) that is retained by the canopy and is evaporated so that it does not reach the ground. Following [33], we assumed that Ec_d asymptotically approaches the canopy retention capacity and can be modeled at daily intervals as:

$$Ec_d = S_t \cdot (1 - e^{-\alpha_h \cdot P_d}) \qquad (4)$$

where S_t is the canopy water retention capacity of the stand at year t and α_h is a parameter describing the slope of the saturation curve. The parameter α_h represents, in a simplified terms, the complex process of water partitioning into throughfall and stem flow [34]. S_t depends on leaf area index of the forest patch at simulated year t (LAI_t) and is calculated by the expression [34,35]:

$$S_t = \frac{LAI_t}{LAI_{max}} \cdot f_h \cdot \log(1 + LAI_t) \qquad (5)$$

where, LAI_{max} is the maximum leaf area index of the forest and f_h is a shape parameter. We avoided unrealistic canopy interception values in the model by setting $Ec_d = P_d$ when $Ec_d > P_d$.

Table 2. Parameter descriptions and parameterization methods used for running simulations in FORMIND-CL v.1.0.

	Description	Value	Units	Method	Reference
Weather generator					
$1/\lambda$	Mean interval time between rainfall events[*]	†	days	a	This study
η	Mean depth of rainfall events[*]	†	mm day^{-1}	a	This study
T_μ	Mean daily temperature[*]	†	°C	a	This study
T_σ	Standard deviation daily temperature[*]	†	°C	a	This study
Rg_μ	Daily global radiation above canopy[*¶]	†	μmol (photons) m^{-2} s^{-1}	a	This study
Rg_σ	Standard deviation of Rg_μ [*¶]	†	μmol (photons) m^{-2} s^{-1}	a	This study
Hydrologic submodel					
n	Vertically averaged porosity of the soil	0.757	-	b	[53]
z	Soil depth	520	mm	b	[14]
k_{soil}	Saturated hydraulic conductivity	4	mm day^{-1}	b	[36,54]
α_h	Slope of the canopy saturation curve	0.7	-	b, c	[34]
f_h	Parameter of the relationship LAI and canopy storage capacity	3	mm day^{-1}	c	[35]
LAI_{max}	Maximum LAI of the studied forest	5.5	m^2 m^{-2}	b	[81]
WUE	Water-use efficiency	9	g CO$_2$ kg^{-1}H$_2$O	e	[14]
γ	Psychometer constant	65	Pa K^{-1}	b	[43,44]
L	Latent heat of vaporization of water	2.56×10^6	J kg^{-1}	e	[43,44]
Δ	Rate of change of saturated vapor pressure with temperature	c	Pa K^{-1}	d	[43,44]
θ_{wp}	Wilting point of the soil	0.125	-	b	[36]
θ_{fc}	Field capacity of the soil	0.3	-	b	[36]

Method refers to a: calculated from daily meteorological data from *Senda Darwin Biological Station*, period 1998–2009, b: from literature, c: calibrated with field data, d: calculated, e: calibrated using literature.
†values indicated in Table 3.
*per season.
¶calculated for dry (Pd<1 mm) and wet days (Pd≥1 mm).

Soil moisture modeling. Drainage out of the root zone ($Q(s,d)$) was modeled according to [1]. When the soil is saturated ($s = 1$), soil water is permitted to percolate at a rate equivalent to the saturated hydraulic conductivity of the soil (k_{soil}, mm day^{-1}, [36]). Runoff occurs when the soil is saturated and no more water can be held in place. The excess of water is assumed to leave the system, which is reasonable to assume given the large rainfall intensity in the study area. When $s < 1$, soil deep percolation rate is calculated using the empirical relationship of Neilson [37],

$$Q(s,d) = k_{soil} \cdot s^2 \qquad (6)$$

Transpiration. Water-use efficiency describes the proportion of water used for the assimilation of a unit of carbon in the photosynthesis [38,39]. This concept can be used to estimate transpiration of trees (Tr, mmol H$_2$O m^{-2} s^{-1}) from:

$$Tr = \frac{PB}{WUE} \qquad (7)$$

where, PB is the gross biomass production of the tree (μmol carbon dioxide m^{-2} s^{-1}), and WUE is a parameter denoting water-use efficiency at stand level. PB is obtained from the rate of single-leaf photosynthesis following [40], which is integrated over

the total LAI of the tree to account for self-shading [13]. The resulting photosynthetic rate is then multiplied by the crown area of the tree to obtain PB (see also equations in Appendix S1 and [13]). Daily transpiration (Tr_d) of trees is obtained from equation 8 and dividing Tr by the length of the active photosynthetic period per year.

The daily potential evapotranspiration (PET_d, mm day^{-1}) describes a physical limit for the amount of water that can be held and transported away from the canopy under given climatic conditions. Evaporation is neglected in the model; therefore, it is assumed that maximum water losses by vegetation are limited by the difference between PET_d and the canopy interception of the day (Ec_d), as follows:

$$PET_d - Ec_d \geq Tr_d \qquad (8)$$

PET_d is calculated using a modified Penman-Monteith expression in case of aerodynamic conductance [41,42] and determined by the variation of the daily net radiation flux (Rn_d, J m^{-2} day^{-1}):

$$PET_d = (\frac{\Delta}{\Delta + \gamma}) Rn_d / L \qquad (9)$$

where γ is the psychometric constant (ca. 65 Pa K^{-1}, slightly depends on temperature), L is the latent heat of vaporization of

Table 3. Parameter values used to run the weather generator under different climatic scenarios.

Rainfall	Current climate			Future scenarios		
	Seasonal sum (mm, average ± sd)	1/λ (days)	η (mm day^{-1})	Seasonal sum (mm, average)	1/λ (days)	η (mm day^{-1})
DJF	284.6±132.7	0.9	8.3	238–131	0.99–1.36	4.2–7.5
MAM	543.6±138.6	0.55	12.8	813.4	0.55	12.8
JJA	813.4±197.2	0.29	16	543.6	0.29	16
SON	424.5±144.7	0.54	9.5	382–212	0.59–0.8	4.7–8.5
Annual sum (mm, average ± sd)	2094.8±353.8					
Temperature	T_μ (°C)	T_σ (°C)			T_μ (°C)	
DJF	12.5	2.6			16.5	
MAM	10.1	3.6			13.1	
JJA	8.4	4.4			10.4	
SON	9.8	3.3			11.8	
Radiation	Rg_μ (Rg_σ)	Rg_μ (Rg_σ)				
	[P_d<1 mm]	[P_d≥1 mm]				
DJF	1413.9 (317.0)	986.9 (378.2)				
MAM	701.2 (294.1)	398.8 (261.6)				
JJA	408.9 (159.1)	229.7 (139.4)				
SON	1065.5 (348.7)	640.8 (320.1)				

Current climate indicate parameters used to run the model under current climate based on instrumental records (weather station at *Senda Darwin Biological Station*, period 1998–2009). Radiation describes parameters daily global radiation Rg_μ and Rg_σ (the latter in brackets, μmol(photons) m^{-2} s^{-1}). Temperature is mean daily air temperature. *Future scenarios* are the range of climatic parameters that were varied to run the model under increased drought (36 scenarios in total, see *Methods* for details). *DJF*: December to February (austral summer, growing season); *MAM*: March to May (austral autumn); *JJA*: June to August (austral winter), *SON*: September to November (spring, growing season). sd: standard deviation.

water (ca. 2.56×10^6 J kg^{-1} slightly depends on temperature). The rate of change of saturated vapor pressure with temperature (Δ, Pa K^{-1}) is calculated as [43,44]:

$$\Delta = \frac{2.503x10^6 e^{17.269\frac{T_d}{237.3+T_d}}}{(237.3+T_d)^2} \tag{10}$$

Rn_d was calculated from latitude, day of the year, sunshine hours and daily air temperature (T_d, °C) following [42,45].

Soil moisture impact on tree biomass production. The dependence of water uptake for tree biomass production on soil moisture is described by a function representing a reduction factor due to water scarcity ($\omega(s)$, $0 \leq \omega(s) \leq 1$ [46]). This factor accounts indirectly for the impact of water demand on potential photosynthetic production (i.e. possible to achieve under competition for light). $\omega(s)$ is implemented as a daily reduction factor due to water scarcity by,

$$\omega(s) = 0 \text{ if } s < \theta_{wp}$$

$$\omega(s) = \frac{s - \theta_{wp}}{\theta_{msw} - \theta_{wp}} \text{ if } \theta_{wp} < s \leq \theta_{msw}$$

$$\omega(s) = 1 \text{ if } s > \theta_{msw} \tag{11}$$

where, θ_{wp} is the wilting point, and θ_{msw} represents a threshold when enough soil moisture is available for potential tree biomass production. We calculated θ_{msw} from:

$$\theta_{msw} = \theta_{wp} + \frac{1}{3}(\theta_{fc} - \theta_{wp}) \tag{12}$$

where, θ_{fc} is the soil field capacity. θ_{wp}, θ_{msw}, and θ_{msw} are expressed as normalized soil moisture. In the model, the wilting point (θ_{wp}) determines the minimum soil moisture content necessary for tree biomass production. Thus, we assumed a linear reduction of biomass production when soil water content was between θ_{msw} and θ_{wp}. The water required for biomass production of trees is completely removed from the soil compartment when soil moisture reaches θ_{msw} (i.e., $\omega(s) = 1$), after the calculation of maximum possible transpiration of trees. Both biomass production and water supply are reduced until the water needed for biomass production corresponds with θ_{wp}. The calculated rate of biomass production influences tree respiration rate through maintenance and growth respiration, which are calculated subsequently in the model (see [13] for equations and Fig. S1 for a diagram). All calculations are performed for every tree and pooled together to calculate the stand-level values.

Weather generator. Rainfall time series, representing the frequency and depth of rainfall events, were constructed as series of random numbers generated by probability distributions. The interval between rainfall events, τ (day) can be expressed as an exponential distribution given by [47].

$$f_T(\tau) = \lambda \cdot e^{-\lambda \tau}, \text{ for } \tau \geq 0 \tag{13}$$

Table 4. Model estimates of water balance components for a young secondary forest (YS, 60 years-old) and an old-growth North Patagonian forest (OG, >500 years-old) located in northern Chiloé Island, Chile, under current climate.

Variable	This study		Literature	
	mm year^{-1} ± sd	%	Value	Source*
Young secondary stand (YS)				
Canopy Interception	381.5±30.3	20.0	20–40%	1
Deep percolation	980.7±156.8	49.9	47%	1
ET	573.6±35.1	30.3	45.2%	1
Net precipitation	1617.2±434.6	80.0	60–80%	1
Runoff	722.9±397.7	33.3		
Soil moisture	-	62.5		
Transpiration	192.1±18.8	10.3	22%	1
Old-growth stand (OG)				
Canopy Interception	378.3±33.6	19.9	17.8%	2, 3
Deep percolation	907.4±142.4	46.5	66.5%	2
ET	648.0±44.1	34.3	19.9–33.3%	2
Net precipitation	1591.8±368.6	80.1		
Runoff	665.2±321.9	31.8	30–55%	4
Soil moisture	-	55.3		
Transpiration	269.7±26.1	14.4		
Potential evapotranspiration	769±4		576–724 mm year^{-1}	5

Model results are the average of 100 simulations per stand (see *Methods* for details), with annual sum of rainfall averaging 1970–2000 mm year^{-1}. Literature refers to values reported by independent studies in Chilean temperate rainforests and comparable forests elsewhere. sd: standard deviation; %: percentage of total annual rainfall; ET: Evapotranspiration (sum of canopy interception and transpiration). sd: standard deviation.
*(1) Data for other broad-leaved evergreen forests, ca. 200 years old. Annual rainfall 2500 mm year^{-1} [82,83]. (2): Mixed deciduous-broad-leaved old-growth forest. Annual rainfall 2400 mm year^{-1} [12]. (3) Mixed broad-leaved and conifer forest, ca. 200 years old in New Zealand. Annual rainfall 3400 mm year^{-1} [84]. (4) Annual rainfall 1700–4500 mm year^{-1}, data from evergreen, broad-leaved forests with 90% cover [85]. (5) Annual rainfall of 2427–3991 mm year^{-1} [86], weather stations of Castro and Punta Corona.

where $1/\lambda$ is the mean time interval between rainfall events (days). Total daily rainfall (P_d) depends on the amount of rain of each event (h, mm day^{-1}), which is also assumed to be an independent random variable, expressed by an exponential probability density function [47]:

$$P_d(h) = \frac{1}{\eta} e^{-\frac{1}{\eta} h}, \text{ for } h \geq 0 \qquad (14)$$

where η is the mean depth of rainfall events (mm day^{-1}). Both $1/\lambda$ and η parameters are calculated for each season of the year. We obtained daily global radiation (Rg_d) from EBSD instrumental records (period from May 1998 to February 2009). Rg_d varied among seasons in relation to daily rainfall (T-test, $p<0.001$). Therefore, in the model, Rg_d was distributed as a Gaussian variable whose mean and standard deviation depended on P_d (mean, Rg_μ, and standard deviation, Rg_σ). Values of Rg_μ and Rg_σ were obtained from instrumental records and varied depending on a threshold value of 1 mm of P_d and the season of the year. Daily temperature (°C, T_d) was simulated by a Gaussian random variable with parameters (mean, T_μ, and standard deviation, T_σ) that varied according to season of the year.

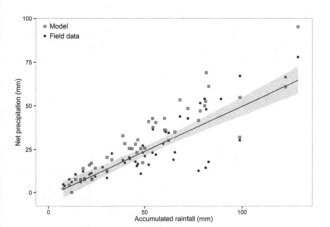

Figure 2. Comparison between measured and modeled net precipitation for 50 rain events recorded in a young secondary stand in northern Chiloé Island, Chile, for the time period 2007–2010. The line represents a linear regression between field measured net precipitation and accumulated rainfall during each event. The gray area represents 0.95 confidence intervals. Model results are for a forest patch are of 400 m^2, with a LAI = 5.

Field data

Stand structure. We characterized stand structure in terms of tree species dominance (basal area, m^2 ha^{-1}) and size (dbh) distribution in a young secondary and an old-growth North Patagonian forest stand (hereafter YS and OG respectively) found in a flat forested area at EBSD (Fig. 1). The YS stand was initiated from a stand-replacing fire about 60 years ago and currently presented no evidence of logging. In 2007, we set up two 20×20 m

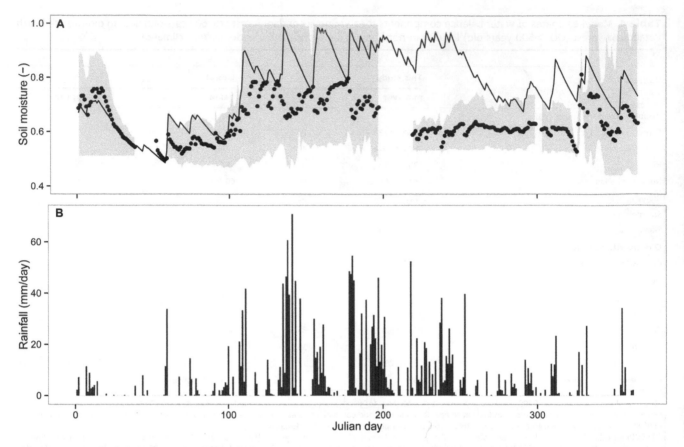

Figure 3. Variation in soil moisture during 2008 in a young secondary, North Patagonian forest stand (YS) in northern Chiloé Island, Chile. a) Comparison between observed and simulated soil moisture (normalized data, dimensionless). Observed data were obtained from five sensors randomly placed inside a 400-m² plot in YS (daily means denoted by filled dots, gray area showing the range of data). Simulated soil moisture (line) is for one-hectare forest with a successional age of 60 years (LAI = 4.5). Soil parameters are the same as described in Table 2 (see also *Methods*). b) Rain events recorded instrumentally during 2008.

plots to measure the hydrologic balance and stand structure. All trees rooted within each plot with stems >1.3 m height and > 5 cm *dbh* were permanently marked with numbered aluminum tags, identified to species, and their *dbh* measured to the nearest cm. Structure and composition of YS is comparable to that described for young-secondary stands elsewhere on northern Chiloé Island (Table 1). OG was an unmanaged forest stand >590 years old, without evidence of recent human disturbance, and representative of old-growth North Patagonian forest on Chiloé Island and elsewhere in the region [26]. Sampling methods, stand history, species composition and structure of OG is described by [28]. OG had total basal area of 72 m² ha⁻¹ and density of 2610 trees ha⁻¹, with a mixed dominance of Podocarpaceae species, *D. winteri*, *N. nitida* and Myrtaceous tree species in the understory [28].

Hydrologic measurements. We estimated net precipitation using throughfall measurements taken in two YS forest plots (i.e. rainfall falling through canopy gaps plus canopy drip), adding stemflow (water running down the stems). We conducted these measurements of volume accumulated during rainfall events occurred between June 2007 and December 2010. During this time period, we also analyzed hourly records of rainfall from the meteorological station at EBSD to obtain daily incident rainfall above the canopy. Rainfall events considered in the analysis occurred with a separation of at least two hours without rain to allow for full drip from the forest canopy. Stemflow collectors

consisted of a 2 mm thick smooth polycarbonate sheet molded around the stem to form a funnel. A hose led from the lowest point of the funnel to a 25 l polythene container, where the stemflow volume was collected after each rain event. Stemflow collectors were placed in 10 randomly selected trees of the two main canopy species, *D. winteri* and *N. nitida* (*dbh* >10 cm), in each plot. We eliminated two trees from our stemflow estimates that died during the study period. We converted the volume of collected water to millimeters of rain assuming that the surface of the collectors equals the projected tree crown area. Crown area was approximated by the area of an ellipse. Throughfall collectors, 0.12×2 m long (0.7 m² total area per plot) gutters were held, with a slight inclination, 0.5 m above the ground at three different locations within each plot. Collectors were connected with a funnel to a 25 l polythene container. Soil matric potential was measured every 30 minutes with four sensors per plot (*WatchDog Data Loggers* 450 and 800) placed approximately in every quarter of each plot, beneath the canopy and at ca. 15 cm soil depth. Continuous soil moisture measurements were obtained for the period January 2007 to March 2009.

Model parameterization

We used a previous model parameterization for North Patagonian forests including all main canopy tree species (11 tree species) occurring in the studied forests. The calibration, validation and robustness of this parameter set to reproduce forest stand

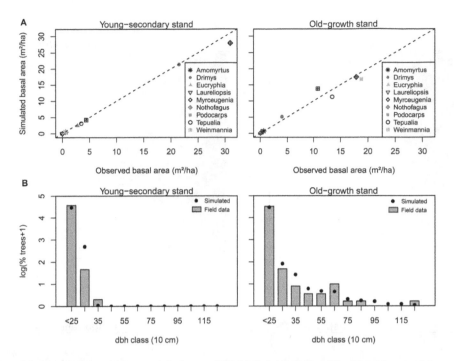

Figure 4. Comparison of forest structure between (a) observed (field data) and simulated basal area of tree species (Spearman's $r^2 > 0.9$, $p < 0.01$ in both cases) and (b) *dbh* distributions for the young secondary (YS) and old-growth (OG) North Patagonian forest stands studied in Chiloé Island, Chile. Simulated OG structure was obtained initializing the model with inventory data.

structure is discussed in detail by [11]. Here, we describe calibration of parameters related to the inclusion of hydrologic balance into the model. New parameters needed to run FORMIND-CL v.1.0 and their values are shown in Table 2.

The parameter f_h describing the relationship between leaf area index and canopy water storage capacity was calibrated following [34] and assuming that storage capacity reaches 4.9 mm day^{-1} at a leaf area index of 5.0 as measured by [14]. LAI_{max} was set to 5.5 following the maximum value observed in other Chilean temperate rainforests [48]. The slope of the saturation curve of the canopy rain retention capacity (α_h) was set according to common values for broad-leaved temperate trees [34]. To the best of our knowledge no estimation exists for water-use efficiency at stand scale in forests of the study area. Therefore we calibrated *WUE* using transpiration estimates of Díaz *et al.* [14] in Chiloé Island and the potential canopy photosynthetic rate estimated by the model for the study area under current climate ($Tr = 296$ mm year^{-1} and $PB = 32.9$ tC ha^{-1}). The selected *WUE* was then confirmed by comparison with reported values from other temperate rainforests [49–52]. Soil characteristics (porosity and depth) followed field descriptions from Chiloé Island [53]. We set water-retention and percolation properties of the soil (parameters θ_{wp}, θ_{fc} and k_{soil}) to average values [36,54] using texture classes (loam to silty loam) described for soils in the study area [53]. Daily records of rainfall from the EBSD weather station (60 m a.s.l, period from May 1998 to February 2009) were directly used to calculate rainfall parameters for the current climate simulations (Table 3). EBSD is the nearest and most representative weather station for the climate at the study site. We calculated the mean time interval between rainfall events from the duration (days) of rain events occurring in each season. The mean depth of rainfall events was calculated by dividing seasonal rainfall sum by the amount of wet days ($P_d > 0$). These calculations were done only for seasons with >85 daily records. We avoided the potential

overestimation of annual rainfall by normalizing predicted seasonal rainfall sum by prescribed seasonal rainfall averages (Table 3). The weather generator reproduced well the seasonal fluctuations in rainfall, temperature and radiation during the year with no significant departures from observed climatic records (Fig. S2, S3). Calibrated climatic parameters were assumed representative for growing conditions of North Patagonian forests on Chiloé Island and neighboring regions in the mainland.

Analyses

Model verification. We compared field measurements of net precipitation with model predictions at daily temporal scale. For this analysis we selected 50 rain events, representing field measurement intervals <20 days long, for which accumulated rainfall during the event (i.e. sum of daily rainfall from first to last day of the event) was correctly measured by EBSD weather station, and for which <50% of containers were filled completely during the rain event. Hydrologic parameters used for model estimation of net precipitation are indicated in Table 2. LAI measurements were unavailable for YS, thus we set LAI to comparable, averaged values reported by Diaz *et al.* [14] for the same forest type. We qualitatively compared daily variation in soil moisture produced by the model with field observations. For this analysis soil matric potential obtained for an entire year was transformed to soil moisture contents using a water retention curve [55]. We set parameters of the water retention curve following [56] for loam to silty loam texture classes and particle density of the soil type under study (ca. 2.0 g/cm^3, [53]). Model simulations were run using rainfall data for the same period of field measurements conducted over a whole year, i.e. 2008. Note that for both analyses spatial scale of model results (>1 ha) differed from the scale of field measurements (400 m^2 plots). We also compared model results with field measurements of hydrologic

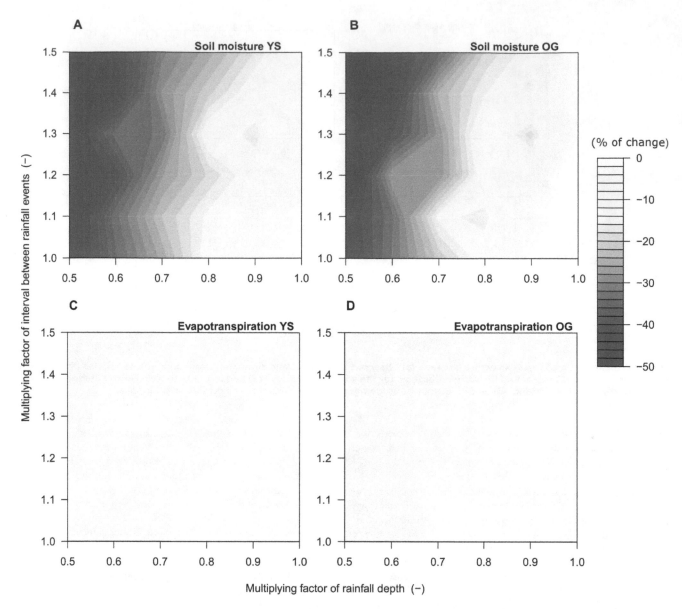

Figure 5. Sensitivity of soil moisture (a, b) and evapotranspiration (c, d) predicted by the model under increased drought in a young-secondary (YS) and an old-growth (OG) North Patagonian forest. Results represent the difference between the average under current climate (indicated in Table 4) and the average under future scenarios. A value of 0% indicates no change. The y-axis represents the variation of mean interval time between rainfall events (parameter $1/\lambda$) and the x-axis represents the variation in the mean depth of rainfall events (η) under increased drought. To represent increased drought scenarios, the parameter $1/\lambda$ was multiplied by a factor ranging from 1 (current climate) to 1.5 whereas the parameter η was multiplied by a factor ranging from 1 (current climate) to 0.5. The axes of the figures correspond to these multiplying factors of rainfall parameters. Results are the averages of 30 simulations per scenario for YS and OG.

balance at yearly temporal scales in temperate rainforests elsewhere in Chile (i.e. independent studies).

We tested model performance to reproduce forest structure under current climate. For this analysis, we compared tree basal areas and stand *dbh* distributions predicted by the model with measured structure of both YS and OG North Patagonian stands. Model comparison for the YS was performed after 60 years of succession, with succession initiated from a treeless state. To compare OG forest structure, we initiated simulations with stand inventory data and run the model for 1000 years to allow the simulated stand to reach dynamic equilibrium. We compared data at the end of the simulations with the known OG structure [28]. For each stands, we ran 100 simulations of 1 ha (i.e. 25×0.04 ha

forest patches, 2500 patches in total) using current climate parameters and model parameters listed in Table 2 and 3. Demographic and species parameters were taken from [11] (site Tepual). To assess the consequences of the hydrologic submodel for the simulated forest composition, we compared species-specific basal areas reported by [11] with data from the simulations using the model version described in this study. Simulations were conducted following methods outlined by [11] and results were compared at the corresponding successional ages of YS and OG.

Simulations under increased drought. We tested the sensitivity of model predictions for forest structure and hydrological balance to changes in rainfall regimes. Climatic scenarios were selected to represent the potential range of expected climate

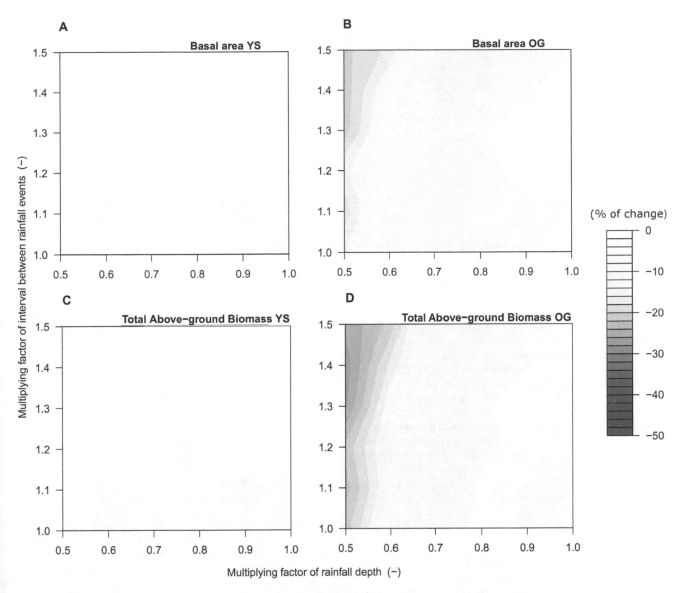

Figure 6. Sensitivity of total basal area (a, b) and above-ground biomass (c, d) predicted by the model under increased drought in a young-secondary (YS) and an old-growth (OG) North Patagonian forest. Results are the percentage of change between the average under current climate (indicated in Table 4) and the average under future scenarios. Results are the average of 30 simulations per scenario for YS and OG. The axes of the figures are as in Figure 5.

change (here increased drought) predicted for this century in the study area. To this end, we used a regional climate model downscaled for Chilean landscapes (PRECIS-DGF model, [17]). The business-as-usual scenario provided by PRECIS-DGF suggests a 50% decrease in rainfall during the growing season by year 2100. We used the daily scale output of the PRECIS-DGF model for year 2100 to calculate seasonal climatic parameters for year 2100. We linearly interpolated this seasonal rate of change between 1998–2009 and 2100. We used this scenario to set the limit of change in rainfall by year 2100 and developed climatic scenarios covering the change from current climate to the business-as-usual scenario. Climate scenarios were developed by gradually changing current climate parameters $1/\lambda$ and η until they reached the estimated value for year 2100, i.e. 50% of the current value. First, we reduced η (mean depth of rainfall events) multiplying current values by 0.9 to 0.5 in steps of 0.1. Then, we increased the parameter $1/\lambda$ (interval between events) multiplying

current values by 1.1 to 1.5 in steps of 0.1. These two parameters were first varied separately (keeping the second parameter constant) and then both together. Parameter variations produced a total of 36 climate scenarios, i.e. six levels of each of the two rainfall parameters, including the current climate scenario (Table 3). In addition, we assessed the impact of warming trends on potential evapotranspiration, and its influence on other hydrologic components and forest structure. To this end, we ran an additional scenario including temperature changes expected for year 2100, with parameters $1/\lambda$ and η kept at their current values (Table 3). We considered four output variables computed at yearly temporal scale: total basal area, above-ground biomass, evapotranspiration (computed as the sum of canopy transpiration and interception) and soil moisture. The latter is dynamically linked to forest processes such as annual gross biomass production of each tree (PB, cf. Eq. 8, see also Fig. S1). We described changes in forest structure at different successional stages based on simulations for

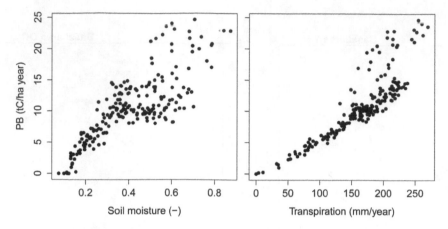

Figure 7. Shifts in annual gross biomass production (*PB*) of the YS North Patagonian forest stand in response to soil moisture and canopy transpiration changes in northern Chiloé Island, Chile. Results are simulations for one hectare of forest with a successional age of 60 years.

both YS and OG forest stands. Simulations were initialized and ran under the same conditions described in section *Model verification*. We ran 30 simulations per scenario and for each forest stand. To assess the impacts of increased drought on forests we compared the differences between means of the studied variable under current climate and means of the same variable under climate change scenarios. Finally, we checked soil moisture influence on *PB* and *Tr* by running simulations under different soil moisture conditions. We ran 216 simulations, at a scale of one ha, by monotonically varying current rainfall depth parameter (*1/λ*) from 0 to 100% of the current value (Table 2). All other parameters were kept constant. We assessed results at the corresponding successional ages of YS and OG, with succession initiated from a treeless state. All statistical analyses were done in R [57].

Results

Model results under current climate

Model predictions for the major components of forest hydrologic balance at a yearly temporal scale were similar to reported values for broad-leaved temperate rainforests in Chile and elsewhere (Table 4, independent studies). At a daily scale, the hydrologic model captured a large portion of net precipitation variability measured in the field (N = 50, r^2 = 0.8, *P*<0.001, Fig. 2). However, for rain events accumulating >100 mm, the model predicted higher net precipitation than recorded in the field (Fig. 2). Modeled daily variation in soil moisture resembled the daily pattern of soil matric potential measured in the field during the year (Fig. 3, r^2 = −0.65, *p*<0.001). Soil moisture increased during austral fall and winter (Fig. 3a, 59< julian day <242) due to a higher frequency and depth of rainfall events and reduced transpiration of trees. In contrast, during the growing season, soil moisture gradually decreased according to the model and field data (Fig. 3). Forest structure simulated by the model with the inclusion of the hydrologic submodel was qualitatively similar to field data (Fig. 4a), but with some departures from observed data for small dbh-classes in YS (<35 cm, Fig. 4b). Total basal area of YS simulated by the model was similar to basal area observed in the field but total basal area of OG was underestimated (Fig. S4). However, model predictions for forest structure including the hydrologic submodel varied in the same manner across species as

in the original model version and resembled field data for both stands (r^2 >0.97, *P*<0.001, Fig. S4).

Net precipitation and runoff predictions under current climate did not differ between YS and OG stands (*P*>0.12, two-sample Wilcoxon test, Table 4). Water loss through deep percolation was lower in OG than in YS (*P*<0.001, two-sample Wilcoxon test, Table 4). Evapotranspiration was higher in OG than YS, mainly due to higher canopy transpiration rates in OG (Table 4, *P*< 0.001, two-sample Wilcoxon test). Modeled soil moisture was lower in OG than in YS (0.55 and 0.63 respectively, *P*<0.001, two-sample Wilcoxon test, Table 4).

Simulations under increased drought

The model predicted changes in hydrological components and forest structure when simulations were run under expected trends of increasing drought (Fig. 5, 6). In both forest stands, YS and OG, we found similar responses of soil moisture to changes in rainfall parameters, with declines of soil moisture up to 50% (Fig. 5a, b). Reducing *η* to values <80% of the current value (Table 2, e.g. multiplying factor of 0.8) consistently reduced soil moisture by 20% (Fig. 5a, b). The influence of *1/λ* (mean interval time between rainfall events) was negligible when *η* was kept constant at its current value (Fig. 5a, b). The model consistently predicted less evapotranspiration (hereafter ET) relative to current values (Fig. 5c, d). In YS, decreases in ET under drier climate were < 50 mm year^{-1} (<8% reduction relative to current values, Fig. 5c). In OG forest, ET was reduced up to 94.4 mm year^{-1} from the current value (15% reduction relative to current values Fig. 5d). Such decreases in ET occurred when current value of *η* was multiplied by 0.6 and *1/λ* was multiplied by 1.2 (Fig. 5d). An increase in PET was predicted by the model when simulations included warming trends (*P*<0.001, two-sample Wilcoxon test, Fig. S5). Only transpiration and ET in YS changed in response to warming trends and increased drought (*P*<0.05, two-sample Wilcoxon test, Fig. S5). Moreover, in both forest stands, changes in PET due to warming trends did not transfer to changes in forest structure (*P*>0.05, two-sample Wilcoxon test, see also Fig. S5).

We did not find distinct differences in basal area and aboveground biomass (AGB) in YS attributable to changes in rainfall parameters (Fig. 6). In contrast, in OG forest increased drought produced decreases in basal area by 21% of the current value (Fig. 6b, current basal area of 63.8 m^2 ha^{-1}) and decreases in AGB by 27% of the current value (Fig. 6d, current AGB of

309.5 tC ha^{-1}). Main changes in basal area and AGB of OG stand were predicted when current η and $1/\lambda$ parameters were multiplied by 0.6 and 1.3, respectively (Fig. 6b, d).

Discussion

Model performance

We developed and evaluated the performance of an individual- and process-based dynamic forest model that incorporates detailed calculations of water cycling for temperate rainforests of southern South America. The model allows for the investigation of dynamic linkages between rainfall trends and forest processes at stand scale. Parameters selected for this study were taken from the literature or calibrated using our own field data (Table 2), thus they can be considered as empirically based. The model incorporates the main known hydrological controls on forest processes in temperate rainforests of southern South America (Table 2, 4).

The accuracy of model predictions regarding forest composition and structure were comparable to previous results (Fig. S4), which were performed using a previous model version without hydrologic balance calculations [11]. Model prediction of total basal area of forests decreased with the inclusion of the hydrologic module, mainly due to fewer large trees (>100 cm dbh) predicted by the model in OG stands (Fig. 4b). However, model predictions of forest structure were considered realistic at small spatial scales (< 0.2 ha) used for field sampling, compared to model results (25 ha, see also discussion in [11]). Some departures between observed and predicted daily values of net precipitation could also be attributed to canopy heterogeneity operating at different spatial scales of model results and empirical data. These discrepancies may appear when comparing forest hydrologic balance estimated for 1 ha against net precipitation measured within considerably smaller areas (<1 m^2). Further, model predictions of daily net precipitation were based on average estimated LAI values for YS because of the lack of field measurements. LAI is a relevant variable to understanding biogeochemical cycles [58], and it should therefore be incorporated in future hydrologic analyses. Additionally, sampling errors during field measurements cannot be disregarded. Despite these limitations, we considered model predictions of forest hydrology acceptable at daily temporal scale for YS (Fig. 2, 3).

At yearly temporal scale, the model reproduced the main hydrologic components reported for similar evergreen, broad-leaved temperate rainforests in southern South America and elsewhere (Table 4, independent studies). Differences in deep percolation predicted by the present model compared to values reported in Table 4 might be accounted for specific physical characteristics of glacially originated soils on northern Chiloé Island. Deep percolation in the model is mainly depending on k_{soil}, which was calibrated specifically for soils from northern Chiloé Island. Applying this model to forests developing on other soil types in SSA (e.g. volcanic originated soils) will require a site-specific calibration of soil parameters (Table 2). Moreover, we compared model results against a different forest type (Table 4), with shared dominance by evergreen and deciduous tree species, and differing in LAI and annual net precipitation. Despite these broad differences, Table 4 suggests that selected hydrologic parameters (e.g. α_{soil}, f_h, and LAI_{max}, see Table 1) yielded reasonable values for canopy water retention capacity of broad-leaved temperate rainforests in SSA.

Forest responses to increased drought

According to predictions from our model, North Patagonian forests are likely to be altered by increased drought predicted for this century by climate change models. In our modeling study, the simulation of 50% reduction in summer rainfall predicted for the study area (business-as-usual scenario, [17]) can induce changes in both hydrological balance (up to ca. 100 mm year^{-1} decrease in ET, Fig. 5) and forest structure (up to 83 tC ha^{-1} decrease in current AGB, Fig. 6), even without considering the potential ecological effects of concomitant global warming. A direct interpretation of changes in rainfall regimes is possible because the model accounts explicitly for changes in frequency and depth of rainfall events [8,32,47]. Decreasing in the depth of rainfall alone can induce some structural changes in the studied forest type (Fig. 6), but simultaneous changes in the frequency and depth of rainfall produced the strongest changes in hydrology and structure of stands (Fig. 6). These results highlight the impact on forest structure and growth of the duration and frequency of water limitation periods.

Reductions of basal area due to increased drought (Fig. 6) conform to empirical findings that drought increases the likelihood of mortality of large trees [59–61]. Our model formulation implies that trees assimilating greater biomass will have increasing demands of water for growth (Eq. 7, see also [62]). For example, annual gross biomass production increased with greater soil moisture availability during the year, a mechanism triggered by increased canopy transpiration (Fig. 7). Consequently, big trees that occur primarily in old-growth forests (Fig. 4b) experience increased stress-induced mortality due to greater hydrologic limitations during dry years. Trees that die under increased drought produce a decrease in both stand basal area and above-ground biomass (AGB, Fig. 6). Moreover, the model predicted that the OG forest has a higher PB than the young secondary stand (31.6±2.3 vs. 18.2±1.2 tC ha^{-1} year^{-1}, both obtained under current climate), and consequently a higher water demand for biomass production. Under increased drought, water demand for biomass production in OG forest is not fully covered by soil moisture supply, which causes the predicted decline of AGB and basal area (Fig. 6). We propose that the contrasting ET and structural patterns between YS and OG predicted for the coming decades under increasing drought are mainly due to significant limitation of available soil moisture for biomass production in OG, with lower impact on YS.

Research needs and model application

Here, we focused on developing an accurate model for assessing the influence of hydrologic processes involved in forest dynamics. Using the model, we quantitatively demonstrated the relative importance of soil moisture on forest structure (Fig. 6). We excluded other hydrologic processes to keep model formulation simple and results tractable with empirical information available. Processes such as water table dynamics, root dynamics, increased run off in slopes, and soil moisture dynamics across multiple soil layers can be incorporated as more empirical data becomes available for model calibration and validation. However, we strongly suggest that future model applications prioritize processes known to have an influence on the system under study [63]. In our study area, variations in the height of the water table may interfere with ecological processes such as tree establishment and mortality [14,24]. Our model provides a convenient starting point to incorporate water table dynamics into the analysis of climate change impacts, and to explore its effects on long-term forest dynamics. Our results also highlight the need of further fieldwork and experimental research on less known, mechanistic soil parameters (e.g. k_{soil}).

The rise of atmospheric carbon dioxide concentration and ensuing climate change are influencing water-use efficiency of forests [64,65]. WUE under future climate is likely to differ among forests [66], local scales [67], and species [68]. Forest water-use efficiency is a sensitive parameter in our model (e.g. Fig. S6) and illustrates the need for detailed studies of the expected variations in this parameter under drought, and their connection to gas exchange capacity of trees in different topographic settings and for a large set of species in SSA. It was beyond the scope of the present study to discuss model behavior along drought-to-moist gradients operating at regional scales because such analyses require an accurate and quantitative assessment of species-specific water-use efficiency [4]. In SSA, WUE variations have been experimentally tested in few study sites and for only three tree species included in our study [49,69]. Additional experimental research should address the complex interaction between photo-synthetic carbon assimilation and water loss via transpiration with declining water supply [70]. Future research can address this question by studying changes in photosynthetic parameters along climatic gradients. Incorporating WUE as a species-specific parameter or a state dependent variable in our model is straightforward based on further empirical information.

Here, we deliberately excluded the influence of expected regional changes in temperature on forest processes (e.g. tree growth) to rather emphasize direct impacts of drought on model results. However, the model suggests that changes in PET due to expected warming trends in the study region are negligible compared to the strong impacts of increased drought (Fig. S5). Our simulated scenarios have been done using the common assumption of vegetation dynamic models that climate-forest interactions under inter-annual variation of climatic conditions can be used as proxy also for impacts of long-term climatic variations. In the later case it would be possible that some tree species would show adaptations effects. However, tree species adaptation to climate variability is still poorly understood in SSA. Moreover, warming can modify individual tree growth by affecting photosynthesis and both plant and soil respiration [38] [71], nutrient dynamics [72] and tree establishment [73]. An undergo-ing study analyzes the combined effect of temperature and rainfall changes on tree demography in the study area [74].

To the best of our knowledge, this is the first application of a forest gap model in temperate rainforests of SSA that integrates dynamic calculations of forest hydrology. The present work uses the best information available to ensure that climate patterns were directly comparable to hydrologic field measurements used for model calibration. However, our results on climate change impacts should be interpreted with caution because our baseline climate is constrained to a short period (1998–2009) within a long-term trend of rainfall (time period 1901–2005). Long-term monitoring of forest hydrology and dynamics can corroborate our results. To date, long-term monitoring (>10 years) of forest hydrology is lacking in forests of SSA. As more empirical data becomes available, the model can be revised and updated. The model developed here allows for the analysis of multiple environmental factors driving forest dynamics. For example, our model can help us understand the long-term responses of regional forest to drought events induced by El Niño Southern Oscillation that amplify background tree mortality rates in *Nothofagus* forest of SSA [75]. Moreover, increased regional drought is likely to interact with other drivers of global change such as changes in fire regimes, massive introduction of exotic forestry species, and forest fragmentation. To date, the implications of these drivers in the context of a changing climate remain poorly understood in SSA. These interacting growing threats demand from ecologists to understand and integrate multiple dimensions of global change on forest functioning. The model presented here is a particularly suitable tool for analyzing broad global change questions in forests of SSA because it also includes logging and fragmentation submodels [13,76]. Model-based experiments can also contribute to develop sound management strategies that anticipate forest responses to increasing drought and other drivers of climate change.

Conclusions

We developed and applied a forest dynamic model to analyze the impact of climate-driven increased drought on ecological and hydrological processes. The developed model was accurate for depicting forest hydrology at stand scales (i.e. <100 ha) and allowed the analysis of the dynamical linkages among rainfall regimes, soil moisture variation, and individual tree growth. Using the model we demonstrated that evergreen, broad-leaved temper-ate rainforests in southern South America are expected to be highly sensitive to future climate change, particularly increases in drought during parts of the year. Increased summer drought predicted for this century will likely impair biomass carbon accumulation, and amplify background tree mortality rates in this region. The developed model expands the range of applicability of gap models to assess climate change impacts in remote and understudied regions of the world, such as temperate forests of the southern hemisphere. It also represents an advance in the development of simple, general models to account for complex and dynamical processes operating at multiple spatial scales in forests.

Supporting Information

Figure S1 A diagram of the hydrologic submodel of FORMIND-CL v.**1.0.** Interaction between processes and variables in the hydrologic submodel, and their respective time scales of calculations. Arrows indicate whether the results of a model calculation influence the calculations of another submodel. Blue boxes represent analyzed variables of this study. All calculations are done in yearly time steps in the model, excepting the ones indicated in the dashed box. Variable notations follow the text. AGB: Above-ground biomass, LAI: leaf area index.

Figure S2 Weather generator results. Density functions of daily rainfall and daily mean temperature predicted by the weather generator compared to observed weather records from EBSD weather station.

Figure S3 Weather generator results. Comparison between simulated and observed climatic patterns during the year. Simulations were run for 100 years using parameters in Table 3. Daily data were averaged by seasons (mean daily temperature and daily radiation). Rainfall is the amount of rainfall during each season. Observed weather data are from EBSD weather station and seasons according to table 3.

Figure S4 Forest composition predictions. Model results for forest composition using different model versions compared to field data. Simulations run under the same conditions detailed in Methods section.

Figure S5 Drought induced simulations with warming included. Changes in hydrologic components and forest structure when warming and increased drought was considered. PET: Potential evapotranspiration (mm year^{-1}), T: transpiration (mm year^{-1}), Ec: canopy interception (mm year^{-1}), ET: evapotranspiration (mm year^{-1}), BAT: total basal area (m^2 ha^{-1}), BT: Total biomass (tC ha^{-1}). Result of a two-sample Wilcoxon test is shown on the upper right of each panel. Pink lines, drought induced simulations with warming included, blue lines drought induced simulations without warming, circles represent the values of simulation results. Note different scales for the axes.

Figure S6　Sensitivity of evaporatranspiration. Changes in evapotranspiration (ET) of the old-growth stand under current climate when using different water-use efficiency values (WUE). Simulations run under the same conditions detailed in Methods section.

References

1. Rodriguez-Iturbe I, Porporato A, Ridolfi L, Isham V, Cox DR (1999) Probabilistic modelling of water balance at a point: the role of climate, soil and vegetation. Proceedings of the Royal Society a-Mathematical Physical and Engineering Sciences 455: 3789–3805.
2. Botkin DB, Wallis JR, Janak JF (1972) Some Ecological Consequences of a Computer Model of Forest Growth. Journal of Ecology 60: 849–872.
3. Bugmann H (2001) A review of forest gap models. Climatic Change 51: 259–305.
4. Bugmann H, Cramer W (1998) Improving the behaviour of forest gap models along drought gradients. Forest Ecology and Management 103: 247–263.
5. Bugmann HKM, Wullschleger SD, Price DT, Ogle K, Clark DF, et al. (2001) Comparing the performance of forest gap models in North America. Climatic Change 51: 349–388.
6. Reynolds JF, Bugmann H, Pitelka LF (2001) How much physiology is needed in forest gap models for simulating long-term vegetation response to global change? Challenges, limitations, and potentials. Climatic Change 51: 541–557.
7. Asbjornsen H, Goldsmith GR, Alvarado-Barrientos MS, Rebel K, Van Osch FP, et al. (2011) Ecohydrological advances and applications in plant-water relations research: a review. Journal of Plant Ecology-Uk 4: 3–22.
8. Porporato A, Daly E, Rodriguez-Iturbe I (2004) Soil water balance and ecosystem response to climate change. American Naturalist 164: 625–632.
9. Vose JV, Sun G, Ford CR, Bredemeier M, Otsuki K, et al. (2011) Forest ecohydrological research in the 21st century: what are the critical needs? Ecohydrology 4: 146–158.
10. Hall GMJ, Hollinger DY (2000) Simulating New Zealand forest dynamics with a generalized temperate forest gap model. Ecological Applications 10: 115–130.
11. Gutierrez AG, Huth A (2012) Successional stages of primary temperate rainforests of Chiloe Island, Chile. Perspectives in Plant Ecology Evolution and Systematics 14: 243–256.
12. Echeverría C, Newton AC, Lara A, Benayas JMR, Coomes DA (2007) Impacts of forest fragmentation on species composition and forest structure in the temperate landscape of southern Chile. Global Ecology and Biogeography 16: 426–439.
13. Rüger N, Gutiérrez AG, Kissling WD, Armesto JJ, Huth A (2007) Ecological impacts of different harvesting scenarios for temperate evergreen rain forest in southern Chile - A simulation experiment. Forest Ecology and Management 252: 52–66.
14. Díaz MF, Bigelow S, Armesto JJ (2007) Alteration of the hydrologic cycle due to forest clearing and its consequences for rainforest succession. Forest Ecology and Management 244: 32–40.
15. Little C, Lara A, McPhee J, Urrutia R (2009) Revealing the impact of forest exotic plantations on water yield in large scale watersheds in South-Central Chile. Journal of Hydrology 374: 162–170.
16. Trenberth KE, Jones PD, Ambenje P, Bojariu R, Easterling D, et al. (2007) Observations: Surface and Atmospheric Climate Change. In: Solomon S, Qin D, Manning M, Chen Z, Marquis M et al., editors. Climate Change 2007: The Physical Science Basis Contribution of Working Group I to the Fourth Assessment Report of the Intergovernmental Panel on Climate Change. Cambridge, United Kingdom and New York, NY, USA.: Cambridge University Press. 336
17. DGF CONAMA (2006) Estudio de la variabilidad climática en Chile para el siglo XXI. Santiago: Departamento de Geofísica. Universidad de Chile. 63 p.
18. Christensen JH, Hewitson B, Busuioc A, Chen A, Gao X, et al. (2007) Regional Climate Projection. In: Solomon S, Qin D, Manning M, Chen Z, Marquis M et al., editors. Climate Change 2007: The Physical Science Basis Contribution of Working Group I to the Fourth Assessment Report of the Intergovernmental Panel on Climate Change. Cambridge, United Kingdom and New York, NY, USA: Cambridge University Press. 94
19. Alaback PB (1991) Comparative Ecology of Temperate Rain-Forests of the America Along Analogous Climatic Gradients. Revista Chilena de Historia Natural 64: 399–412.
20. Armesto JJ, Smith-Ramírez C, Carmona MR, Celis-Diez JL, Díaz I, et al. (2009) Old-growth temperate rain forests of South America: Conservation, plant-animal interactions, and baseline biogeochemical processes. In: Wirth C, Gleixner G, Heimann M, editors. Old-growth forests: Function, fate and value. Berlin, Heidelberg: Springer New York. 367–390
21. Armesto JJ, Rozzi R, Smith-Ramírez C, Arroyo MTK (1998) Conservation targets in South American temperate forests. Science 282: 1271–1272.
22. Willson MF, Armesto JJ (1996) The natural history of Chiloé: on Darwin's trail. Revista Chilena de Historia Natural 69: 149–161.
23. Holdgate MW (1961) Vegetation and soils in the South Chilean Islands. Journal of Ecology 49: 559–580.
24. Díaz MF, Armesto JJ (2007) Physical and biotic constraints on tree regeneration in secondary shrublands of Chiloe Island, Chile. Revista Chilena De Historia Natural 80: 13–26.
25. di Castri F, Hajek E (1976) Bioclimatología de Chile. Santiago: Universidad Católica de Chile.
26. Gutiérrez AG, Armesto JJ, Aravena JC, Carmona M, Carrasco NV, et al. (2009) Structural and environmental characterization of old-growth temperate rain-forests of northern Chiloe Island, Chile: Regional and global relevance. Forest Ecology and Management 258: 376–388.
27. Veblen TT, Schlegel FM, Oltremari JV (1983) Temperate broad-leaved evergreen forest of South America. In: Ovington JD, editor. Temperate Broad-Leaved Evergreen Forest. Amsterdam: Elsevier Science Publishers. 5–31
28. Gutiérrez AG, Armesto JJ, Aravena JC (2004) Disturbance and regeneration dynamics of an old-growth North Patagonian rain forest in Chiloe Island, Chile. Journal of Ecology 92: 598–608.
29. Köhler P (2000) Modelling anthropogenic impacts on the growth of tropical rain forests-using an individual-oriented forest growth model for the analyses of logging and fragmentation in three case studies [PhD thesis http://hdl.handle.net/10013/epic.15101].Osnabrück, Germany: Center for Environmental Systems Research and Department of Physics. University of Kassel. Der Andere Verlag.
30. Köhler P, Chave J, Riera B, Huth A (2003) Simulating the long-term response of tropical wet forests to fragmentation. Ecosystems 6: 114–128.
31. Köhler P, Huth A (1998) The effects of tree species grouping in tropical rainforest modelling: Simulations with the individual-based model FORMIND. Ecological Modelling 109: 301–321.
32. Kumagai T, Katul GG, Saitoh TM, Sato Y, Manfroi OJ, et al. (2004) Water cycling in a Bornean tropical rain forest under current and projected precipitation scenarios. Water Resources Research 40: W01104.
33. Rutter AJ, Morton AJ (1977) Predictive Model of Rainfall Interception in Forests.3. Sensitivity of Model to Stand Parameters and Meteorological Variables. Journal of Applied Ecology 14: 567–588.
34. Wattenbach M, Hattermann F, Weng R, Wechsung F, Krysanova V, et al. (2005) A simplified approach to implement forest eco-hydrological properties in regional hydrological modelling. Ecological Modelling 187: 40–59.
35. Rey JM (1999) Modelling potential evapotranspiration of potential vegetation. Ecological Modelling 123: 141–159.
36. Maidment DR (1993) Handbook of Hydrology; McGraw-Hill, editor: McGraw-Hill.
37. Neilson RP (1995) A Model for Predicting Continental-Scale Vegetation Distribution and Water-Balance. Ecological Applications 5: 362–385.
38. Lambers H, Chapin FS, Pons TL (1998) Plant physiological ecology. New York: Springer. 540 p.
39. Bazzaz FA (1979) Physiological ecology of plant succession. Annual Review of Ecology and Systematics 10: 351–371.

Acknowledgments

We appreciate the useful comments offered by Hans Pretzsch, and anonymous reviewers in previous versions of the manuscript. This is a contribution to the Research Program of the Chilean LTSER network.

Author Contributions

Conceived and designed the experiments: AGG AH JJA MFD. Performed the experiments: AGG JJA MFD. Analyzed the data: AGG AH MFD. Contributed reagents/materials/analysis tools: AGG JJA AH. Wrote the paper: AGG.

40. Thornley HMJ, Johnson IR (1990) Plant and Crop Modelling – A mathematical approach to plant and crop physiology. Oxford, UK.: Clarendon Press.

41. Gerten D, Schaphoff S, Haberlandt U, Lucht W, Sitch S (2004) Terrestrial vegetation and water balance - hydrological evaluation of a dynamic global vegetation model. Journal of Hydrology 286: 249–270.

42. Venevsky S, Maksyutov S (2007) SEVER: A modification of the LPJ global dynamic vegetation model for daily time step and parallel computation. Environmental Modelling & Software 22: 104–109.

43. Prentice IC, Sykes MT, Cramer W (1993) A Simulation-Model for the Transient Effects of Climate Change on Forest Landscapes. Ecological Modelling 65: 51–70.

44. Haxeltine A, Prentice IC (1996) BIOME3: An equilibrium terrestrial biosphere model based on ecophysiological constraints, resource availability, and competition among plant functional types. Global Biogeochemical Cycles 10: 693–709.

45. Prentice IC, Cramer W, Harrison SP, Leemans R, Monserud RA, et al. (1992) A Global Biome Model Based on Plant Physiology and Dominance, Soil Properties and Climate. Journal of Biogeography 19: 117–134.

46. Dingman SL (2002) Physical hydrology. New Jersey: Prentice Hall. 646 p.

47. Laio F, Porporato A, Ridolfi L, Rodriguez-Iturbe I (2001) Plants in water-controlled ecosystems: active role in hydrologic processes and response to water stress - II. Probabilistic soil moisture dynamics. Advances in Water Resources 24: 707–723.

48. Lusk CH (2001) When is a gap not a gap? Light levels and leaf area index in bamboo-filled gaps in a Chilean rain forest. Gayana Botanica 58: 25–30.

49. Piper FI, Corcuera LJ, Alberdi M, Lusk C (2007) Differential photosynthetic and survival responses to soil drought in two evergreen Nothofagus species. Annals of Forest Science 64: 447–452.

50. Zuñiga R, Alberdi M, Reyes-Diaz M, Olivares E, Hess S, et al. (2006) Seasonal changes in the photosynthetic performance of two evergreen Nothofagus species in south central Chile. Revista Chilena de Historia Natural 79: 489–504.

51. Cunningham SC (2005) Photosynthetic responses to vapour pressure deficit in temperate and tropical evergreen rainforest trees of Australia. Oecologia 142: 521–528.

52. Brodribb T, Hill RS (1998) The photosynthetic drought physiology of a diverse group of southern hemisphere conifer species is correlated with minimum seasonal rainfall. Functional Ecology 12: 465–471.

53. Janssen I, Kruemmelbein J, Horn R, Ellies AS (2004) Physical and hydraulic properties of the ñadi soils in south Chile - Comparison between untilled and tilled soil. Revista de la Ciencia del Suelo y Nutrición Vegetal 4: 14–28.

54. Sitch S, Smith B, Prentice IC, Arneth A, Bondeau A, et al. (2003) Evaluation of ecosystem dynamics, plant geography and terrestrial carbon cycling in the LPJ dynamic global vegetation model. Global Change Biology 9: 161–185.

55. van Genuchten MT (1980) A closed-form equation for predicting the hydraulic conductivity of unsaturated soils. Soil Science Society of America Journal 44: 892–898.

56. Seki K (2007) SWRC fit-a nonlinear fitting program with a water retention curve for soils having unimodal and bimodal pore structure. Hydrology and Earth System Sciences Discussions 4: 407–437.

57. R-Development-Core-Team (2012) R: A language and environment for statistical computing. Vienna, Austria.: R Foundation for Statistical Computing.

58. Breda NJJ (2003) Ground-based measurements of leaf area index: a review of methods, instruments and current controversies. Journal of Experimental Botany 54: 2403–2417.

59. Condit R, Hubbell SP, Foster RB (1995) Mortality-Rates of 205 Neotropical Tree and Shrub Species and the Impact of a Severe Drought. Ecological Monographs 65: 419–439.

60. Allen CD, Macalady AK, Chenchouni H, Bachelet D, McDowell N, et al. (2010) A global overview of drought and heat-induced tree mortality reveals emerging climate change risks for forests. Forest Ecology and Management 259: 660–684.

61. Phillips OL, van der Heijden G, Lewis SL, Lopez-Gonzalez G, Aragao LEOC, et al. (2010) Drought-mortality relationships for tropical forests. New Phytologist 187: 631–646.

62. Larcher W (2001) Ökophysiologie der Pflanzen. Stuttgart: 6th edition, Verlag Eugen Ullmer.

63. Grimm V, Railsback SF (2005) Individual-based modeling and ecology. New Jersey: Princeton University Press.

64. Penuelas J, Canadell JG, Ogaya R (2011) Increased water-use efficiency during the 20th century did not translate into enhanced tree growth. Global Ecology and Biogeography 20: 597–608.

65. Keenan TF, Hollinger DY, Bohrer G, Dragoni D, Munger JW, et al. (2013) Increase in forest water-use efficiency as atmospheric carbon dioxide concentrations rise. Nature 499: 324–+.

66. Silva LCR, Anand M (2013) Probing for the influence of atmospheric CO_2 and climate change on forest ecosystems across biomes. Global Ecology and Biogeography 22: 83–92.

67. Penuelas J, Hunt JM, Ogaya R, Jump AS (2008) Twentieth century changes of tree-ring delta C-13 at the southern range-edge of Fagus sylvatica: increasing water-use efficiency does not avoid the growth decline induced by warming at low altitudes. Global Change Biology 14: 1076–1088.

68. Levesque M, Saurer M, Siegwolf R, Eilmann B, Brang P, et al. (2013) Drought response of five conifer species under contrasting water availability suggests high vulnerability of Norway spruce and European larch. Global Change Biology 19: 3184–3199.

69. Figueroa JA, Cabrera HM, Queirolo C, Hinojosa LF (2010) Variability of water relations and photosynthesis in *Eucryphia cordifolia* Cav. (Cunoniaceae) over the range of its latitudinal and altitudinal distribution in Chile. Tree Physiology 30: 574–585.

70. Chaves MM, Pereira JS, Maroco J, Rodrigues ML, Ricardo CPP, et al. (2002) How plants cope with water stress in the field. Photosynthesis and growth. Annals of Botany 89: 907–916.

71. Ryan MG, Law BE (2005) Interpreting, measuring, and modeling soil respiration. Biogeochemistry 73: 3–27.

72. Aerts R, Chapin FS (2000) The mineral nutrition of wild plants revisited: A re-evaluation of processes and patterns. Advances in Ecological Research, Vol 30 30: 1–67.

73. Hobbie SE, Chapin FS (1998) An experimental test of limits to tree establishment in Arctic tundra. Journal of Ecology 86: 449–461.

74. Gutiérrez AG (2010) Long-term dynamics and the response of temperate rainforests of Chiloé Island (Chile) to climate change [Phd Thesis]. Freising, Germany: Technische Universität München. 170 p.

75. Suarez ML, Ghermandi L, Kitzberger T (2004) Factors predisposing episodic drought-induced tree mortality in Nothofagus-site, climatic sensitivity and growth trends. Journal of Ecology 92: 954–966.

76. Groeneveld J, Alves LF, Bernacci LC, Catharino ELM, Knogge C, et al. (2009) The impact of fragmentation and density regulation on forest succession in the Atlantic rain forest. Ecological Modelling 220: 2450–2459.

77. Donoso PJ, Soto DP, Bertin RA (2007) Size-density relationships in *Drimys winteri* secondary forests of the Chiloe Island, Chile: Effects of physiography and species composition. Forest Ecology and Management 239: 120–127.

78. Navarro C, Donoso C, Sandoval V (1999) Los renovales de Canelo. In: Donoso C, Lara A, editors. Silvicultura de los bosques nativos de Chile. Santiago, Chile: Editorial Universitaria.

79. Naulin P (2002) Estimacion de la biomasa en un renoval de canelo (Drimys winteri J.R. et Foster) en la Comuna de Ancud. Santiago: Universidad de Chile. 45 p.

80. Aravena JC, Carmona MR, Perez CA, Armesto JJ (2002) Changes in tree species richness, stand structure and soil properties in a successional chronosequence in northern Chiloe Island, Chile. Revista Chilena De Historia Natural 75: 339–360.

81. Lusk CH (2001) Leaf life spans of some conifers of the temperate forests of South America. Revista Chilena De Historia Natural 74: 711–718.

82. Huber A, Iroume A (2001) Variability of annual rainfall partitioning for different sites and forest covers in Chile. Journal of Hydrology 248: 78–92.

83. Oyarzun CE, Godoy R, Staelens J, Donoso PJ, Verhoest NEC (2011) Seasonal and annual throughfall and stemflow in Andean temperate rainforests. Hydrological Processes 25: 623–633.

84. Barbour MM, Hunt JE, Walcroft AS, Rogers GND, McSeveny TM, et al. (2005) Components of ecosystem evaporation in a temperate coniferous rainforest, with canopy transpiration scaled using sapwood density. New Phytologist 165: 549–558.

85. Lara A, Little C, Urrutia R, McPhee J, Alvarez-Garreton C, et al. (2009) Assessment of ecosystem services as an opportunity for the conservation and management of native forests in Chile. Forest Ecology and Management 258: 415–424.

86. CIREN (1994) Delimitación y descripción de microregiones para la transferencia tecnológica de INDAP. Santiago: CORFO. 99 p.

11

Tree Diversity Mediates the Distribution of Longhorn Beetles (Coleoptera: Cerambycidae) in a Changing Tropical Landscape (Southern Yunnan, SW China)

Ling-Zeng Meng[1,2]*, Konrad Martin[2], Andreas Weigel[3], Xiao-Dong Yang[1]

1 Key Laboratory of Tropical Forest Ecology, Xishuangbanna Tropical Botanical Garden, Chinese Academy of Sciences, Yunnan, China, **2** University of Hohenheim, Agroecology in the Tropics and Subtropics (380b), Stuttgart, Germany, **3** Rosalia Umweltmanagement, Am Schloßgarten 6, Wernburg, Germany

Abstract

Longhorn beetles (Coleoptera : Cerambycidae) have been used to identify sites of high biological diversity and conservation value in cultivated landscapes, but were rarely studied in changing landscapes of humid tropics. This study was conducted in a region of southern Yunnan, China, which was dominated by natural rainforest until 30 years ago, but is successively transformed into commercial rubber monoculture plantations since that time. The objectives were to investigate longhorn beetle species diversity and distribution in the major land use types of this landscape and to estimate the effects of an expected expansion of rubber plantations on the longhorn beetle assemblages. The results showed that tree species diversity (181 species in total) and longhorn beetle diversity (220 species in total) were closely related with no significant differences between the tree and longhorn beetles assemblages shown by similarity distance analysis. There was a highly positive relationship between the estimated species richness of longhorn beetles and the number of tree species. Individual numbers of longhorn beetles and trees were also highly positive related at the sampling sites. Non-metric multidimensional scaling revealed that the degree of canopy coverage, succession age and tree diversity explained 78.5% of the total variation in longhorn beetle assemblage composition. Natural forest sites had significantly higher numbers of species and individuals than any other type of habitat. Although young rubber plantations bear the highest longhorn beetle diversity outside forests (half of the total number of longhorn beetle species recorded in total), they can not provide permanent habitats for most of these species, because they develop into closed canopy plantations with less suitable habitat conditions. Therefore, along with an expected expansion of rubber cultivation which largely proceeds at the expense of forest areas, the habitat conditions for longhorn beetles in this region might decrease dramatically in future.

Editor: Christopher J. Lortie, York U, Canada

Funding: This study was supported by National Natural Science Foundation of China (NSFC-31200322). Environment friendly rubber plantation project of Xishuangbanna Tropical Botanical Garden, Chinese Academy of Sciences also partly funded this study. The work of the second author (K. Martin) was supported by the project SURUMER (Sustainable Rubber Cultivation in the Mekong Region), funded by the German Federal Ministry of Education and Research (BMBF) program "Sustainable Land Management". The funders had no role in study design, data collection and analysis, decision to publish, or preparation of the manuscript.

Competing Interests: AW is an employee of Rosalia Umweltmanagement. There are no patents, products in development or marketed products to declare.

* E-mail: mlz@xtbg.org.cn

Introduction

Longhorn beetles (Coleoptera : Cerambycidae) have been used to identify sites of high biological diversity and conservation value in cultivated landscapes which are usually composed of heterogeneous mosaics of different land use [1,2,3]. Longhorn beetles almost exclusively feed on living, dying or dead woody plants in the larval stage. Relationships between longhorn beetles and host plants are often quite specific, but there is a great range in the breadth of host tree species that may used by the larvae of different species [4]. Longhorn beetles can play an important role in the decomposition of dead wood and therefore were also considered as "ecosystem engineers" [5]. Furthermore, many longhorn beetle adults visit flowers to feed on nectar and/or pollen and therefore act as pollinators.

Longhorn beetle diversity and distribution was shown to be affected by forest management practices [6,7,8,9,10], invasive tree species [11], habitat destruction and degradation [12,13], habitat fragmentation [14], environmental gradients [15], disturbances of fire, drought and windstorm [16,17,18,19,20,21], spatial heterogeneity [22] and effects of host species preferences [23]. Conclusions drawn from those studies are that most longhorn beetle species are concentrated on undisturbed or primary forest, whereas secondary forest and artificial plantations support less species. Furthermore, increasing intensification and disturbance tends to reduce specialist longhorn beetle species and to homogenize the beetle assemblages between the various habitat types of a landscape. Most studies were conducted in temperate regions, where the original vegetation has disappeared or been strongly modified in the course of a usually long history of land cultivation. However, very little research has been done to analyze the effects of land use change on beetle assemblages in relatively young cultivated landscapes of tropical rainforest regions (but see [24] for Mexico).

Table 1. Overview and description of the 13 sampling sites in the study area of the Naban River valley within Naban River Watershed National Nature Reserve (NRWNNR).

Location	Site code	Site descriptions
Forest		
Mandian	MD-FO	Primary forest, closed canopy 35 m high
Naban	NB-FO	Secondary forest, closed canopy 35 m high
Anmaxinzhai	AM-FO	Secondary forest, closed canopy 35 m high
Guomenshan	GMS-FO	Primary forest, closed canopy 35 m high
Rubber plantations		
Mandian	MD-RU	5 years, trees 7 m high, open canopy
Naban	NB-RU	8 years, trees 12 m high, open canopy
Anmaxinzhai	AM-RU	20 years, trees 20 m high, closed canopy
Shiyidui	SYU-RU	40 years, trees 30 m high, closed canopy
Open land		
Naban	NB-OP	Forest clearfell between NB-FO and NB-RU
Anmaxinzhai	AM-OP	Grassland on a ridge
Guomenshan	GMS-OP	Shrubland succession
Fallow		
Mandian	MD-FA	Rice field fallow
Guomenshan	GMS-FA	Rice field fallow

This study was conducted in the tropical landscape of southern Yunnan Province, China. This region is part of the 'Indo-Burma hotspot', one of the 34 global hotspots exceptionally rich in biodiversity [25]. The specific study area represents a tributary valley of the Mekong River. There, traditional land use systems are irrigated rice fields along the river courses and shifting cultivation systems on the slopes, but the largest proportion of the land area was covered with primary and secondary forest until about 30 years ago. Since then, large areas of forest have been, and still are, successively transformed into commercial rubber (*Hevea brasiliensis*) monoculture plantations. The predominant habitat types of the investigated landscape therefore include natural forest plots, open shrubland and grassland and agricultural fields as well as rubber plantations of different age. This land use pattern is representative of the development of tropical southern Yunnan. Detailed data from a typical subregion (Xishuangbanna Prefecture) showed that between 1998 and 2006, rubber plantations increased from 12% of the total land coverage to 46%, whereas forested areas dropped from 49 to 28% [26]. Tropical seasonal rainforest was the type of land most affected by the expansion of rubber plantations [27].

The objective of this study was to investigate the longhorn beetle species diversity and distribution in this fragmented landscape in relation to land use type and to assess the response of the longhorn beetle guilds to the recent changes in land use. Due to the importance of woody plants as resources for longhorn beetles, the land use types were characterized by their tree species inventory, which largely corresponds with the successional stage of the predominant land use types, including rice field fallows, grassland and shrubland, young and old rubber plantations and natural forest plots. These data are used to estimate the effects on longhorn beetle diversity and distribution as a consequence of the expected increase of rubber plantations, which is related to the loss and reduction of forest cover and the increasing fragmentation of

the landscape. Based on the idea that land use changes by human beings affect the species richness and composition in longhorn beetles, we hypothesis that some specialist species will be more sensitive than others to forest coverage decreased. We also expect that artificial rubber plantation at young ages especially those closing to forest patches which were mixed with some young tree species can be served as a temporary suitable habitat for some generalist species.

Materials and Methods

Ethics Statement

The Naban River Watershed National Natural Reserve (NRWNNR) provided the permission for the field studies.

Study area and Sampling Localities

The study was carried out in the Naban River valley (ca. 11.000 ha) within the Naban River Watershed National Nature Reserve (NRWNNR) in Xishuangbanna, southern Yunnan province, south-west China (22° 10′ N and 100° 38′ E). The region represents the northernmost part of the humid tropics in Asia with a climate influenced by Monsoon and three distinct seasons: cool-dry (October-January, with the lowest monthly temperature of 15°C in December), hot-dry (February-April, with the highest monthly temperature of 25°C in April) and a rainy season (May-September) with annual precipitation of almost 1600 mm. The natural vegetation of the study region is tropical rainforest, consisting of different types of evergreen and seasonal forest depends on topography and elevation [28,29]. Secondary and primary forest plots and fragments are widespread in the study area, but most cultivated land is covered by rubber plantations. Valley bottoms are dominated by rice fields, and there are various fruit and vegetable crops around the small villages. Shrubland and grassland areas are found along the slopes. To represent the most typical habitat types of this landscape, we selected 13 sampling localities including forest plots, rubber plantations, fallows and open lands (Table S1). Further descriptions are given in Table 1.

Field Methods

Beetle sampling was carried out by using a combined trap system including bowl traps and Malaise traps (3.5×2.0×1.5 m, length×width×height) [30] at all sites, and aerial collectors in the canopy area of trees in forests and in rubber plantations. Bowl traps were plastic pots with a diameter of 35 cm and a depth of 15 cm put on the soil surface, one third height filled with a mixture of liquid of blue colored anti-freeze (ethanol-glycol). At each site, one yellow bowl trap was arranged at a distance of ca. 5 m from one side of Malaise trap, and a red bowl trap was established at the other side with a same distance. Malaise traps had two separate collection ports and were installed in directions that cover the typical land use type of the trap site. Aerial collectors were constructed of two pieces of transparent plastic plates (50×30 cm, height×width) which were arranged crosswise and fixed upon a red plastic bowl of 30 cm in diameter. These traps were installed on canopy tree branches using ropes. The collecting bottles of the Malaise traps and the bowls of the aerial collectors were also filled with a mixture of liquid of blue colored anti-freeze (ethanol-glycol).

Traps collections were conducted in different seasonal periods covering (a) the beginning of the rainy season (May-July 2008), (b) the beginning of the cool-dry season (September-November 2008), and (c) the transition period from the hot-dry to the rainy season (March-June 2009). At all sites, traps were emptied every 10 days during the collecting periods (with few exceptions where traps were destroyed or collection was impossible due to heavy rains).

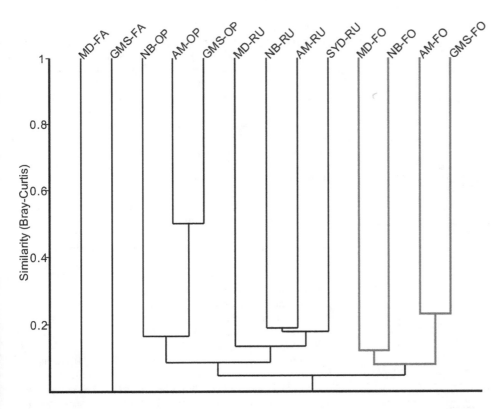

Figure 1. Quantitative similarity cluster analysis of the tree species at the sampling localities (site codes see Table 1), generated from the Bray-Curtis index using UPGMA through stratigraphically constrained clustering. The dendrogram shows four subgroups clearly classified by habitat categories.

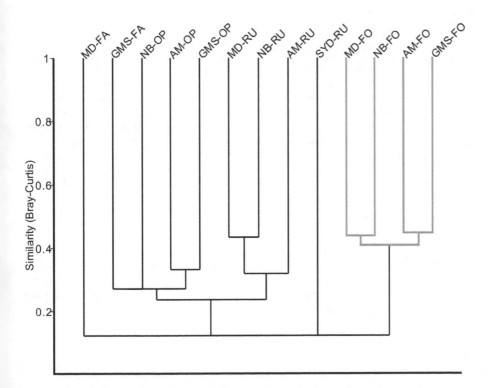

Figure 2. Quantitative similarity cluster analysis of the longhorn beetles at the sampling localities (site codes see Table 1), generated from the Bray-Curtis index using UPGMA through stratigraphically constrained clustering. The dendrogram shows three subgroups clearly classified by habitat categories.

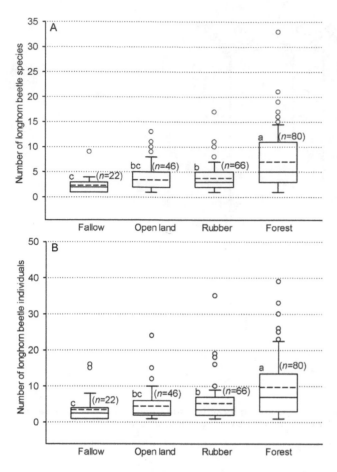

Figure 3. Number of (A) species and (B) individuals of longhorn beetles in the four subgroup produced by habitat categories. Box and whisker plots illustrate the 5th, 25th, 50th (median), 75th, and 95th percentiles, and the means as the dashed line. Different letters indicate significant differences ($P<0.05$; Mann-Whitney U-Test). Small circle indicate outliners.

The beetle specimens were preserved in 70% ethanol for further identification to species level. Data analyses are based on numbers of species and individuals combined from all traps and trap types per site and the total counts from all collecting periods. Voucher specimen of the collected beetles are kept at the National Zoological Museum of China, Institute of Zoology, CAS, Beijing.

Vascular plant species inventories from each of the 13 trap sites were recorded in March 2009. At the four natural forest and four rubber plantation sites, four 20×20 m^2 plots were established around the trap locations to record the numbers of tree and liana species. Other plant species including young trees (representing the groundcover vegetation <2 m) in the plots were recorded from four 5×5 m^2 subplots within each of the four large plots. Total species numbers of the four small and the four large plots were used for further calculations. In the three open land and two fallow sites, records from all four 20×20 m^2 plots per site provided the plant species numbers used for calculations. Voucher specimens of the recorded and identified plant species were deposited at the Herbarium of the Xishuangbanna Tropical Botanical Garden (XTBG), CAS, Yunnan, China.

Data Analysis

Cluster analyses through stratigraphically constrained clustering were performed to identify quantitatively similar groups of tree species (including young trees but without shrub and liana) and longhorn beetle assemblages among the different sites and habitats using Bray-Curtis index of similarity. Among the algorithms for hierarchical clustering, we selected the unweighted pair-group method using averages (UPGMA) which is conventionally used in ecology [31,32]. One-way ANOSIM (analysis of similarities) global tests were then applied to test for differences between subgroups produced by the cluster analysis. The abundance data of tree and longhorn beetle were log normalized through $\log_{10}(X+1)$ transformation to bring out category features of original data and downweight those a few very dominant taxa. Differences in total species numbers and abundances of longhorn beetles between the subgroups were compared by the Mann-Whitney U-test using Minitab 15.0 software (Minitab Inc. State College PA, USA). This non-parametric test was applied because the assumptions of homogeneity of variances and normality (tested with the Shapiro-Wilk normality test) were not met according to [33]. Furthermore, we used a one-way ANOVA to test whether the Bray-Curtis similarity distance indices among the 13 sampling sites produced by tree and longhorn beetle data assemblages were different from each other.

Non-metric multidimensional scaling (NMS; [34]) using the Bray–Curtis index for abundance data was applied to display and test for differences in longhorn beetle assemblage composition across the land use types. This ordination score was performed with PC-ORD software [35] with the following parameters employed in the NMS procedure: Sorensen distance measure; a maximum number of 500 iterations; random starting coordinates; 100 runs with real data; step down in dimensionality (initial step length = 0.2); 100 runs with randomized data. A total of 12 vegetation and land use variables after log transformation were included in the NMS analysis to test their effects on the longhorn beetle assemblages. Two variables that refer to number of tree species and individuals, respectively, were included. Further 5 variables refer to plant species richness, including the total number of vascular plant species per site and the species numbers of different life forms, i.e. grasses, forbs (non-woody plants other than grasses), and lianas. The remaining variables are the degree of tree canopy cover and the degree of ground vegetation cover, the maximum vegetation height, the successional age of the study site (years after establishment of the present vegetation or age of the trees), and the discrimination between four categories of land use type, represented by rice field fallows, early successions (forest clear fell, grassland, scanty shrubland), rubber plantations and natural forest (Table 1). Correlations between the ordination and the environmental variables were calculated with the Pearson coefficient. Indicator Species Analysis based on the combined values of relative abundance and relative frequency of species [36] was used to identify longhorn beetle species affiliated with specific land use types. The indicator value of each of the recorded species was calculated with PC-ORD software [35] using 4999 runs in a Monte Carlo test considering values at $P<0.05$.

The Chao 1 estimator was used to estimate total species richness of longhorn beetles per site. This estimator is the sum of the observed number of species and the quotient $a^2/2b$, where a and b equal the number of species represented by one (singletons) and two (doubletons) individuals, respectively. Calculations were conducted using the software package EstimateS (Version 8.2.0; [37]). To examine the relationship between Chao 1 estimated species richness of longhorn beetles and the variables of tree species and individuals at sampling location, these were included

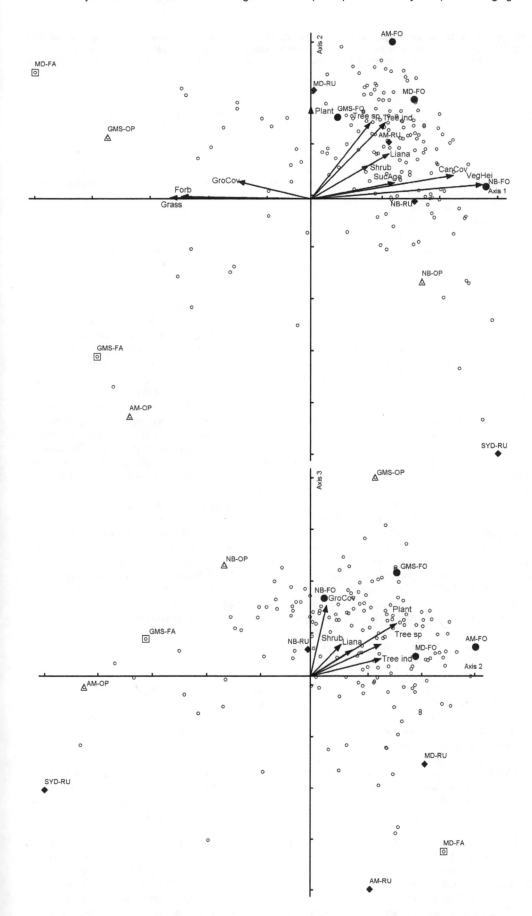

Figure 4. NMS ordination of longhorn beetle assemblages at the 13 sample sites: (♦) rubber plantations; (△) grassland and shrubland; (□) rice field fallows; (•) forest. Blue small circle means longhorn beetle species. Variables with Pearson correlation coefficient at $P<0.05$ are shown. Cumulative variation in the original dataset explained by ordination is 78.5% (Axis 1 = 41.0%, Axis 2 = 19.2%, Axis 3 = 18.4%, Final stress = 9.58, Final instability = 0.00001).

in the NMS analysis, using polynomial regression and Spearman's Rho non-parametric correlation.

Results

A total of 1434 tree individuals representing 181 species were recorded across all study sites (Table S2). The cluster analysis dendrogram based on the quantitative similarities of the tree inventories is shown in Fig. 1. As supported by global one-way ANOSIM tests, meaningful differences between assemblages occurred at a similarity of almost 0.1 and generated 4 subgroups (global $R=0.622$; $P<0.0001$). The dendrogram shows two major subgroups which are clearly distinguished by land use categories. Subgroup 1 includes all four mature forest sites. Subgroup 2 represents the four rubber plantation sites of different age (2a) and three open land sites (2b). The two fallow sites clearly diverge from the other groups due to their extremely low presence of trees.

A total of 1404 specimens representing 220 longhorn beetle species were recorded across all study sites and recording periods. Consistent with the highest tree species diversity, more than half (780) of all individuals were recorded from forest sites (Table S3). Meaningful differences between communities occurred at a similarity of about 0.25 and generated three main subgroups (global $R=0.4093$; $P<0.01$; Fig. 2). Subgroup 1 is composed of the four forest sites. Subgroup 2 is represented by three of the four rubber plantation sites. Subgroup 3 includes three open land sites and one fallow site. Another fallow site and the youngest rubber plantation (5 years) diverge from these subgroups. The Bray-Curtis similarity distance index among 13 sampling sites showed no significant differences between the tree and longhorn beetles assemblages (one-way ANOVA, $F=2.864$, df = 1,181, $P=0.092$).

Based on the analysis of the a priori selected habitat types, Fig. 3 shows that forest sites had the highest and fallows had the lowest mean value of both species and individuals. Rubber plantations and open land sites showed intermediate values with no significant differences between each other. The mean number of species and individuals was significantly different comparing all habitat types ($P<0.001$; Mann-Whitney U-Test).

The NMS ordination of the longhorn beetle assemblages represented 78.5% of the variation with a recommended three dimensional solution (Fig. 4) at a final stress of 9.58. Axis 1 accounts for 41.0% of this variation while axes 2 and 3 represent 19.2% and 18.4%, respectively. All environmental variables except the land use category showed significant effects at $P<0.05$ (Pearson correlation coefficient) for longhorn beetles (Fig. 4). Axis 1 was not only positively related with vegetation height, canopy coverage and succession age, but also with the plant variables referring to the number of tree species and individuals and also to the numbers of shrubs and lianas. Contrary to this, Axis 1 was negatively related with the ground coverage and number of grass and forb species at the sampling sites. Six longhorn beetle species affiliated with forest sites with indicator values at $P<0.05$ were identified. No indicator species affiliated with the other three land use types were found (see Table S3).

There was a highly positive relationship between the estimated species richness of longhorn beetles and the number of tree species at the sampling sites ($rs=0.7718$; $P<0.001$) (Fig. 5a). The species richness of longhorn beetles and the number of tree species also showed a highly positive relationship between each other ($rs=0.8527$; $P<0.001$) (Fig. 5b). Furthermore, longhorn individual numbers was also highly positive related to the number of tree individuals at the sampling sites ($rs=0.8560$; $P<0.001$) (Fig. 5c). The polynomial regression between the diversity index of trees and longhorn beetles at the sampling sites showed a highly positive relationship ($P<0.01$).

Discussion

The overall results on longhorn beetle species richness and abundance in the different habitat types of the study landscape clearly indicate that only one land use type, natural forest, possesses a degree of uniqueness in species diversity. The highest total numbers of species and individuals were recorded from the forest sites at which 780 individuals represented 193 species, from 220 species recorded overall. Furthermore, longhorn beetle species numbers at the forest sites were related to tree species diversity at the respective sites. However, only 6 forest habitat specialist species were identified. The number of forest specialists might indeed be higher than the estimated species number through the Chao estimator, because most of the species were only represented by few individuals which were not enough to show statistically significant results. For example, from the 193 species recorded from forest sites, 166 species were only represented with numbers between one and 6 individuals. Although these species were not denoted as indicator species, most of them are probably forest specialists, but they are merely rare.

The positive relationships between the numbers of longhorn beetle species and individuals on the one hand and the number of tree species and individuals on the other clearly indicate the importance of forest diversity for the longhorn beetle assemblages of the study landscape. Tree diversity may also indirectly reflect the availability of dead and decaying wood for various specialists. The quantity of this resource was not analyzed explicitly in the present study, but was found to be of high importance for saproxylic beetles in studies of northern temperate and boreal forests [38,39].

The highest number of longhorn beetle species outside forest habitats were recorded from rubber plantations (109 species, 344 individuals), with higher numbers in young plantations (5 and 8 years) than in older stands (20 and 40 years). It can be assumed that most of the species recorded there originate from natural forest sites and temporarily colonized rubber plantations due to the structural similarity of these plantations with forest. Most longhorn beetle species including the most common ones (*Chlorophorus arciferus*, *Pterolophia annulata* and *Xylotrechus buqueti*) were recorded from both forest and rubber plantations. However, for a few species such as *Nupserha nigriceps*, young rubber plantations might represent a suitable habitat because they were most common in this type of habitat and rare in forest sites. Young rubber plantations established after the clear felling of natural forest have open canopies and exhibit a rich undergrowth vegetation with regrown young forest trees and dead wood that remained from the destroyed forest. These conditions might explain the higher longhorn beetle diversity in young compared to old closed rubber stands. The latter only show a poor undergrowth vegetation, because most of the plants occurring in young

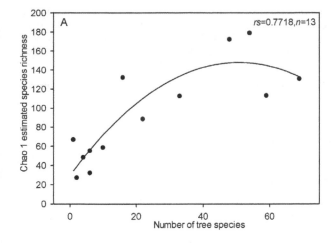

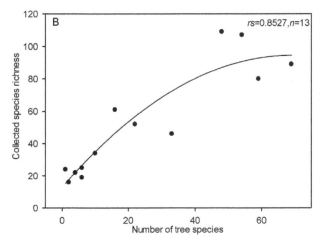

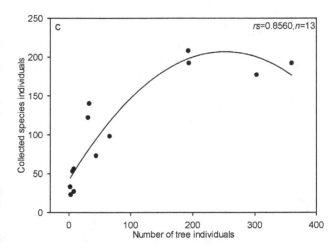

Figure 5. The polynomial regression between the diversity index of trees and longhorn beetles at 13 sampling location within Naban River Watershed National Nature Reserve. (A) Number of tree species and estimated Chao1 species richness of longhorn beetles; (B) Number of tree species and collected species richness of longhorn beetles; (C) Number of tree individuals and collected species individuals of longhorn beetles.

plantations became shaded out after the closing of the rubber tree canopy. Similarly, a study conducted in Japan [40] showed that in

different in types of plantation forests (broad leaf and conifer), the species richness of longhorn beetles was highest in young stands and decreased with the age of the stand. Possible causes of the high species richness in young stands stated by these authors include the availability of large amounts of coarse wood debris and flowers, which are resources for oviposition and nutrition for adults, respectively. The same study showed that longhorn beetle diversity was generally lower in conifer stands than in broad-leaved stands, and almost all species that occurred in conifer stands were also collected in broad-leaved stands. Another study conducted in Japan [6] showed that species richness of longhorn beetles was lower in second-growth forests and conifer plantations than in old-growth forests. These results are principally consistent with the findings of the present study referring to the longhorn beetle diversity patterns of forest, young rubber plantations and old rubber plantations, which decreases in this sequence. Although young rubber plantations bear the highest longhorn beetle diversity outside forests (almost half of total number of longhorn beetle species recorded in this study), they can not be considered as alternative habitats for most of the longhorn beetle species originating from forest, because they develop into closed canopy plantations with less suitable habitat conditions. This stage is approximately reached 20 years after planting. Currently, most of the rubber plantations in the study area are less than 20 years old, but will reach an age of about 40 years before the latex productivity declines. Then, the plantations will be clear felled for starting a new plantation cycle with young trees. Therefore, along with an expected expansion of rubber cultivation which largely proceeds at the expense of forest areas, the habitat conditions for longhorn beetles might decrease dramatically in future. Nonetheless, effects of rubber cultivation on longhorn beetles could be mitigated by preserving diverse natural forest plots and in addition, by a management of the rotation cycles of rubber plantations that allows the steady presence of young plantation stages. As recommended in various studies [10,38,41], the leaving of old trees, snags and dead wood in clear-cuts could also contribute to the preservation of longhorn beetles in newly established rubber plantations.

Similar to rubber plantations, most longhorn beetle species recorded from fallow, grassland and shrubland sites were also found in forests. The relatively low species and individual numbers from these habitat types might indicate that most species appeared there incidentally, reflecting the mobility of species which are able to move within the landscape. Overall, the highest species richness is clearly concentrated on the forest plots of the study landscape. On the other hand, a study from a temperate landscape of Switzerland [42] indicates that ecotones with a gradual transition from open land into the forest provide an important habitat for longhorn beetles, and the authors conclude that many of the so-called forest species are in fact forest edge species. They argue that additional resources such as pollen, sunlight and heat are necessary to meet the demands of many adult longhorn beetles, while in the forest interior a sufficient amount of dead wood as a substrate for larval development may be available. However, besides of the differences in climate and overall species diversity, these two landscapes differ in their land use history. The research area of the present study is characterized by recent and extreme changes through rubber cultivation, whereas many cultivated landscapes of temperate Europe have developed over hundreds or thousands of years and now provide habitats for species originating from other (warmer) regions. We therefore assume that the largest portion of the recorded longhorn beetle assemblage in the tropical landscape of the present study originates from the local natural forest, and that with few exceptions (such as *Nupserha nigriceps*) the large majority of the species can not benefit from wood provided by rubber trees and from the creation of open habitats. This, however, was not the case in certain other groups of insects recorded from the same region. For example, Meng et al. [43] found that the species richness and densities of ground beetles (Carabidae) were significantly higher in open habitats and rice field fallows than in other habitat types of this landscape, including forest, and are probably dominated by species originating from other regions with natural open vegetation. Furthermore, rice field fallows, early natural successions and natural forest each possess a degree of uniqueness in ground beetle species composition. Total variation in ground beetle assemblage composition was explained by the degree of vegetational openness and the groundcover plant species diversity. That is, grass and forb species showed to be the most important plant life types accounting for ground beetle distribution, and not trees. Compared to the ground beetles, longhorn beetles therefore seem to be more suitable indicators for the ecological state of landscapes in tropical forest regions.

Supporting Information

Table S1 Habitat characteristics of the 13 sampling sites and the environmental variables and the environmental variables used in the NMS analysis: total number of vascular plant species (Plant) and species of different life forms; percentage of ground vegetation cover (GroCov) and canopy cover (CanCov); the successional age of the site or age of the trees (SucAge), vegetation height (VegHei) and four categories of land use.

Table S2 List of tree species and numbers of individuals recorded from the 13 study localities compiled by habitat types. Abbreviation code FA, OP, RU and FO means rice fallow, open land, rubber plantation and forest respectively.

Table S3 List of longhorn beetle species and numbers of individuals recorded from the 13 study localities compiled by habitat types (*means indicator species). Abbreviation code FA, OP, RU and FO means rice fallow, open land, rubber plantation and forest respectively.

Acknowledgments

We would like to say thanks to Mr. C. Holzschuh and Dr. Mei-Ying Lin for their comprehensive support within the identification of many species. We are grateful to Mr. Gong Ai from the local Dai people for the numerous contributions in field work. The voucher specimen of the collected beetles were handed to Dr. Mei-Ying Lin at the National Zoological Museum of China, Institute of Zoology, Chinese Academy of Sciences (CAS), Beijing. We also thank Mr. Hong Wang and Mr. Guo-Da Tao for their help in the identification of the plant species. Mr. Shi-Shun Zhou cared for the voucher specimen at the Herbarium of the Xishuangbanna Tropical Botanical Garden (XTBG), Mengla, Yunnan.

Author Contributions

Conceived and designed the experiments: LZM KM. Performed the experiments: LZM AW. Analyzed the data: LZM AW. Contributed reagents/materials/analysis tools: AW XDY. Wrote the paper: LZM KM XDY.

References

1. Holland JD (2007) Sensitivity of Cerambycid biodiversity indicators to definition of high diversity. Biodiversity and Conservation 16: 2599–2609.
2. Jansson N, Bergman KO, Jonsell M, Milberg P (2009) An indicator system for identification of sites of high conservation value for saproxylic oak (Quercus spp.) beetles in southern Sweden. Journal of Insect Conservation 13: 399–412.
3. Sugiura S, Tsuru T, Yamaura Y, Makihara H (2009) Small off-shore islands can serve as important refuges for endemic beetle conservation. Journal of Insect Conservation 13: 377–385.
4. Hanks LM (1999) Influence of the larval host plant on reproductive strategies of cerambycid beetles. Annual Review of Entomology 44: 483–505.
5. Buse J, Ranius T, Assmann T (2008) An endangered longhorn beetle associated with old oaks and its possible role as an ecosystem engineer. Conservation Biology 22: 329–337.
6. Maeto K, Sato S, Miyata H (2002) Species diversity of longicorn beetles in humid warm-temperate forests: the impact of forest management practices on old-growth forest species in southwestern Japan. Biodiversity and Conservation 11: 1919–1937.
7. Taki H, Inoue T, Tanaka H, Makihara H, Sueyoshi M, et al. (2010) Responses of community structure, diversity, and abundance of understory plants and insect assemblages to thinning in plantations. Forest Ecology and Management 259: 607–613.
8. Ulyshen MD, Hanula JL, Horn S, Kilgo JC, Moorman CE (2004) Spatial and temporal patterns of beetles associated with coarse woody debris in managed bottomland hardwood forests. Forest Ecology and Management 199: 259–272.
9. Vodka S, Konvicka M, Cizek L (2009) Habitat preferences of oak-feeding xylophagous beetles in a temperate woodland: implications for forest history and management. Journal of Insect Conservation 13: 553–562.
10. Hjältén J, Stenbacka F, Pettersson RB, Gibb H, Johansson T, et al. (2012) Micro and macro-habitat associations in saproxylic beetles: implications for biodiversity management. Plos One 7: e41100–e41100.
11. Sugiura S, Yamaura Y, Tsuru T, Goto H, Hasegawa M, et al. (2009) Beetle responses to artificial gaps in an oceanic island forest: implications to invasive tree management to conserve endemic species diversity. Biodiversity and Conservation 18: 2101–2118.
12. Baur B, Zschokke S, Coray A, Schlapfer M, Erhardt A (2002) Habitat characteristics of the endangered flightless beetle Dorcadion fuliginator (Coleoptera : Cerambycidae): implications for conservation. Biological Conservation 105: 133–142.
13. Baur B, Coray A, Minoretti N, Zschokke S (2005) Dispersal of the endangered flightless beetle Dorcadion fuliginator (Coleoptera : Cerambycidae) in spatially realistic landscapes. Biological Conservation 124: 49–61.
14. Collinge SK, Holyoak M, Barr CB, Marty JT (2001) Riparian habitat fragmentation and population persistence of the threatened valley elderberry longhorn beetle in central California. Biological Conservation 100: 103–113.
15. Baselga A (2008) Determinants of species richness, endemism and turnover in European longhorn beetles. Ecography 31: 263–271.
16. Muona J, Rutanen I (1994) The short-term impact of fire on the beetle fauna in boreal coniferous forest. Annales Zoologici Fennici 31: 109–121.
17. Moretti M, Obrist MK, Duelli P (2004) Arthropod biodiversity after forest fires: winners and losers in the winter fire regime of the southern Alps. Ecography 27: 173–186.
18. Moretti M, Barbalat S (2004) The effects of wildfires on wood-eating beetles in deciduous forests on the southern slope of the Swiss Alps. Forest Ecology and Management 187: 85–103.
19. Bouget C (2005) Short-term effect of windstorm disturbance on saproxylic beetles in broadleaved temperate forests - Part I: Do environmental changes induce a gap effect? Forest Ecology and Management 216: 1–14.
20. Campbell JW, Hanula JL, Outcalt KW (2008) Effects of prescribed fire and other plant community restoration treatments on tree mortality, bark beetles, and other saproxylic Coleoptera of longleaf pine, Pinus palustris Mill., on the Coastal Plain of Alabama. Forest Ecology and Management 254: 134–144.
21. Moretti M, De Caceres M, Pradella C, Obrist MK, Wermelinger B, et al. (2010) Fire-induced taxonomic and functional changes in saproxylic beetle communities in fire sensitive regions. Ecography 33: 760–771.
22. Talley TS (2007) Which spatial heterogeneity framework? Consequences for conclusions about patchy population distributions. Ecology 88: 1476–1489.
23. Hanks LM, Paine TD, Millar JG (1993) Host species preference and larval performance in the wood-boring beetle Phoracantha semipunctata F. Oecologia 95: 22–29.
24. Corona AM, Toledo VH, Morrone JJ (2007) Does the Trans-mexican Volcanic Belt represent a natural biogeographical unit? An analysis of the distributional patterns of Coleoptera. Journal of Biogeography 34: 1008–1015.
25. Myers N, Mittermeier RA, Mittermeier CG, da Fonseca GAB, Kent J (2000) Biodiversity hotspots for conservation priorities. Nature 403: 853–858.
26. Hu H, Liu W, Cao M (2008) Impact of land use and land cover changes on ecosystem services in Menglun, Xishuangbanna, Southwest China. Environmental Monitoring and Assessment 146: 147–156.
27. Li H, Aide TM, Ma Y, Liu W, Cao M (2007) Demand for rubber is causing the loss of high diversity rain forest in SW China. Biodiversity and Conservation 16: 1731–1745.
28. Cao M, Zou XM, Warren M, Zhu H (2006) Tropical forests of Xishuangbanna, China. Biotropica 38: 306–309.

29. Lue XT, Yin JX, Tang JW (2010) Structure, tree species diversity and composition of tropical seasonal rainforests in Xishuangbanna, south-west China. Journal of Tropical Forest Science 22: 260–270.

30. Townes H, Zoology M, The, Arbor A (1962) Design for a malaise trap. Proceeding of the Entomological Society Washington 64: 253–262.

31. James FC, McCulloch CE (1990) Multivariate-analysis in ecology and systematics - panacea or pandora box. Annual Review of Ecology and Systematics 21: 129–166.

32. Wolda H (1981) Similarity indices, sample size and diversity. Oecologia 50: 296–302.

33. Zar JH (1999) Biostatistical analysis: Prentice Hall.

34. Kruskal J (1964) Nonmetric multidimensional scaling: A numerical method. Psychometrika 29: 115–129.

35. McCune B, Mefford J (2006) PC-ORD: Multivariate Analysis of Ecological Data : Version 6 for Windows: MjM Software Design.

36. Dufrêne M, Legendre P (1997) Species assemblages and indicator species: The need for a flexible asymmetrical approach. Ecological Monographs 67: 345–366.

37. Colwell RK (2006) EstimateS: Statistical estimation of species richness and shared species from samples. Version 8. Persistent. Available: purl.oclc.org/estimates.

38. Martikainen P, Siitonen J, Punttila P, Kaila L, Rauh J (2000) Species richness of Coleoptera in mature managed and old-growth boreal forests in southern Finland. Biological Conservation 94: 199–209.

39. Økland B, Bakke A, Hågvar S, Kvamme T (1996) What factors influence the diversity of saproxylic beetles? A multiscaled study from a spruce forest in southern Norway. Biodiversity and Conservation 5: 75–100.

40. Makino S, Goto H, Hasegawa M, Okabe K, Tanaka H, et al. (2007) Degradation of longicorn beetle (Coleoptera, Cerambycidae, Disteniidae) fauna caused by conversion from broad-leaved to man-made conifer stands of Cryptomeria japonica (Taxodiaceae) in central Japan. Ecological Research 22: 372–381.

41. Hansen AJ, Spies TA, Swanson FJ, Ohmann JL (1991) Conserving biodiversity in managed forests - lessons from natural forests. Bioscience 41: 382–392.

42. Wermelinger B, Flückiger PF, Obrist MK, Duelli P (2007) Horizontal and vertical distribution of saproxylic beetles (Col., Buprestidae, Cerambycidae, Scolytinae) across sections of forest edges. Journal of Applied Entomology 131: 104–114.

43. Meng L-Z, Martin K, Weigel A, Liu J-X (2012) Impact of rubber plantation on carabid beetle communities and species distribution in a changing tropical landscape (southern Yunnan, China). Journal of Insect Conservation 16: 423–432.

12

Complex Spatiotemporal Responses of Global Terrestrial Primary Production to Climate Change and Increasing Atmospheric CO$_2$ in the 21st Century

Shufen Pan[1,2], Hanqin Tian[1]*, Shree R. S. Dangal[1], Chi Zhang[3], Jia Yang[1], Bo Tao[1], Zhiyun Ouyang[2], Xiaoke Wang[2], Chaoqun Lu[1], Wei Ren[1], Kamaljit Banger[1], Qichun Yang[1], Bowen Zhang[1], Xia Li[1]

1 International Center for Climate and Global Change Research, School of Forestry and Wildlife Sciences, Auburn University, Auburn, Alabama, United States of America, 2 State Key Laboratory of Urban and Regional Ecology, Research Center for Eco-Environmental Sciences, Chinese Academy of Sciences, Beijing, China, 3 State Key Laboratory of Desert and Oasis Ecology, Xinjian Institute of Ecology and Geography, Chinese Academy of Sciences, Urumqi, China

Abstract

Quantitative information on the response of global terrestrial net primary production (NPP) to climate change and increasing atmospheric CO$_2$ is essential for climate change adaptation and mitigation in the 21st century. Using a process-based ecosystem model (the Dynamic Land Ecosystem Model, DLEM), we quantified the magnitude and spatiotemporal variations of contemporary (2000s) global NPP, and projected its potential responses to climate and CO$_2$ changes in the 21st century under the Special Report on Emission Scenarios (SRES) A2 and B1 of Intergovernmental Panel on Climate Change (IPCC). We estimated a global terrestrial NPP of 54.6 (52.8–56.4) PgC yr^{-1} as a result of multiple factors during 2000–2009. Climate change would either reduce global NPP (4.6%) under the A2 scenario or slightly enhance NPP (2.2%) under the B1 scenario during 2010–2099. In response to climate change, global NPP would first increase until surface air temperature increases by 1.5°C (until the 2030s) and then level-off or decline after it increases by more than 1.5°C (after the 2030s). This result supports the Copenhagen Accord Acknowledgement, which states that staying below 2°C may not be sufficient and the need to potentially aim for staying below 1.5°C. The CO$_2$ fertilization effect would result in a 12%–13.9% increase in global NPP during the 21st century. The relative CO$_2$ fertilization effect, i.e. change in NPP on per CO$_2$ (ppm) bases, is projected to first increase quickly then level off in the 2070s and even decline by the end of the 2080s, possibly due to CO$_2$ saturation and nutrient limitation. Terrestrial NPP responses to climate change and elevated atmospheric CO$_2$ largely varied among biomes, with the largest increases in the tundra and boreal needleleaf deciduous forest. Compared to the low emission scenario (B1), the high emission scenario (A2) would lead to larger spatiotemporal variations in NPP, and more dramatic and counteracting impacts from climate and increasing atmospheric CO$_2$.

Editor: Benjamin Poulter, Montana State University, United States of America

Funding: This study was supported by NSF Decadal and Regional Climate Prediction using Earth System Models (AGS-1243220), NSF Dynamics of Coupled Natural and Human Systems (1210360), NASA Interdisciplinary Science Program (NNX10AU06G, NNG04GM39C), US Department of Energy NICCR Program (DUKE-UN-07-SC-NICCR-1014). The funders had no role in study design, data collection and analysis, decision to publish, or preparation of the manuscript.

Competing Interests: The authors have declared that no competing interests exist.

* Email: tianhan@auburn.edu

Introduction

Net Primary Productivity (NPP), a balance between photosynthetic carbon (C) uptake (Gross Primary Productivity; GPP) and losses due to plant respiration, represents the net C retained by terrestrial vegetation. It is of particular importance to humans since the largest portion of the food supply comes from terrestrial NPP [1]. NPP is also an important indicator of ecosystem health and services [2,3], and is an essential component of the global C cycle [4]. Terrestrial NPP is sensitive to multiple environmental changes including climate and atmospheric changes [5]. The IPCC Fourth Assessment (AR4) assessment indicated that global average temperature has increased by 0.74°C since the pre-industrial times and that this trend is expected to continue through the 21st century [6]. In addition, atmospheric CO$_2$ concentration have increased from the pre-industrial level of 280 ppm to the

2005 level of 379 ppm [6]. Comparing the 2090s with the 2000s, under the high emission scenario (A2), global mean temperature would increase by 4.6°C, while global annual precipitation would increase by 16.8%. However, the Representative Concentration Pathways (RCP's) scenarios used in the IPCC Fifth Assessment Report (AR5) IPCC report [7] were created in a different way and span a wider range of 21st century projections. There are notable differences among the IPCC SRES scenarios and the IPCC RCP scenarios. The B1 scenario is very close to the RCP 4.5 by 2100, but there is lower emissions at the middle of 21st century [8]. Similarly, the A2 scenario is between the RCP 6.0 and RCP 8.5 scenarios. Projected changes in atmospheric CO$_2$ concentration showed large increases under the A2 scenario, from 379.6 ppm in the 2000s to 809 ppm in the 2090s, which is equivalent to an overall increase of 113.8%. Such dramatic changes in climate and atmospheric composition would profoundly affect the NPP of terrestrial ecosystems. It is of critical importance to quantitatively analyze the contemporary pattern of global NPP and project to what extent climate change and increasing atmospheric CO_2 in the 21st century would alter the magnitude and spatiotemporal patterns of NPP across the terrestrial ecosystems [9].

Previous studies reported that climate and increasing atmospheric CO_2 are the primary drivers for changes in global terrestrial NPP [4,10–12]. Enhanced terrestrial NPP in response to increase in temperature [13] and atmospheric CO_2 concentration [14,15] have been suggested across a range of terrestrial ecosystems. Over the recent three decades, climate change has been the major driver of terrestrial NPP [10,16,17], with further benefits from CO_2 fertilization [4,18]. Nemani et al. [10] reported that climate change resulted in a 6% increase in global terrestrial NPP from 1982 to1999, with the largest increase in low-latitude ecosystems. Zhao and Running et al [16] reported that in the recent decade (2000–2009), high temperature has increased water stresses and autotrophic respiration in the Southern Hemisphere resulting in a decline in global NPP by 0.55 PgC. Potter et al. [17], however, reported that rapid climate warming alleviated temperature limitations in high-latitude ecosystems which led to an increase in global terrestrial NPP by 0.14 PgC during 2000–2009. Higher temperature affect plant phenology, promoting an early growth and increasing the C assimilation in temperature-limited regions [13,19,20] due to the acceleration of enzymatic processes. Also, increasing CO_2 has been found to reduce stomatal conductance [14] and increase water use efficiency [21,22]. However, the stimulated effects of temperature on NPP may also be mitigated by increasing soil water stress and respiration rates induced by temperature rise [18,23,24]. In addition, increased water stress reduces nutrient uptake [25] which could potentially lead to a decline in productivity [26].While there is little doubt that climate change and increasing atmospheric CO_2 are the primary drivers of terrestrial NPP for the recent decade, the relative contribution of different drivers in the future is still unclear.

Process-based ecosystem models are effective tools for future projection of terrestrial NPP in response to global change [27]. Various process-based ecosystem models have been developed to estimate NPP response to changes in climate and increasing atmospheric CO_2 concentration at several scales from continental to global for both contemporary and future climatic conditions [4,28]. Previous modeling studies found that climate change resulted in an overall decline in global NPP, but doubling atmospheric CO_2 concentration resulted in an increase in global NPP by 16–25% [4,18]. In a process-based model comparison study, Cramer et al. [29] found differences in global NPP among 17 models (ranging from 44.4 to 66.3 PgC yr^{-1}) due largely to how the water balance was represented in models. In a similar 17-model comparison study, Friedlingstein et al. [30] found large uncertainties associated with belowground processes that resulted in different responses of NPP to global change factors across models. Thus, realistic historical assessments and future projections of global terrestrial NPP in a rapidly changing climate require more comprehensive models that include ecological, physiological and biogeochemical processes such as changes in phenology, length of growing seasons, nutrient dynamics, and ecohydrological processes.

The purpose of this study is to understand complex responses of terrestrial NPP at latitudinal, biome and global levels to projected climate change and increasing atmospheric CO_2 in the 21st century. To accomplish this task, we first established the baseline estimate of global terrestrial NPP for the first decade of the 21st century by using a well-evaluated process-based ecosystem model (the Dynamic Land Ecosystem Model, DLEM [22]) driven by multiple environmental factors. Then we used the DLEM model to examine responses of terrestrial NPP to projected climate change and increasing atmospheric CO_2 during the rest of the 21st century under the IPCC Special Report on Emissions Scenarios (A2 and B1).The major objectives of this study are: (1) to estimate the contemporary global terrestrial NPP, (2) to project its changing trend in the 21st century, (3) to attribute the relative contribution of climate, elevated CO_2, and their interaction; and (4) to investigate the spatiotemporal pattern of global NPP as well as the response of different biomes to climate and CO_2 changes.

Methods

2.1 Model description

The DLEM is a highly integrated, process-based terrestrial ecosystem model that aims at simulating the structural and functional dynamics of land ecosystems affected by multiple factors including climate, atmospheric compositions (CO_2, nitrogen deposition, and tropospheric ozone), land use and land cover change, and land management practices (harvest, rotation, fertilization etc). The DLEM has five core components (Figure 1): 1) biophysics, 2) plant physiology, 3) soil biogeochemistry, 4) dynamic vegetation, and 5) land use and management [22]. This model has been extensively calibrated against various field data covering forest, grassland, and cropland from the Chinese Ecological Research Network, the US Long Term Ecological Research (LTER) sites, the AmeriFlux network and other field sites. Detailed information on how DLEM simulates these processes is available in our published papers [31–34]. Recently, we updated the model to DLEM 2.0 version, which is characterized by cohort structure, multiple soil layer processes, coupled C, water and nitrogen cycles, multiple greenhouse (GHG) emissions simulation, enhanced land surface processes, and dynamic linkages between terrestrial and riverine ecosystems [34,35]. Below, we briefly describe the simulation of GPP and NPP, calculation of relative CO_2 fertilization effect, input datasets used to drive the DLEM, and global-level evaluation of simulated NPP against satellite data.

2.2 Modeling gross primary productivity (GPP) in the DLEM

The DLEM uses a modified Farquhar's model to simulate GPP [36–39]. The canopy is divided into sunlit and shaded layers. GPP (gC m^{-2} day^{-1}) is calculated by scaling leaf assimilation rates (μmol CO_2 m^{-2} s^{-1}) up to the whole canopy:

$$GPP_{sun} = 12.01 \times 10^{-6} \times A_{sun} \times plai_{sun} \times dayl \times 3600$$

$$GPP_{shade} = 12.01 \times 10^{-6} \times A_{shade} \times plai_{shade} \times dayl \times 3600$$

$$GPP = GPP_{sun} + GPP_{shade}$$

Where GPP_{sun} and GPP_{shade} are the GPP of sunlit and shaded canopy, respectively; A_{sun} and A_{shade} are assimilation rates of sunlit and shaded canopy; $plai_{sun}$ and $plai_{shade}$ are sunlit and shaded leaf area indices; dayl is daytime length (second) in a day. 12.01×10^{-6} is a constant to change the unit from μmol CO_2 to gram C.

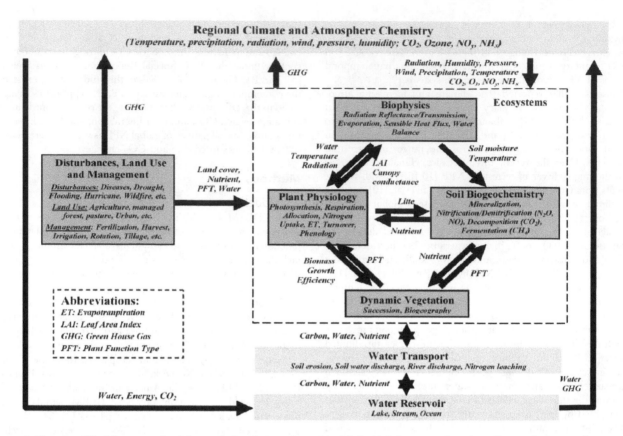

Figure 1. The simplified framework of Dynamic Land Ecosystem Model (DLEM) for assessing the effects of climate change and increasing atmospheric CO₂ concentration on global terrestrial net primary production (NPP).

The $plai_{sun}$ and $plai_{shade}$ are estimated as:

$$plai_{sun} = 1 - EXP(-proj_{LAI})$$

$$plai_{shade} = proj_{LAI} - plai_{sun}$$

Where, $proj_{LAI}$ is the projected leaf area index. Using methods similar to Collatz et al. [37], DLEM determines the C assimilation rate as the minimum of three limiting rates, w_c, w_j, w_e, which are functions that represents the assimilation rates as limited by the efficiency of photosynthetic enzymes system (Rubisco-limited), the amount of Photosynthetically Active Radiation (PAR) captured by leaf chlorophyll (light-limited), and the capacity of leaves to export or utilize photosynthesis products (export-limited) for C₃ species, respectively. For C₄ species, w_e refer to the Phosphoenolpyruvate (PEP) carboxylase limited rate of carboxylation. The sunlit and the shaded canopy C assimilation rate can be estimated as:

$$A = \min(w_c, w_j, w_e) \times Index_{gs}$$

$$w_c = \begin{cases} \dfrac{(c_i - \Gamma_*)V_{\max}}{c_i + K_c(1 + o_i/K_0)} & \text{for } C_3 \text{ plants} \\ V_{\max} & \text{for } C_4 \text{ plants} \end{cases}$$

$$w_j = \begin{cases} \dfrac{(c_i - \Gamma_*)4.6\phi\alpha}{c_i + 2\Gamma_*} & \text{for } C_3 \text{ plants} \\ 4.6\phi\alpha & \text{for } C_4 \text{ plants} \end{cases}$$

$$w_e = \begin{cases} 0.5 V_{\max} & \text{for } C_3 \text{ plants} \\ 4000 V_{\max} \dfrac{c_i}{P_{atm}} & \text{for } C_4 \text{ plants} \end{cases}$$

where c_i is the internal leaf CO₂ concentration (Pa); O_i is the O₂ concentration (Pa); Γ_* is the CO₂ compensation point (Pa); K_c and K_o are the Michaelis-Menten constants for CO₂ and O₂, respectively; α is the quantum efficiency; ϕ is the absorbed photosynthetically active radiation (W·M⁻²); V_{max} is the maximum rate of carboxylation varies with temperature, foliage nitrogen concentration, and soil moisture [38]:

$$V_{\max} = V_{\max 25} a_{v\max}^{\frac{T_{day} - 25}{10}} f(N) f(T_{day}) \beta_t$$

where V_{max25} is the value at 25°C and a_{vmax} is the temperature sensitivity parameter; $f(Tday)$ is a function of temperature-related metabolic processes [36,37].

$$f(T_{day}) = \left[1 + \exp\left(\frac{-220000 + 710(T_{day} + 273.16)}{8.314(T_{day} + 273.16)}\right)\right]^{-1}$$

$f(N)$ adjusts the rate of photosynthesis for foliage nitrogen

$$f(N) = \min\left(\frac{N_{leaf}}{N_{leaf\_opt}}, 1\right)$$

Where N_{leaf} is the nitrogen concentration and N_{leaf_opt} is the optimal leaf nitrogen concentration for photosynthesis.

β_t is a function, ranging from one to zero that represents the soil moisture and the lower temperature effects on stomatal resistance and photosynthesis.

$$\beta_t = \beta(T_{\min}) \times \beta(w)$$

$$\beta(T_{\min}) = \begin{cases} 0 & \text{for } T_{\min} < -8\,^{\circ}C \\ 1 + 0.125 T_{\min} & \text{for } -8\,^{\circ}C \leq T_{\min} \leq 0\,^{\circ}C \\ 1 & \text{for } T_{\min} > 0\,^{\circ}C \end{cases}$$

$$\beta(w) = \sum_{i=1}^{2} w_i r_i$$

$$w_i = \begin{cases} 0 & \text{for } ps_i > psi\_close \\ \frac{psi\_close - ps_i}{psi\_close - psi\_open} & \text{for } psi\_open \leq ps_i \leq psi\_close \\ 1 & \text{for } ps_i < psi\_open \end{cases}$$

where T_{min} is the daily minimum temperature; w_i is the soil water stress of soil layer i; ps_i is the soil water potential of soil layer i, which is estimated from the soil volume water content based on equations from Saxton and Rawls [40]; r_i is the root fractions distributed in soil layer i; psi_close and psi_open are the plant functional specific tolerance of the soil water potential for stomata overall close and open. The water stress in plants is a function of soil water potential which ranges from 0 to 1. Under no water limitations, the soil water stress of soil layer i (w_i) is equal to 1 where the soil water potential is at its maximum i.e., soil water potential when the stomata is opened (psi_open). Under frequent water stress, however, wi is calculated based on wilting point potential of specific plant functional types and depends on the balance between psi_open and psi_close.

Leaf stomatal resistance and leaf photosynthesis are coupled together through the following [37–39].

$$\frac{1}{r_s} = m \frac{A}{c_s} \frac{e_s}{e_i} P_{atm} + b$$

where r_s is the leaf stomatal resistance, m is an empirical parameter, A is the leaf photosynthesis, c_s is the leaf surface CO$_2$ concentration, e_s is the leaf surface vapor pressure, e_i is the saturated vapor pressure inside leaf, b is the minimum stomata conductance with $A = 0$, and P_{atm} is the atmospheric pressure. Together in the following equations:

$$c_s = c_a - 1.37 r_b P_{atm} A$$

$$e_s = \frac{e_a' + e_i r_b}{r_b + r_s}$$

where c_a is the atmospheric CO$_2$ concentration, rb is the boundary resistance, e_a is the vapor pressure of air, and stomatal resistance is the larger of the two roots of this quadratic [38].

$$\left(\frac{mAP_{atm}e_a'}{c_s e_i} + b\right) r_s^2 + \left(\frac{mAP_{atm}r_b}{c_s} + br_b - 1\right) r_s - r_b = 0$$

2.3 Modeling net primary productivity (NPP) in the DLEM

NPP is the net C gain by vegetation and equals the difference between GPP and plant respiration, which is calculated as:

$$Gr = 0.25 \times GPP$$

$$NPP = GPP - Mr - Gr$$

The DLEM estimates maintenance respiration (Mr, unit: gC m^{-2} day^{-1}) and growth respiration (Gr, unit: gC m^{-2} day^{-1}) as a function of assimilated C, surface air temperature and biomass nitrogen content. Gr is calculated by assuming that the fixed part of assimilated C will be used to construct new tissue (for turnover or plant growth). During these processes, 25% of assimilated C is supposed to be used as growth respiration [41]. Maintenance respiration is related to surface temperature and biomass nitrogen content. The following is used to calculate the maintenance respiration of leaves, sapwood, fine roots, and coarse roots:

$$Mr_i = rf \times R_{coeff} \times N_i \times f(T)$$

Where i denotes the C pool of different plant parts (leaf, sapwood, fine root, or coarse root); Mr_i (gC m^{-2} day^{-1}) is the maintenance respiration of different pools; rf is a parameter indicating growing phase, which is set at 0.5 for the non-growing season and 1.0 for the growing season; R_{coeff} is a plant functional type-specific respiration coefficient; N_i (gN m^{-2}) is the nitrogen content of pool i; $f(T)$ is the temperature factor and is calculated as follows:

$$f(T) = e^{308.56 \times \left(\frac{1}{56.02} - \frac{1}{T + 46.02}\right)}$$

Where T is the daily average air temperature for modeling aboveground C pools such as leaves, sapwood, and heartwood or soil temperature for modeling belowground pools such as coarse roots and fine roots.

2.4 The effect of CO_2 fertilization

In this study, we further quantified the effects of direct CO_2 fertilization on terrestrial NPP by calculating the 'beta' (β) effect. β effect measures the strength of changes in terrestrial NPP in response to increasing CO_2 concentration as follows:

$$NPP_{co_2} = \frac{NPP_{clm+co_2} - NPP_{clm}}{CO_2\ concentration(ppm)}$$

Where, NPP_{CO2} is the relative contribution of direct CO_2 fertilization on terrestrial NPP under the A2 and B1 scenarios, $NPP_{clm+CO2}$ is the terrestrial NPP under the climate plus CO_2 experiment and NPP_{clm} is the terrestrial NPP under the climate only experiment. CO_2 concentration (ppm) is the concentration of atmospheric CO_2 under the A2 and B1 scenarios.

2.5 Input datasets

The spatially-explicit data sets for driving the DLEM model include time series of daily climate, CO_2 concentration, annual land cover and land use (LCLU), nitrogen deposition, tropospheric ozone, and land management practices (irrigation and nitrogen fertilizer use). Other ancillary data include river network, cropping system, soil property, and topography maps. Contemporary vegetation map include 18 plant functional types (Figure 2). Cropland and urban distribution datasets were developed by aggregating the 5-arc minute resolution HYDE v3.1 global cropland distribution data [42]. The vegetation map was developed based on global land-cover data derived from Landsat imageries [43], the National Land Cover Dataset 2000 (www.usgs. gov), and the global database of lakes, reservoirs, and wetlands [44]. The vegetation is transient and does not include any disturbance during the course of simulation. Half degree daily climate data (including average, maximum, minimum air temperature, precipitation, relative humidity, and shortwave radiation) were derived from newly available CRU-NCEP climate forcing data (1900–2009, 6-hour, half degree spatial resolution) [45]. The annual nitrogen deposition dataset for the historical period were based on Dentener [46]. Ozone AOT_{40} data sets were based on the global AOT_{40} index developed by Felzer et al. [47]. The gridded monthly CO_2 concentration data were derived from Multi-scale Synthesis and Terrestrial Model Intercomparison Project (MSTIMP) (http://nacp.ornl.gov/MsTMIP.shtml). Consumption of nitrogen fertilizers from 1961 to 2008 were derived from country level Food and Agriculture Organization of the United Nations (FAO) statistic database (http://faostat.fao.org). We then calculated the annual fertilization rate (gN m^{-2}) as the ratio of national fertilizer application amount to total cropland area in each country [48]. The contemporary irrigation map was developed based on LCLU data and global irrigatied fraction map [49,50]. Long-term average climate datasets from 1900 to 1930 were used to represent the initial climate state in 1900.

For future projections, we used two IPCC emission secenarios (A2 and B1) datasets containing atmospheric CO_2 concentration and climate (precipitation and temperature) from the Community Climate System Model (CCSM3) (Figure 3–4). The A2 scenario (high emission scenario) is characterized by rapid population growth and low per capita income, with regionally oriented economic development. The B1 scenario (low emission scenario) describes the same global population as the A2 storyline, but it is less materially intensive in its service and information, economic structure, with emerging clean and resource-efficient technology [6]. The climate datasets were downloaded from the World Climate Research Programme's Coupled Model Intercomparison Project phase 3 (CMIP3) multi-model database (Meehl et al. [51]; www.engr.scu.edu/~emaurer/global_data). These datasets were downscaled as described by Maurer et al. [52] using the bias-correction/spatial downscaling method [53] to a 0.5 degree resolution, based on the 1950–1999 gridded observations of Adam and Lettenmaier [54]. For the future projections (2010–2099), we assumed that nitrogen deposition, ozone pollution, and LCLU remains unchanged from 2009.

Temperature and precipitation have been projected to increase substantially during 2010 to 2099 (Figure 3A–C) with large spatial variations under the A2 and B1 scenarios (Figure 4A–D). Under the A2 scenario, air temperature would increase by 4.6°C (Temperature = 0.008×Year; p-value<0.01), while precipitation would increase by 16.8% (Precipitation = 0.41×Year; p-value< 0.01) by the 2090s compared to the 2000s. Similarly, under the B1 scenario, air temperature would increase by 1.5°C (Temperature = 0.007×Year; p-value<0.01), while precipitation would increase by 7.5% (Precipitation = 0.039×Year; p-value<0.01) during the 2090s compared to the 2000s. Across latitudes, the largest increase in surface air temperature (>5°C) would occur under the A2 scenario in mid- and high-latitude regions of the Northern Hemisphere, while the smallest increase would occur in low latitude regions. In the Southern Hemisphere, a large increase in surface air temperature (>6°C) would occur in parts of Australia. The largest increase in precipitation would occur in high-latitude regions under the A2 and B1 scenarios, while large variations in mid- and high-latitude regions. For instance, there would be no change in precipitation in Africa, Southwestern US, Northwestern China and Australia under both scenarios, while precipitation would increase by >350 mm and >250–350 mm in monsoon Asia under the A2 and B1 scenarios, respectively. In addition, large variation in total precipitation between the Southern and the Northern Hemisphere have been observed due to physical distribution of more landmass resulting in a greater thermal effect in the Northern Hemisphere than in the Southern Hemisphere [55].

2.6 Model parameterization, calibration and evaluation

The DLEM has been parameterized and applied across several regional and continental studies including Asia [31,33,56–60], the United States [22,34,61] and North America [62,63] using long-term observational data for all defined plant functional types. The calibrated parameter values have been used to drive the model for specific plant functional types (Figure 5). In this study, we compared DLEM-simulated global estimates of terrestrial NPP with MODIS NPP to evaluate model performance during 2000–2009. We first evaluated DLEM performance at the global level in simulating spatial pattern of terrestrial NPP by comparing DLEM-simulated NPP with MODIS NPP (Figure 5). We then carried out a grid-to-grid comparison of DLEM-simulated NPP with MODIS product by randomly selecting 6000 grids from MODIS product (10% of the total sampling units). For bare land such as part of Africa, China and Mongolia (Taklamakan desert in West China, Gobi desert in Mongolia), there is no NPP data available from MODIS and we excluded those areas from analysis. At the global scale, the spatial pattern of DLEM-simulated NPP is consistent with that of MODIS NPP (Figure 5A–B). In addition, we found a good agreement between MODIS and DLEM-simulated NPP for randomly selected 6000 grid points (Figure 5C).The fitted line between DLEM-simulated and MODIS derived NPP had a slope of 0.73 and a high correlation coefficient ($R^2 = 0.68$). Our model evaluation for the effect of CO_2 fertilization is available in Lu et al. [60].

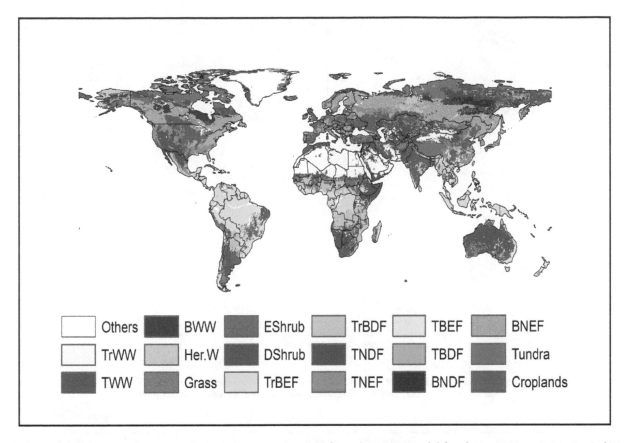

Figure 2. Contemporary vegetation map of the world as observed from the DLEM model for the year 2010. TrWW: Tropical Woody Wetlands; TWW: Temperate Woody Wetlands, BWW: Boreal Woody Wetlands, Her.W: Herbaceous Wetlands, EShrub: Evergreen Shrubland; DShrub: Deciduous Shrubland; TrBEF: Tropical Broadleaf Evergreen Forest; TrBDF: Tropical Broadleaf Deciduous Forest; TNDF: Temperate Needleleaf Deciduous Forest; TNEF: Temperate Needleleaf Evergreen Forest; TBEF: Temperate Broadleaf Evergreen Forest; TBDF: Temperate Broadleaf Deciduous Forest; BNDF: Boreal Needleleaf Deciduous Forest; BNEF: Boreal Needleleaf Evergreen Forest; Others: Desert & Ice.

2.7 Experimental design and model implementation

We designed 2×2 factorial simulation experiments (2 simulations×2 scenarios). The two simulation experiments include: (1) climate change only: only climate changes with time and other environmental factors are held constant during the study period (2010–2099); and (2) climate plus CO_2: both climate and atmospheric CO_2 concentration change during the study period while other environmental factors including LCLU, nitrogen deposition, and tropospheric ozone are held constant at 2009 levels. The A2 and B1 climate scenarios were used to drive these two simulations.

The model simulation follows three important stages: an equilibrium simulation stage, a 3000-year spin-up stage, and a transient simulation stage. The model simulation begins with an equilibrium run with long-term average climate data for the period 1901–1930, with the 1900 levels of atmospheric CO_2 concentration, nitrogen deposition, and LCLU map to develop simulation baselines for C, nitrogen, and water pools. However, the tropospheric ozone data is kept at the 1935 level during the equilibrium. The simulation baseline (equilibrium) is approached when the net C exchange between the atmosphere and terrestrial ecosystems is less than 0.1 gC m^{-2}, the change in soil water pool is less than 0.1 mm, and the difference in soil mineral nitrogen content and nitrogen uptake is less than 0.1 gN m^{-2} among consecutive years. After the equilibrium run, a 3000-year spin up is carried out using transient climate data and LCLU distribution in the 1900 to eliminate system fluctuations caused by simulation

mode shift from equilibrium to transient mode. Finally, a transient simulation is set up, driven by changes in climate, atmospheric chemistry (nitrogen deposition, tropospheric ozone, and atmospheric CO_2 concentration), and LCLU distribution during 1900–2009. In this study, we focused our analysis on global terrestrial NPP during two time periods: 2000–2009 and 2010–2099. For the first period (2000–2009), the simulated results (NPP$_{2000s}$) represent contemporary patterns of global terrestrial NPP. For the second period (2010–2099), simulated results (NPP$_{2090s}$) reflect the evolution of terrestrial NPP by the end of the 21st century. The difference between the two decadal mean NPP (NPP$_{2090s}$ - NPP$_{2000s}$) indicates the overall effects of climate change and CO_2 increase on terrestrial ecosystems in the 21st century.

Results

3.1 Terrestrial NPP in the first decade of 21st century

The DLEM simulation results show a global terrestrial NPP of about 54.57 (52.8–56.4) PgC yr^{-1} during the first decade of the 21st century, with substantial inter-annual variations due to precipitation ($R^2 = 0.63$; $P<0.01$) (Figure 6; left panel). In a specific year, drought or wet climate could substantially alter the magnitude of global terrestrial NPP. For instance, the dry year of 2005 resulted in a decline in global terrestrial NPP by 1.33 Pg C, while the wet year of 2008 increased global terrestrial NPP by 1.82 Pg C compared to the decadal mean (2000–2009). In addition, there are substantial variation in contemporary

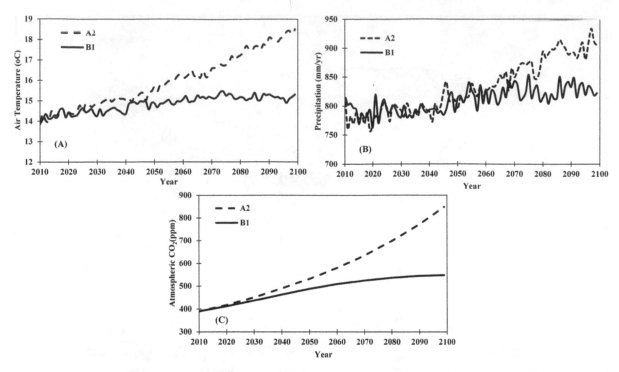

Figure 3. Input datasets used for driving the DLEM model based on CRUNCEP analysis. Temperature and precipitation change for A2 (A) and B1 (B) scenario and changes in CO_2 concentration between A2 and B1 emission scenario.

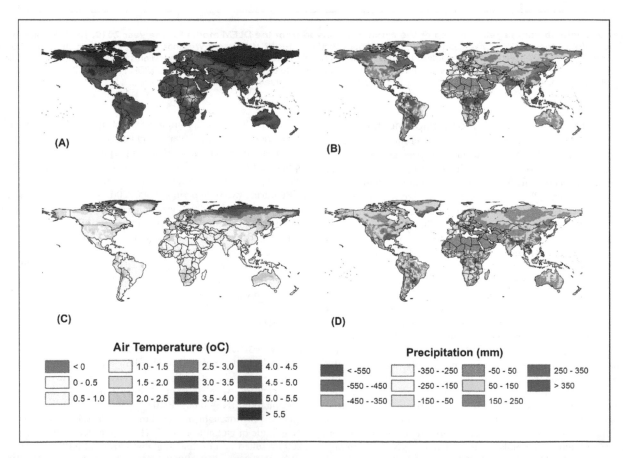

Figure 4. Spatial pattern of temperature and precipitation estimated as an average difference between 2099–2090 and 2000–2009: temperature (A) and precipitation (B) under A2 scenario and temperature (C) and precipitation (D) under B1 scenario.

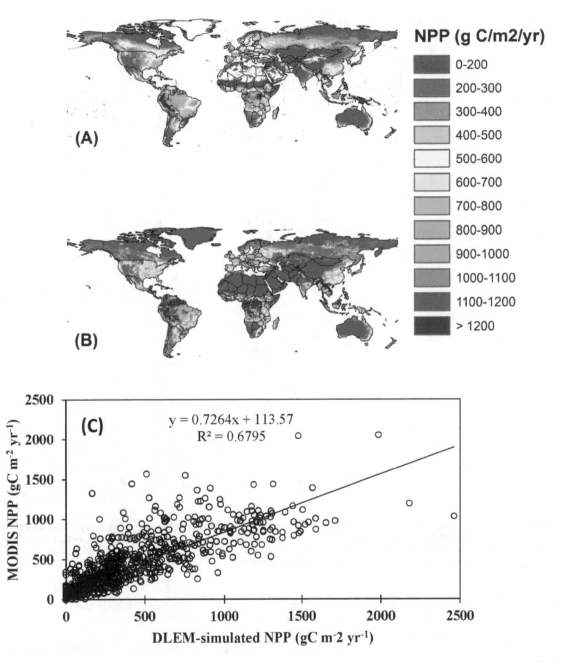

Figure 5. Spatial patterns of MODIS-NPP (A) and DLEM-simulated NPP (B) during 2000–2009 and comparison of the DLEM-simulated NPP with MODIS-NPP (C) for 6000 randomly selected grids.

productivity across different biomes, with the highest NPP of 1122.43 gC m^{-2} yr^{-1} for tropical broadleaf evergreen forest and the lowest NPP of 70.27 gC m^{-2} for tundra vegetation (Figure 6; right panel). It should be noted that boreal needleleaf deciduous forests have experienced fast increase in NPP by 7.42 gC m^{-2} yr^{-1} (p-value<0.05) during 2000–2009, due to substantial warming in the high latitudes [64]. To examine whether the contemporary trend will continue, we further projected climate change effects on ecosystem productivity during the rest of the century (2010–2099).

3.2 Changes in terrestrial NPP induced by climate change and increasing atmospheric CO$_2$ during 2010–2099

3.2.1. Temporal responses of terrestrial NPP to climate change. Our DLEM simulations show that climate change would increase terrestrial NPP by 3.0% until the 2030s under the A2 scenario and by 2.7% until the 2060s under the B1 scenario (Figure 7A; Table 1), but there would be a declining trend afterwards. Climate change under the A2 scenario would result in an overall decline in terrestrial NPP by 2.51 PgC (4.6%) in the 2090s compared to the 2000s (Figure 8; left panel). The B1 scenario shows an increasing trend with the highest increase in NPP by 1.57 PgC in the 2050s; however, NPP would level off after

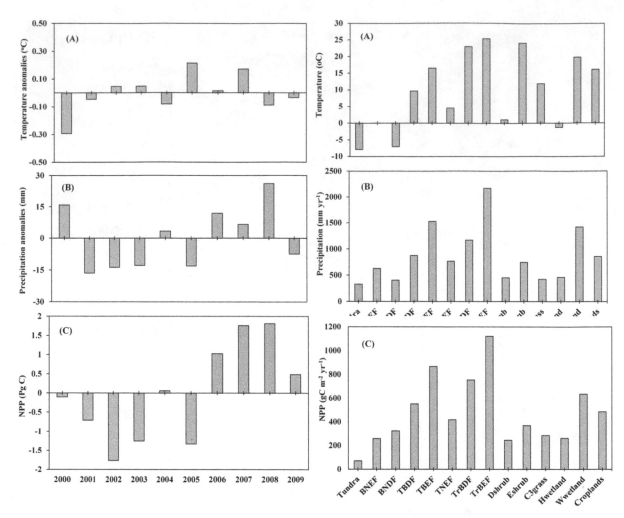

Figure 6. Effect of inter-annual variation in precipitation and temperature on global net primary productivity during the contemporary period (2000–2009) (left panel) and changes in mean annual NPP of major biomes as a function of temperature and precipitation (right panel). Left Panel: Mean annual temperature anomalies (a), annual precipitation anomalies (b), and net primary productivity (c) and right panel: average (2000–2009) temperature (a) average precipitation (b) and average net primary productivity (c).

the 2050s with an overall increase by 1.2 PgC (2.2%) in the 2090s compared to the 2000s (Figure 8; right panel). Under both A2 and B1 scenarios, we found a levelling off of NPP when temperature reaches ~15°C (i.e. a 1.5°C rise compared to the contemporary period (2000–2009) global mean temperature). Interestingly, NPP would decline below the contemporary level after the 2060s when temperature exceeds approximately 16.5°C under the A2 scenario (Figure 8; left panel). Global terrestrial NPP, however, shows a complex temporal response to changes in precipitation during the 21st century. NPP increases with increasing precipitation through the 21st century under the B1 scenario (P-value<0.01) while under the A2 scenario, it increases until the 2060s (P-value<0.2).

3.2.2. Temporal response of terrestrial NPP to climate change and increasing CO_2 concentration. Climate change coupled with increasing atmospheric CO_2 concentration would result in higher global terrestrial NPP than climate change alone under both the A2 and B1 scenarios (Figure 7A). The largest increase in terrestrial NPP would occur under the A2 scenario with an overall increase in the 2060s by 7.73 PgC (14.2%) compared to the 2000s; however, NPP would show a declining trend after the 2060s with a net decrease of 0.12 PgC (compared to the 2060s) in the 2090s. The B1 scenario shows an overall increasing trend

through the 21st century with the largest increase in terrestrial NPP by 6.55 PgC (12.0%) (Table 1).

3.3 Spatial variation of terrestrial NPP induced by climate change and increasing atmospheric CO_2 during 2010–2099

3.3.1. Latitudinal and spatial responses of terrestrial NPP to climate change. Our results show substantial variation in climate-induced NPP change along latitudes during the 21st century (Figure 7B–D; Table 1). The magnitude of terrestrial NPP would be highest in low latitude regions (30.1–42.1 PgC yr^{-1}), and lowest in high latitude regions (2.23–3.85 PgC yr^{-1}). Compared to the 2000 s, the A2 climate would increase terrestrial NPP in mid- and high-latitudes by 4.6% and 34.8%, respectively, but would decrease terrestrial NPP in low latitude by 11.3% in the 2090s. The B1 climate scenario shows similar temporal pattern for mid- and high-latitide regions with an increase in terrestrial NPP by 1.6% and 13.6%, respectively. The low latitude regions, however, show a declining trend with an increase in terrestrial NPP by 2.0% in the 2060s and 1.7% in the 2090s when compared to the 2000s. Under the A2 climate-only scenario, terrestrial NPP would increase in high latitude region by 0.82 PgC while decrease in

Table 1. Decadal changes in global terrestrial net primary production (NPP) and across low-, mid-, and high-latitude regions under A2 and B1 scenario for climate and climate plus CO_2 experiments.

Global NPP (PgC yr⁻¹)

2000s	54.57			
	A2 (Climate Only)	A2 (Climate + CO_2)	B1(Climate Only)	B1 (Climate + CO_2)
2030s	56.20(3.0%)	59.13(8.4%)	55.66(2.0%)	57.68(5.7%)
2060s	55.17(1.1%)	62.30(14.2%)	56.05(2.7%)	60.27(10.5%)
2090s	52.06(−4.6%)	62.18(13.9%)	55.77(2.2%)	61.12(12.0%)

Low Latitude NPP (PgC yr⁻¹)

2000s	36.09			
	A2 (Climate Only)	A2 (Climate + CO_2)	B1(Climate Only)	B1 (Climate + CO_2)
2030s	37.24(3.2%)	39.32(8.9%)	37.10(2.8%)	38.51(6.7%)
2060s	35.90(−0.5%)	40.96(13.5%)	36.83(2.0%)	39.83(10.4%)
2090s	32.02(−11.3%)	38.92(7.8%)	36.70(1.7%)	40.57(12.4%)

Middle Latitude NPP (PgC yr⁻¹)

2000s	16.09			
	A2 (Climate Only)	A2 (Climate + CO_2)	B1(Climate Only)	B1 (Climate + CO_2)
2030s	16.32(1.4%)	17.08(6.2%)	15.94(−0.9%)	16.49(2.5%)
2060s	16.43(2.1%)	18.28(13.6%)	16.53(2.7%)	17.62(9.5%)
2090s	16.83(4.6%)	19.64(22.0%)	16.36(1.6%)	17.69(9.9%)

High Latitude NPP (PgC yr⁻¹)

2000s	2.39			
	A2 (Climate Only)	A2 (Climate + CO_2)	B1(Climate Only)	B1 (Climate + CO_2)
2030s	2.65(10.9%)	2.73(14.5%)	2.62(9.7%)	2.68(12.4%)
2060s	2.84(19.1%)	3.06(28.1%)	2.70(13.0%)	2.82(18.2%)
2090s	3.21(34.8%)	3.62(51.9%)	2.71(13.6%)	2.86(19.9%)

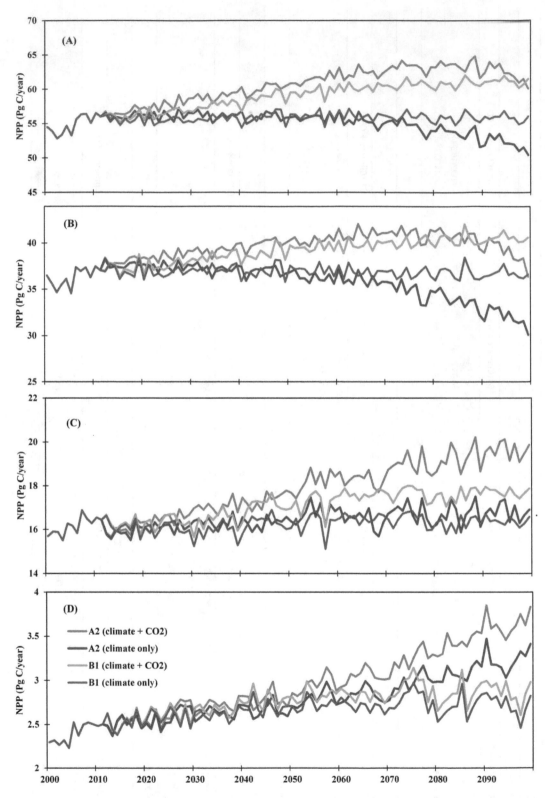

Figure 7. Temporal pattern of change in terrestrial NPP: Global (A), low-latitude (B), mid-latitude (C) and high-latitude (D) as a function of climate and increasing atmospheric CO$_2$ under A2 and B1.

low latitude terrestrial NPP by 4.07 PgC during the 2090s compared to the 2000s.

Our climate-only experiment shows an increase in terrestrial NPP by 50–100 gC m^{-2} yr^{-1} in boreal and arctic regions

(Figure 9) under both the A2 and B1 scenarios. However, the largest decline in terrestrial NPP would occur in tropical regions (>250 gC m^{-2} yr^{-1}) such as parts of South America, Africa, South Asia and Australia under the A2 scenario due to rapidly

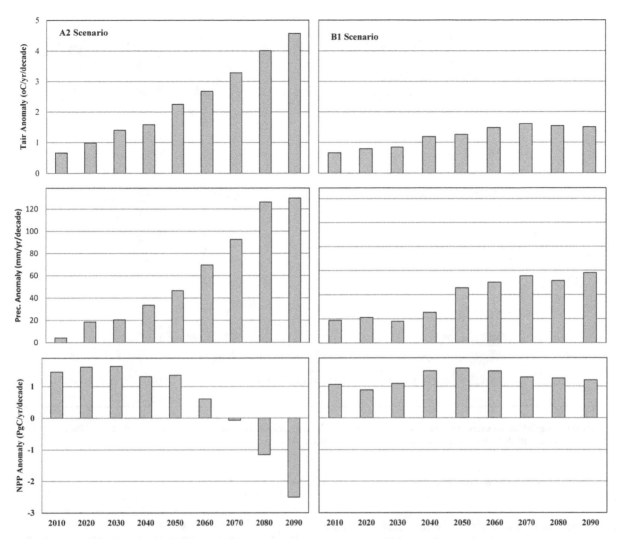

Figure 8. Effect of temperature and precipitation on global net primary productivity during the rest of the 21ˢᵗ century (2010–2099) under A2 (left panel) and B1 (right panel) climate change scenarios.

warming temperature and likely drought events. The climate-only experiment under the B1 scenario, however, shows no substantial decrease in terrestrial NPP in tropical regions. Our results further show that in a particular year, drought or rainfall deficits combined with increasing temperature can potentially reduce terrestrial NPP. Compared to the average precipitation of the 21ˢᵗ century, terrestrial ecosystems experienced the largest reduction in global precipitation (Figure 10) of about 68.97 and 43.07 mm yr^{-1} in 2019 and 2020 under the A2 and B1 scenario, respectively. This annual precipitation deficit resulted in a reduction in global terrestrial NPP by 375 gC m^{-2} in the tropics and low latitude regions (Figure 10).

3.3.2. Latitudinal and spatial responses of terrestrial NPP to climate change and increasing CO_2 concentration. Climate plus CO_2 experiment under both the A2 and B1 scenarios show substantial difference in the magnitude and temporal pattern of terrestrial NPP along latitudes (Figure 7B–D). The A2 climate plus CO_2 experiment shows a substantial increase in terrestrial NPP in low latitude regions until the 2060s where NPP would increase by 4.87 PgC which is equivalent to an increase of 13.5%. However, terrestrial NPP would start to decline in the 2090s (Table 1). In mid- and high-latitude regions,

terrestrial NPP would continue to increase through the 21ˢᵗ century, with an increase of about 22.0% and 51.9%, respectively by the end of the century (2090s vs. 2000s). The climate plus CO_2 experiment under the B1 scenario shows an increasing trend in terrestrial NPP across all latitudes where NPP would increase by 12.4%, 9.9%, and 19.9% in low-, mid-, and high-latitude regions, respectively. During the 2090 s, the largest magnitude (4.48 PgC) of increase in terrestrial NPP would occur in low latitude regions under the B1 scenario, while the largest rate (51.9%) of increase would occur in high latitude regions, under the A2 scenario.

Large increase in terrestrial NPP would occur in tropical regions especially in central and southern Africa, and the Amazon basin under the A2 scenario, while modest increase under the B1 scenario (Figure 9B, 9D). The A2 scenario shows large increase in terrestrial NPP by >200 gC m^{-2} yr^{-1}, while the B1 scenario shows an increase of 100–200 gC m^{-2} yr^{-1} in tropical regions. Large increase in NPP in the tropical regions due to increasing atmospheric CO_2 concentration is primarily because DLEM uses a Farquhar model that shows a higher NPP enhancement at high temperatures under elevated CO_2. However, the mid- and high-latitude regions show lower NPP enhancement in response to

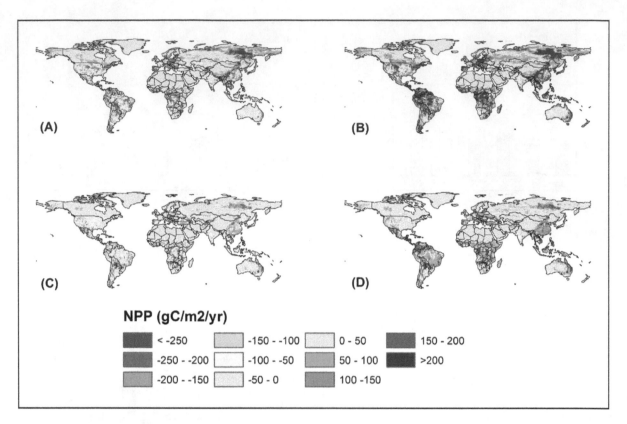

Figure 9. Spatial variation in terrestrial NPP as influenced by climate-only and climate with CO₂. Climate only (A) and climate with CO₂ (B) under A2 scenario, and climate only (C) and climate plus CO₂ (D) under B1 scenario.

increasing CO_2 concentration possibly because of nitrogen limitations.

3.4 Biome variation of terrestrial NPP response to climate change and increasing CO_2 during 2010–2099

3.4.1. Biome NPP response to climate change. Our results show large variation in the response of various biomes to future climate change (Table 2). The A2 climate scenario would result in a NPP decrease for major biomes such as tropical broadleaf deciduous and tropical evergreen forest, evergreen shrubs, grasses, woody wetland and croplands in the 2090s. The largest percent decline would occur in tropical broadleaf evergreen forest where NPP would decrease by 104.0 gC m^{-2}, equivalent to a decrease of 9.3% compared to the contemporary NPP. The largest percent increase under the A2 scenario would occur in tundra and boreal needleleaf deciduous forest by 41.5 gC m^{-2} and 176 gC m^{-2} equivalent to an increase of 59.1% and 54.3%; respectively. The B1 climate scenario, however, shows no substantial change in terrestrial NPP across different biomes compared to the A2 scenario. We found the largest percent decline in NPP in woody wetland by 2.4%, and the largest percent increase in NPP in boreal needleleaf deciduous forest by 16.6% in the climate-only simulation under the B1 scenario. Croplands, in particular, show a decline in NPP due to climate variability under both the A2 and B1 scenarios with the largest decline of 9.9% under the A2 scenario.

3.4.2 Biome NPP response to climate change and increasing CO_2 concentration during 2010–2099. Climate change and increasing atmospheric CO_2 concentration would result in an overall increase in terrestrial NPP across all biomes through the 21st century under both the A2 and B1 scenarios (Table 2). The strength of CO_2 fertilization effect would be largest in tundra and boreal needleleaf deciduous forest under the A2 scenario where NPP would increase by 55.1 gC m^{-2} and 257.6 gC m^{-2} equivalent to an increase of 78.4% and 79.5%; respectively. Simulations with climate change and increasing atmospheric CO_2 would result in an increase in tropical broadleaf evergreen forest NPP by 129.6 gC m^{-2} (which decreases with climate alone), equivalent to a percent increase of 11.5%, indicating that CO_2 ferilization could have a substantial effect on terrestrial NPP. The B1 scenario, however, shows a relatively small increase in terrestrial NPP (7.5–22.8%) across all biomes. Interestingly, croplands show the largest increase in NPP by 45.9 gC m^{-2} (9.4%) under B1 scenario when climate change and increasing atmospheric CO_2 concentration are included in an experiment. This increase is higher than that under the A2 scenario when climate change is coupled with increasing atmospheric CO_2 concentration.

3.5 The CO_2 fertilization effects on terrestrial NPP

Our results show that terrestrial NPP during the 2090s would increase due to increasing atmospheric CO_2 across the globe, but the magnitude of increase varied substantially across different latitudes (Figure 11A–B). The largest increase in terrestrial NPP due to increasing atmospheric CO_2 would be prevalent in the tropical regions where NPP would increase by >225 gC m^{-2} under the A2 scenario and 75–100 gC m^{-2} under the B1 scenario. A closer look at the effect of atmospheric CO_2 elevation under the A2 scenario shows large increase in global NPP of about 10.1 PgC, equivalent to an increase of 18.5% during the 2090s compared to the 2000s.

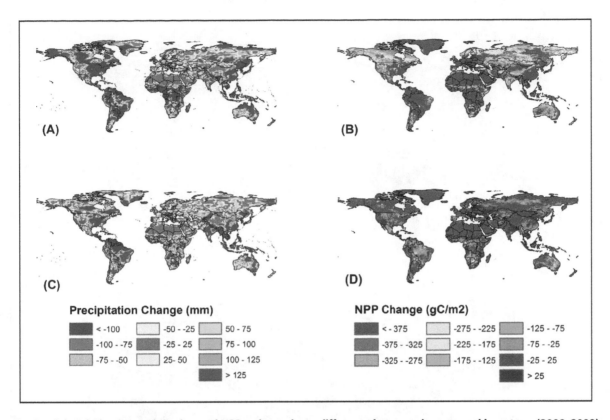

Figure 10. Spatial variation in precipitation and NPP estimated as a difference between dry year and long term (2000–2099) mean: precipitation difference between 2019 and long term mean (A) for A2 scenario, NPP difference between 2019 and long term mean (B) for A2 scenario climate-only simulation, precipitation difference between 2020 and long term mean (C) for B1 scenario and NPP difference between 2020 and long term mean (D) for B1 climate-only simulation.

Table 2. Decadal mean of terrestrial net primary production (NPP) in the contemporary period (2000–2009) and change in NPP between the 2090s and the 2000s among major biomes.

	Decadal Mean	A2		B1	
	2000s	Net Change (% change)			
PFTs	(gC m^{-2} yr^{-1})	climate only	Climate + CO$_2$	climate only	Climate + CO$_2$
Tundra	70.27	41.5 (59.1)	55.1 (78.4)	8.2 (11.7)	13.0 (18.5)
BNEF	258.64	2.4 (0.9)	34.3 (13.3)	4.1 (1.6)	19.3 (7.5)
BNDF	324.14	176.0 (54.3)	257.6 (79.5)	53.8 (16.6)	74.0 (22.8)
TBDF	551.64	48.7 (8.8)	116.2 (21.1)	20.6 (3.7)	55.0 (10.0)
TBEF	867.63	18.2 (2.1)	187.2 (21.6)	15.8 (1.8)	78.8 (9.1)
TNEF	418.12	25.6 (6.1)	87.3 (20.9)	8.9 (2.1)	36.1 (8.6)
TrBDF	753.62	−76.1 (−10.1)	56.1 (7.4)	35.4 (4.7)	105.1(13.9)
TrBEF	1122.43	−104.0 (−9.3)	129.6 (11.5)	−1.6 (−0.1)	106.1 (9.4)
Deciduous Shrub	245.39	38.8 (15.8)	85.6 (34.9)	20.7 (8.4)	40.1 (16.3)
Evergreen Shrub	369.19	−55.8 (−15.1)	22.2 (6.0)	26.1 (7.1)	77.0 (20.9)
C3 grass	284.8	−0.6 (−0.2)	65.0 (22.8)	9.9 (3.5)	40.2 (14.1)
Herbaceous wetland	261.8	36.3 (13.9)	79.5 (30.4)	14.1 (5.4)	33.6 (12.8)
Woody wetland	635.90	−86.2 (−13.5)	29.5 (4.6)	−15.8 (−2.4)	52.3 (8.2)
Cropland	487.02	−48.2 (−9.9)	21.6 (4.4)	−4.3 (−0.9)	45.9 (9.4)

Note: BNEF, Boreal Needleleaf Evergreen Forest; BNDF, Boreal Needleleaf Deciduous Forest; TBDF, Temperate Broadleaf Deciduous Forest; TBEF, Temperate Broadleaf Evergreen Forest; TNEF, Temperate Needleleaf Evergreen Forest; TrBDF, Tropical Broadleaf Deciduous Forest; TrBEF, Tropical Broadleaf Evergreen Forest. Numbers in parenthesis represents percentage change between the 2090s and the 2000s.

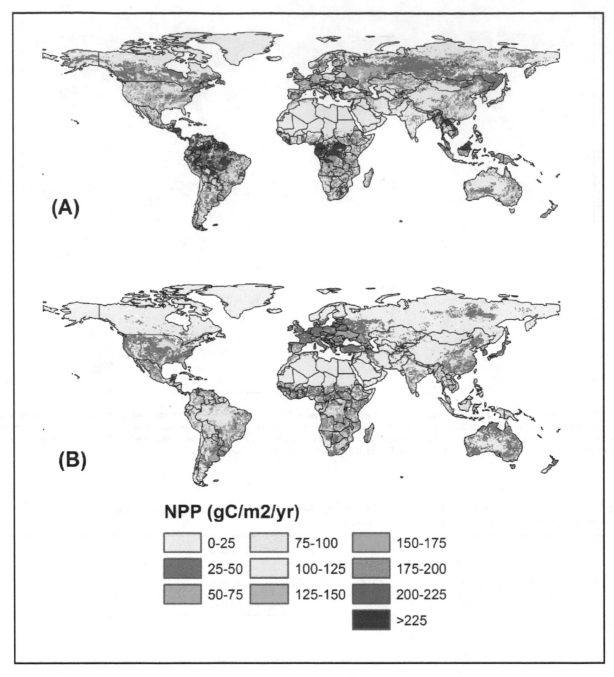

Figure 11. Contribution of increasing atmospheric CO₂ concentration to NPP during the 2090s calculated as a difference between climate plus CO₂ and climate-only experiments: A2 scenario (A) and B1 scenario (B).

We further calculated the 'beta' (β) effect that measures the strength of changes in terrestrial NPP in response to increasing atmospheric CO_2 concentration to examine the relative contribution of direct CO_2 fertilization on terrestrial NPP (Figure 12). By the end of 21st century, large increase in global terrestrial NPP would occur under the A2 scenario, while a small increase would occur under the B1 scenario. At the global scale, the effect of direct CO_2 fertilization was 91.4 mgC m^{-2}/CO_2 (ppm) under the A2 scenario, implying that about 91.4 mgC m^{-2} would be fixed by plants as NPP by using 1 ppm of atmospheric CO_2 by the 2090s. The effect of CO_2 fertilization, however, would start to decline after the 2080 s, with a net reduction in stimulative effect by 8%.

Across different latitudes, the effect of CO_2 fertilization is highest in low latitudes (127.87 mgC m^{-2}/CO_2 (ppm)), and lowest in high latitude (15.52 mgC m^{-2}/CO_2 (ppm)). Although a higher effect of CO_2 feritlization is found under the A2 scenario in low latitude regions, this effect would decline after the 2070s with a net reduction in strength by 12.63 mgC m^{-2} during the 2090s compared to the 2070s. Interestingly, the strength of CO_2 fertilization would increase continuously under the B1 scenario in low latitude regions through the 21st century. The strength of CO_2 fertilization is lower in mid- and high-latitude regions compared to low latitude regions where NPP shows a continuous increase in response to CO_2 fertilization through the 21st century.

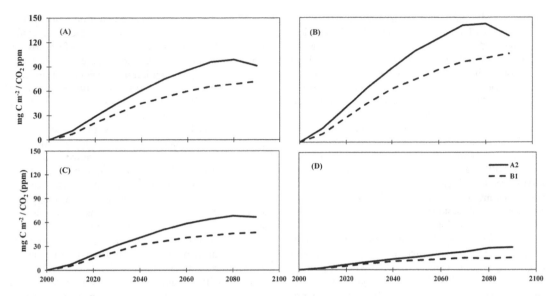

Figure 12. The effect of CO_2 ferilization on terrestrial NPP across the globe (A), low-latitude (B), mid-latitude (C), and high-latitude (D) under A2 and B1 scenarios. For each unit of CO_2 (ppm), the A2 scenario show a highest rate of increase in NPP (mgC m^{-2}) compared to B1 scenario.

Discussion

4.1 Comparison of DLEM-simulated NPP with previous estimates

The DLEM-simulated global terrestrial NPP is 54.57 PgC yr^{-1} for the period 2000–2009, which is comparable to MODIS-based estimate of 53.5 PgC yr^{-1} during the 2000s [16], and also falls in the range of 44–66 PgC yr^{-1} as estimated by 17 global terrestrial biosphere models [29]. The large variation in global NPP estimates among these models are primarily caused by different model representations of nutrient and water constraints on NPP in various terrestrial ecosystems. For instance, both field- and model-based studies indicated that carbon and nitrogen are closely interacted in many terrestrial ecosystems [60]. Some of these models do not simulate carbon-nitrogen interaction, which generally result in an overestimate of global NPP. In addition, moisture in root zone is critically important for plants especially in areas under frequent water stress [65,66]. Remote sensing algorithms, uses a nonlinear function of increasing temperature to estimate production responses to water limitation, which is not capable of accurately accounting for environmental stresses such as rooting depths. Such function can enhance heat stress greatly in low latitude regions, introducing large uncertainties in NPP estimates.

Our results indicate that high latitude biomes have the greatest potential to increase NPP in response to future climate and CO_2 changes. Mainly due to climate change effect, the boreal needleleaf deciduous forest and tundra would have the largest NPP increase by 79.5% and 78.4% in the 21st century, respectively (Table 2). Hill and Henry [67] found that aboveground biomass of tundra sedge community increased on average by 158% in 2005 compared to the 1980s primarily due to climate warming. Our climate dataset indicates boreal forests and tundra ecosystems would experience 3–6°C temperature increase in the 21st century (Figure 4). In contrast, the model predicted a decline in NPP in tropical forests in the climate-only experiment under the A2 scenario that predicted rapid temperature increase in the 21st century. This prediction is supported by Clark et al. [68]'s long-term study in tropical rain forest, which suggests that temperature increase in tropical regions would suppress forests NPP.

4.2 Climate change effects on global NPP in the 21st century

Our climate-only scenario simulations indicate that temperature rise in the 21st century would first promote and then reduce global terrestrial NPP. We found that under the A2 scenario the turning point of global NPP would occur when the global mean temperature reaches about 16.5°C. As temperature crosses 16.5°C, the terrestrial biosphere would show a net reduction in NPP compared to the contemporary period (Figure 8; left panel). Our results also show a levelling-off of terrestrial NPP when the global mean temperature reaches about 15°C under both scenarios. Compared with the contemporary global temperature of ~13.5°C, our finding justifies the goal of maintaining global warming rate under 2°C by the Copenhagen Accord [69]. However, the temperature value of 15°C should not be treated as a threshold to judge climate effects on ecosystem NPP at regional scale, due to the complex responses of ecosystems to climate change in different regions [70]. Our analysis shows that NPP would increase rapidly in high latitude region (e.g., by 34% under the A2 scenario) while decrease in low latitude region in response to climate warming (Figure 9A). The decline in terrestrial NPP in low latitudes is possibly the result of enhanced autotrophic respiration and increased moisture stress caused by rising temperature [27,29,30]. Warming may cause increasing moisture stress resulting in a decline in net nitrogen mineralization, therefore lead to reduction in NPP [25]. In contrast, temperature increase in mid- and high-latitudes can stimulate NPP by alleviating low temperature constraints to plant growth [10], lengthening growing season [13,71], and enhancing nitrogen mineralization and availability [72].

More extreme precipitation regimes are expected to have a substantial effect on terrestrial NPP during the 21st century, and therefore ecological implications of greater intra-annual variability and extremes in rainfall events have received much attention from the scientific community [73,74]. Our study suggests that precipitation deficit would result in reduction in terrestrial NPP by 375 gC m^{-2} across eastern United States, South America, Africa, Europe and Southeast Asia under the A2 scenario. The B1 scenario, however, shows a reduction of up to 175 gC m^{-2} with no net change in most of the areas across the globe. Such reduction in terrestrial NPP due to decrease in precipitation has been reported worldwide [16,75,76]. Although precipitation is projected to increase during the 21st century under the A2 and B1 scenarios, their spatio-temporal variability is a major factor that determines the magnitude of terrestrial NPP. For instance, we observed a reduction in global precipitation by 8% in 2019 that led to a decline in terrestrial NPP of about 375 gC m^{-2} dominated largely in the tropics. Mohamed et al. [77] found larger NPP decline in tropical regions indicating greater impacts of drought and high sensitivity and weaker physiological adjustment to climate variability particularly precipitation. Such variability in precipitation will induce more decline of NPP in the tropics. In addition, frequent extreme drought may counteract the effects of anticipated warming and growing season extension and reduce terrestrial production [76] in mid- and high-latitude reigons. Increase in drought events is primarily the result of less frequent but more intense precipitation events which would likely decrease NPP in wetter ecosystems but increase NPP in drier ecosystems [73]. Therefore, a deeper understanding of the importance of more extreme precipitation patterns relative to other global change drivers such as increasing atmospheric CO_2 concentration, elevated temperature and atmospheric pollution is critical to project future NPP in a changing global environment.

4.3 CO_2 fertilization effect and its interaction with climate change

While climate change may have negative (-4.6% under the A2 scenario) to small positive (2.2% under the B1 scenario) effect on global terrestrial NPP, our Climate plus CO_2 simulations project notable increase (12% under the B1 scenario to 13.9% under the A2 scenario) in NPP by the end of 21st century. Majority of laboratory studies of tree growth under high CO_2 concentration show enhanced growth rates, on average, with a 60% increase in productivity as a result of doubling of atmospheric CO_2 [18]. This explains the prominent CO_2 fertilization effect found in low latitude, especially under the A2 scenario (Figure 12; Table 1 and 2), which would more than compensate the negative effect (-76.1--104 gC m^{-2} yr^{-1}) of climate change and result in an increase in NPP (132.2–233.6 gC m^{-2} yr^{-1}) in tropical forest during the 21st century. Elevated CO_2 was also found to reduce stomatal conductance and increase water use efficiency in water-limited environments [78]. Our study indicates that CO_2 fertilization effect might help the Mediteranian evergreen shrubland to overcome severe NPP loss (-15.1%) in response to the projected future drought (Figure 4) due to the poleward shift of subtropical dry zones [79], and resulted increase in shrubland NPP (6%) by the end of the 21st century under the A2 scenario (Table 2).

However, field studies also found large uncertainties related to terrestrial ecosystems' responses to CO_2 enrichment. Plant response to elevated atmospheric CO_2 can be modified by increasing temperature [80], soil nutrient deficiency [81], and tropospheric ozone pollution [82]. Temperature rise in the 21st century could enhance mineralization rate of soil nutrients, which would support higher ecosystem productivity. In low latitudes, however, nutrient leaching rate will also increase due to intensive and increasing precipitation. Our analyses indicate that at the beginning of 21st century, relative effect of CO_2 fertilization on terrestrial NPP would increase fast; then as a result of CO_2

saturation and gradually intensified water and nutrient limitations, the effect of CO_2 fertilization would reach its maximum potential in the 2070s, level off and even decline afterwards (Figure 12).

While climate change would result in large increase in terrestrial NPP in mid- and high-latitude regions due to greater growing season extension compared to low-latitude region [83], the effect of increasing CO_2 concentration on plant growth will be stronger in low-latitude region during the 21st century. This is primarily because of much stronger photosynthetic response under elevated CO_2 at high temperatures [84]. The DLEM uses a Farquhar model to examine the response of terrestrial primary production to climate change and increasing atmospheric CO_2 concentration. The Farquhar model may cause a much stronger CO_2 enhancement at high temperature compared to low temperatures, resulting in higher NPP response to increasing CO_2 concentration in low latitude compared to mid- and high-latitude regions. Hickler et al. [85] evaluated the Farquhar photosynthesis model and found that the optimum temperature for primary production shifts to higher temperatures under elevated CO_2. The DLEM simulates a stronger NPP response to elevated CO_2 under drier environments because elevated CO_2 reduces the negative effect of drought on plant growth resulting in higher NPP response in low-latitude region. However, mid- and high-latitude regions are primarily thought to be nitrogen limited. Increase in temperature and precipitation in mid- and high-latitude regions would possibly enhance decomposition of soil organic matter resulting in more nitrogen available for plant uptake. The resultant increase in plant nitrogen uptake with further benefits from elevated CO_2 would result in an increase in NPP in mid- and high-latitude ecosystems such as eastern Europe, eastern Russia, and parts of China (Figure 9B, 9D). Our regional variation in the response of terrestrial NPP to rising CO_2 concentration are similar to Kirschbaum [86] who used a modified Farquhar model to simulate the response of photosynthesis to elevated CO_2.

4.4 Uncertainty and future research needs

Several limitations and uncertainties are inherent in this study regarding input data, model parameterization and simulations. Our goal is to investigate impacts of future climate change and elevated CO_2 concentration on the global terrestrial NPP. We applied one GCMs (CCSM3) climate output together with atmospheric CO_2 data under two emission scenarios (A2 and B1). Discrepancies existing in different global climate models would lead to different estimation of NPP response as projected climate variables change [87]. Further analysis is helpful to explore the uncertainty ranges by adopting climate projections derived from multiple climate models [29]. In addition, terrestrial NPP response to increasing CO_2 concentration in the model is based on calibration and parameterization at several Free-Air Concentration Enrichment (FACE) sites based on Ainsworth and Long [88]. Because current long-running FACE experiments are all located in the temperate zone [89], we have very little empirical information available on the response of NPP to elevated CO_2 in the tropics and high-latitude ecosystems. While parameters were well calibrated based on existing field observations, some processes such as responses of C assimilation/allocation and stomatal conductance to elevated temperature and CO_2 may change due to plant acclimation [90] and dynamic responses of phenology and growing season length [91], which have not been included in the current DLEM simulations. In addition, we only attempted to quantify the terrestrial ecosystem response to seasonal and interannual climatic variability, and increasing atmospheric CO_2 concentration, but did not consider how climate induced functional change in ecosystem processes [92] may affect terrestrial NPP. We should be aware of how other environmental factors (e.g. nitrogen deposition, tropospheric ozone, and LCLU change and land management, fertilization and irrigation of croplands) may work together with climate change and elevated CO_2 to influence terrestrial NPP. Also, disturbance such as timber harvest and cropland abandonment may have a substantial effect on C dynamics at the regional scale [93]. Therefore, future research must take into account additional environmental and human factors such as nitrogen deposition, ozone pollution and land use/land cover change. Furthermore, model representations of ecosystem processes associated with human activities must be improved.

Conclusion and Implications for Climate Change Policy

The DLEM simulations estimate a mean global terrestrial NPP of 54.57 Pg C yr^{-1} in the first decade of the 21st century, resulting from multiple environmental factors including climate, atmospheric CO_2 concentration, nitrogen deposition, tropospheric ozone, and LCLU change. Climate change in the 21st century would either reduce global NPP by 4.6% under the A2 scenario or slightly enhance NPP (2.2%) under the B1 scenario. In response to climate change, global NPP would first increase and then decline after global warming exceeds 1.5°C, which justifies the goal of keeping the global warming rate under 2°C by the Copenhagen Accord (2009). These results also support the Accord Acknowledgement, which states that staying below 2°C may not be sufficient and a review in 2015 will be conducted to assess the need to potentially aim for staying below 1.5°C. Terrestrial NPP at high latitude (60°N–90°N) in the Northern Hemisphere would benefit from climate change, but decline at low latitude (30°S–30°N) and the central USA. The CO_2 fertilization effect would improve global productivity as simulated by the DLEM forced by climate plus CO_2, showing notable increase in NPP by the end of 21st century (12% under the B1 scenario to 13.9% under the A2 scenario). Ecosystem responses to increasing CO_2, however, are complex and involve large uncertainties, especially in the tropics.

The relative CO_2 fertilization effect, i.e. change in NPP on per CO_2 (ppm) bases, was projected to increase quickly, level off in the 2070s, and then even decline by the end of the 2080s, possibly due to CO_2 saturation and nutrient limitation. Compared to the low emission scenario (B1), earth ecosystems would have a less predictable and thus unfavorable future under the high emission scenario (A2) due to large uncertainties related to the global NPP, which are characterized by stronger spatial variation, more complex temporal dynamics, larger differences in biomes' responses to global changes, and more dramatic and counteracting impacts from climate and elevated CO_2.

Our model projections indicate that high emission scenario (A2) will not necessarily negatively affect global terrestrial NPP in the future. However, the A2 scenario describes a future with high ecological uncertainty, the rest of 21st century may experience frequent climate extreme events and increasing ecological risks, and a global ecosystem whose behavior and trend are difficult to predict and therefore can hardly be protected with efficiency. The world under the A2 scenario would likely face more serious challenges in food security and water scarcity given a continually growing world human population. To avoid this scenario, we should shift to the low emission scenario.

Acknowledgments

This study has been supported by NSF Decadal and Regional Climate Prediction using Earth System Models (AGS-1243220), NSF Dynamics of Coupled Natural and Human Systems (1210360), NASA Interdisciplinary Science Program (NNX10AU06G, NNG04GM39C), US Department of Energy NICCR Program (DUKE-UN-07-SC-NICCR-1014).

Author Contributions

Conceived and designed the experiments: HT SP. Performed the experiments: SP JY BT CL WR. Analyzed the data: SP SD JY BT. Contributed reagents/materials/analysis tools: ZO XW KB QY BZ XL. Contributed to the writing of the manuscript: HT SP DS JY BT CL WR ZO XW KB QY BZ XL.

References

1. Vitousek PM, Ehrlich PR, Ehrlich AH, Matson PA (1986) Human appropriation of the products of photosynthesis. Bioscience: 368–373.
2. Costanza R, d'Arge R, De Groot R, Farber S, Grasso M, et al. (1997) The value of the world's ecosystem services and natural capital. Ecological economics 25: 3–15.
3. Running S, Nemani RR, Heinsch FA, Zhao M, Reeves M, et al. (2004) A continuous satellite-derived measure of global terrestrial primary production. Bioscience 54: 547–560. doi:10.1641/0006-3568%282004%29054%5B0547:acsmog%5D2.0.co;2.
4. Melillo J, McGuire AD, Kicklighter DW, Moore B, Vorosmarty CJ, et al. (1993) Global climate change and terrestrial net primary production. Nature 363: 234–240. doi:10.1038/363234a0.
5. Chapin III F, Woodwell G, Randerson JT, Rastetter EB, Lovett G, et al. (2006) Reconciling carbon-cycle concepts, terminology, and methods. Ecosystems 9: 1041–1050. doi:10.1007/s10021-005-0105-7.
6. IPCC (2007) The Physical Science Basis. Contribution of Working Group I to the Fourth Assessment Report of the Intergovernmental Panel on Climate Change. Cambridge University Press, Cambridge, UK.
7. IPCC (2014) The Physical Science Basis: Working Group I Contribution to the IPCC Fifth Assessment Report Cambridge University Press, Cambridge, UK.
8. Snover A, Mauger GS, Whitely Binder LC, Krosby M a, I T (2013) Climate Change Impacts and Adaptation in Washington State: Technical Summaries for Decision Makers. State of Knowledge Report prepared for the Washington State Department of Ecology Climate Impacts Group, University of Washington, Seattle: 130.
9. Pan S, Tian H, Dangal SR, Ouyang Z, Tao B, et al. (2014) Modeling and Monitoring Terrestrial Primary Production in a Changing Global Environment: Toward a Multiscale Synthesis of Observation and Simulation. Advances in Meteorology 2014: 17. doi:10.1155/2014/965936.
10. Nemani RR, Keeling CD, Hashimoto H, Jolly WM, Piper SC, et al. (2003) Climate-driven increases in global terrestrial net primary production from 1982 to 1999. Science 300: 1560–1563. doi:10.1126/science.1082750.
11. Chapin S, McFarland J, David McGuire A, Euskirchen ES, Ruess RW, et al. (2009) The changing global carbon cycle: linking plant–soil carbon dynamics to global consequences. Journal of Ecology 97: 840–850. doi:10.1111/j.1365-2745.2009.01529.x.
12. Friedlingstein P, Bopp L, Ciais P, Dufresne JL, Fairhead L, et al. (2001) Positive feedback between future climate change and the carbon cycle. Geophysical Research Letters 28: 1543–1546. doi:10.1029/2000GL012015.
13. Myneni RB, Keeling C, Tucker C, Asrar G, Nemani R (1997) Increased plant growth in the northern high latitudes from 1981 to 1991. Nature 386: 698–702. doi:10.1038/386698a0.
14. Leakey AD, Ainsworth EA, Bernacchi CJ, Rogers A, Long SP, et al. (2009) Elevated CO_2 effects on plant carbon, nitrogen, and water relations: six important lessons from FACE. Journal of Experimental Botany 60: 2859–2876. doi:10.1093/jxb/erp096.
15. Norby RJ, DeLucia EH, Gielen B, Calfapietra C, Giardina CP, et al. (2005) Forest response to elevated CO_2 is conserved across a broad range of productivity. Proceedings of the National Academy of Sciences of the United States of America 102: 18052–18056. doi:10.1073/pnas.0509478102.
16. Zhao M, Running SW (2010) Drought-induced reduction in global terrestrial net primary production from 2000 through 2009. Science 329: 940–943. doi:10.1126/science.1192666.
17. Potter CS, Klooster S, Genovese V (2012) Net primary production of terrestrial ecosystems from 2000 to 2009. Climatic Change 115: 365–378. doi:10.1007/s10584-012-0460-2.
18. Cao M, Woodward FI (1998) Dynamic responses of terrestrial ecosystem carbon cycling to global climate change. Nature 393: 249–252. doi:10.1038/30460.
19. Black T, Chen W, Barr A, Arain M, Chen Z, et al. (2000) Increased carbon sequestration by a boreal deciduous forest in years with a warm spring. Geophysical Research Letters 27: 1271–1274. doi:10.1029/1999GL011234.
20. Hyvönen R, Ågren GI, Linder S, Persson T, Cotrufo MF, et al. (2007) The likely impact of elevated [CO_2], nitrogen deposition, increased temperature and management on carbon sequestration in temperate and boreal forest ecosystems:

a literature review. New Phytologist 173: 463–480. doi:10.1111/j.1469-8137. 2007.01967.x.

21. Feng X (1999) Trends in intrinsic water-use efficiency of natural trees for the past 100–200 years: a response to atmospheric CO_2 concentration. Geochimica et Cosmochimica Acta 63: 1891–1903. doi:10.1016/S0016-7037(99)00088-5.

22. Tian, Chen G, Liu M, Zhang C, Sun G, et al. (2010) Model estimates of net primary productivity, evapotranspiration, and water use efficiency in the terrestrial ecosystems of the southern United States during 1895–2007. Forest ecology and management 259: 1311–1327. doi:10.1016/j.foreco.2009.10.009.

23. Bonan GB (2008) Forests and climate change: forcings, feedbacks, and the climate benefits of forests. Science 320: 1444–1449. doi:10.1126/science. 1155121.

24. Ge Z-M, Kellomäki S, Peltola H, Zhou X, Väisänen H, et al. (2013) Impacts of climate change on primary production and carbon sequestration of boreal Norway spruce forests: Finland as a model. Climatic Change: 1–15. doi:10.1007/s10584-012-0607-1.

25. Felzer BS, Cronin TW, Melillo JM, Kicklighter DW, Schlosser CA, et al. (2011) Nitrogen effect on carbon-water coupling in forests, grasslands, and shrublands in the arid western United States. Journal of Geophysical Research: Biogeosciences (2005–2012) 116. doi:10.1029/2010JG001621.

26. Tian H, Melillo J, Kicklighter D, McGuire A, Helfrich Iii J, et al. (2000) Climatic and biotic controls on annual carbon storage in Amazonian ecosystems. Global Ecology and Biogeography 9: 315–335.

27. Sitch S, Huntingford C, Gedney N, Levy P, Lomas M, et al. (2008) Evaluation of the terrestrial carbon cycle, future plant geography and climate-carbon cycle feedbacks using five Dynamic Global Vegetation Models (DGVMs). Global Change Biology 14: 2015–2039. doi:10.1111/j.1365-2486.2008.01626.x.

28. Sitch S, Smith B, Prentice IC, Arneth A, Bondeau A, et al. (2003) Evaluation of ecosystem dynamics, plant geography and terrestrial carbon cycling in the LPJ dynamic global vegetation model. Global Change Biology 9: 161–185. doi:10.1046/j.1365-2486.2003.00569.x.

29. Cramer, Kicklighter D, Bondeau A, Iii BM, Churkina G, et al. (1999) Comparing global models of terrestrial net primary productivity (NPP): overview and key results. Global change biology 5: 1–15. doi:10.1046/j.1365-2486. 1999.00009.x.

30. Friedlingstein P, Cox P, Betts R, Bopp L, Von Bloh W, et al. (2006) Climate-carbon cycle feedback analysis: Results from the (CMIP)-M-4 model intercomparison. Journal of Climate 19: 3337–3353. doi:10.1175/Jcli3800.1.

31. Tian H, Xu X, Lu C, Liu M, Ren W, et al. (2011) Net exchanges of CO_2, CH_4, and N_2O between China's terrestrial ecosystems and the atmosphere and their contributions to global climate warming. Journal of Geophysical Research: Biogeosciences (2005–2012) 116. doi:10.1029/2010JG001393.

32. Lu C, Tian H (2013) Net greenhouse gas balance in response to nitrogen enrichment: perspectives from a coupled biogeochemical model. Global Change Biology 19: 571–588. doi:10.1111/gcb.12049.

33. Ren W, Tian H, Tao B, Huang Y, Pan S (2012) China's crop productivity and soil carbon storage as influenced by multifactor global change. Global Change Biology 18: 2945–2957. doi:10.1111/j.1365-2486.2012.02741.x.

34. Tian H, Chen GS, Zhang C, Liu ML, Sun G, et al. (2012) Century-Scale Responses of Ecosystem Carbon Storage and Flux to Multiple Environmental Changes in the Southern United States. Ecosystems 15: 674–694. doi:10.1007/s10021-012-9539-x.

35. Liu M, Tian H, Yang Q, Yang J, Song X, et al. (2013) Long-term trends in evapotranspiration and runoff over the drainage basins of the Gulf of Mexico during 1901–2008. Water Resources Research 49. doi:10.1002/wrcr.20180.

36. Farquhar GD, Caemmerer SV, Berry JA (1980) A Biochemical-Model of Photosynthetic Co_2 Assimilation in Leaves of C-3 Species. Planta 149: 78–90. doi:10.1007/Bf00386231.

37. Collatz GJ, Ball JT, Grivet C, Berry JA (1991) Physiological and environmental regulation of stomatal conductance, photosynthesis and transpiration: a model that includes a laminar boundary layer. Agricultural and Forest Meteorology 54: 107–136. doi:10.1016/0168-1923(91)90002-8.

38. Bonan GB (1996) Land surface model (LSM version 1.0) for ecological, hydrological, and atmospheric studies: Technical description and users guide. Technical note. National Center for Atmospheric Research, Boulder, CO (United States). Climate and Global Dynamics Div.

39. Sellers P, Randall D, Collatz G, Berry J, Field C, et al. (1996) A revised land surface parameterization (SiB2) for atmospheric GCMs. Part I: Model formulation. Journal of Climate 9: 676–705. doi:10.1175/1520-0442 (1996)009<0676:ARLSPF>2.0.CO;2.

40. Saxton K, Rawls W (2006) Soil water characteristic estimates by texture and organic matter for hydrologic solutions. Soil Science Society of America Journal 70: 1569–1578. doi:10.2136/sssaj2005.0117.

41. Ryan MG (1991) Effects of climate change on plant respiration. Ecological Applications 1: 157–167.

42. Klein Goldewijk K, Beusen A, Van Drecht G, De Vos M (2011) The HYDE 3.1 spatially explicit database of human-induced global land-use change over the past 12,000 years. Global Ecology and Biogeography 20: 73–86. doi:10.1111/j.1466-8238.2010.00587.x.

43. DeFries R (2008) Terrestrial Vegetation in the Coupled Human-Earth System: Contributions of Remote Sensing. Annual Review of Environment and Resources 33: 369–390. doi:10.1146/annurev.environ.33.020107.113339.

44. Lehner B, Döll P (2004) Development and validation of a global database of lakes, reservoirs and wetlands. Journal of Hydrology 296: 1–22. doi:10.1016/j.jhydrol.2004.03.028.

45. Wei Y, Liu S, Huntzinger D, Michalak A, Viovy N, et al. (2013) The North American Carbon Program Multi-scale Synthesis and Terrestrial Model Intercomparison Project–Part 2: Environmental driver data. Geoscientific Model Development Discussions 6: 5375–5422. doi:10.5194/gmdd-6-5375-2013.

46. Dentener F (2006) Global maps of atmospheric nitrogen deposition, 1860, 1993, and 2050. Data set Available on-line (http://wwwdaacornlgov) from Oak Ridge National Laboratory Distributed Active Archive Center, Oak Ridge, TN, USA.

47. Felzer B, Kicklighter D, Melillo J, Wang C, Zhuang Q, et al. (2004) Effects of ozone on net primary production and carbon sequestration in the conterminous United States using a biogeochemistry model. Tellus B 56: 230–248. doi:10.1007/s10584-005-6776-4.

48. Tian H, Lu C, Chen G, Xu X, Liu M, et al. (2011) Climate and land use controls over terrestrial water use efficiency in monsoon Asia. Ecohydrology 4: 322–340. doi:10.1002/eco.216.

49. Leff B, Ramankutty N, Foley JA (2004) Geographic distribution of major crops across the world. Global Biogeochemical Cycles 18. doi:10.1029/2003 GB002108.

50. Siebert S, Döll P, Feick S, Hoogeveen J, Frenken K (2007) Global map of irrigation areas version 4.0. 1. Johann Wolfgang Goethe University, Frankfurt am Main, Germany/Food and Agriculture Organization of the United Nations, Rome, Italy.

51. Meehl GA, Covey C, Taylor KE, Delworth T, Stouffer RJ, et al. (2007) The WCRP CMIP3 multimodel dataset: A new era in climate change research. Bulletin of the American Meteorological Society 88: 1383–1394. doi:10.1175/BAMS-88-9-1383.

52. Maurer E, Adam J, Wood A (2008) Climate model based consensus on the hydrologic impacts of climate change to the Rio Lempa basin of Central America. Hydrology and Earth System Sciences Discussions 5: 3099–3128. doi:10.5194/hessd-5-3099-2008.

53. Wood AW, Leung LR, Sridhar V, Lettenmaier D (2004) Hydrologic implications of dynamical and statistical approaches to downscaling climate model outputs. Climatic Change 62: 189–216. doi:10.1023/b:clim. 0000013685.99609.9e.

54. Adam JC, Lettenmaier DP (2003) Adjustment of global gridded precipitation for systematic bias. Journal of Geophysical Research: Atmospheres (1984–2012) 108. doi:10.1029/2002jd002499.

55. Dore MH (2005) Climate change and changes in global precipitation patterns: What do we know? Environment international 31: 1167–1181. doi:10.1016/j.envint.2005.03.004.

56. Ren W, Tian H, Liu M, Zhang C, Chen G, et al. (2007) Effects of tropospheric ozone pollution on net primary productivity and carbon storage in terrestrial ecosystems of China. Journal of Geophysical Research: Atmospheres (1984–2012) 112. doi:10.1029/2007jd008521.

57. Ren W, Tian H, Tao B, Chappelka A, Sun G, et al. (2011) Impacts of tropospheric ozone and climate change on net primary productivity and net carbon exchange of China's forest ecosystems. Global Ecology and Biogeography 20: 391–406. doi:10.1111/j.1466-8238.2010.00606.x.

58. Tian H, Melillo JM, Lu C, Kicklighter D, Liu M, et al. (2011) China's terrestrial carbon balance: Contributions from multiple global change factors. Global Biogeochemical Cycles 25. doi:10.1029/2010GB003838.

59. Tian H, Xu X, Zhang C, Ren W, Chen G, et al. (2009) Forecasting and assessing the large-scale and long-term impacts of global environmental change on terrestrial ecosystems in the United States and China. Real World Ecology: Springer. 235–266. doi:10.1007/978-0-387-77942-3_9.

60. Lu C, Tian H, Liu M, Ren W, Xu X, et al. (2012) Effect of nitrogen deposition on China's terrestrial carbon uptake in the context of multifactor environmental changes. Ecological Applications 22: 53–75. doi:10.1890/10-1685.1.

61. Song X, Tian H, Xu X, Hui D, Chen G, et al. (2013) Projecting terrestrial carbon sequestration of the southeastern United States in the 21st century. Ecosphere 4: art88. doi:10.1890/es12-00398.1.

62. Tian, Xu X, Liu M, Ren W, Zhang C, et al. (2010) Spatial and temporal patterns of CH_4 and N_2O fluxes in terrestrial ecosystems of North America during 1979–2008: application of a global biogeochemistry model. Biogeosciences 7: 2673–2694. doi:10.5194/bg-7-2673-2010.

63. Xu X, Tian H, Zhang C, Liu M, Ren W, et al. (2010) Attribution of spatial and temporal variations in terrestrial methane flux over North America. Biogeosciences Discussions 7: 5383–5428. doi:10.5194/bgd-7-5383-2010.

64. Kaufman DS, Schneider DP, McKay NP, Ammann CM, Bradley RS, et al. (2009) Recent warming reverses long-term Arctic cooling. Science 325: 1236–1239. doi:10.1126/science.1173983.

65. Irvine J, Law B, Kurpius M, Anthoni P, Moore D, et al. (2004) Age-related changes in ecosystem structure and function and effects on water and carbon exchange in ponderosa pine. Tree Physiology 24: 753–763. doi:10.1093/treephys/24.7.753.

66. Zhang X, Friedl MA, Schaaf CB (2006) Global vegetation phenology from Moderate Resolution Imaging Spectroradiometer (MODIS): Evaluation of global patterns and comparison with in situ measurements. Journal of Geophysical Research: Biogeosciences (2005–2012) 111. doi:10.1029/2006 jg000217.

67. Hill GB, Henry GH (2011) Responses of High Arctic wet sedge tundra to climate warming since 1980. Global Change Biology 17: 276–287. doi:10.1111/j.1365-2486.2010.02244.x.

68. Clark DA, Piper S, Keeling C, Clark D (2003) Tropical rain forest tree growth and atmospheric carbon dynamics linked to interannual temperature variation during 1984–2000. Proceedings of the National Academy of Sciences 100: 5852–5857. doi:10.1073/pnas.0935903100.

69. Accord C (2009) U.N. Framework Convention on Climate Change. United Nations.

70. Luo Y, Gerten D, Le Maire G, Parton WJ, Weng E, et al. (2008) Modeled interactive effects of precipitation, temperature, and [CO$_2$] on ecosystem carbon and water dynamics in different climatic zones. Global Change Biology 14: 1986–1999. doi:10.1111/j.1365-2486.2008.01629.x.

71. Piao S, Friedlingstein P, Ciais P, Viovy N, Demarty J (2007) Growing season extension and its impact on terrestrial carbon cycle in the Northern Hemisphere over the past 2 decades. Global Biogeochemical Cycles 21. doi:10.1029/2006gb002888.

72. Tian H, Melillo JM, Kicklighter D, McGuire A, Helfrich J (1999) The sensitivity of terrestrial carbon storage to historical climate variability and atmospheric CO$_2$ in the United States. Tellus Series B-Chemical and Physical Meteorology 51: 414–452. doi:10.1034/j.1600-0889.1999.00021.x.

73. Knapp AK, Beier C, Briske DD, Classen AT, Luo Y, et al. (2008) Consequences of more extreme precipitation regimes for terrestrial ecosystems. Bioscience 58: 811–821. doi:10.1641/b580908.

74. Jentsch A, Kreyling J, Beierkuhnlein C (2007) A new generation of climate-change experiments: events, not trends. Frontiers in Ecology and the Environment 5: 365–374. doi:10.1890/060097.

75. Hanson PJ, Weltzin JF (2000) Drought disturbance from climate change: response of United States forests. Science of the Total Environment 262: 205–220. doi:10.1016/S0048-9697(00)00523-4.

76. Ciais P, Reichstein M, Viovy N, Granier A, Ogée J, et al. (2005) Europe-wide reduction in primary productivity caused by the heat and drought in 2003. Nature 437: 529–533. doi:10.1038/nature03972.

77. Mohamed M, Babiker IS, Chen Z, Ikeda K, Ohta K, et al. (2004) The role of climate variability in the inter-annual variation of terrestrial net primary production (NPP). Science of the Total Environment 332: 123–137. doi:10.1016/j.scitotenv.2004.03.009.

78. Morison J, Gifford R (1984) Plant growth and water use with limited water supply in high CO$_2$ concentrations. I. Leaf area, water use and transpiration. Functional Plant Biology 11: 361–374. doi:10.1071/pp9840361.

79. Seidel DJ, Fu Q, Randel WJ, Reichler TJ (2007) Widening of the tropical belt in a changing climate. Nature geoscience 1: 21–24. doi:10.1038/ngeo.2007.38.

80. Boisvenue C, Running SW (2006) Impacts of climate change on natural forest productivity–evidence since the middle of the 20th century. Global Change Biology 12: 862–882. doi:10.1111/j.1365-2486.2006.01134.x.

81. Shaw MR, Zavaleta ES, Chiariello NR, Cleland EE, Mooney HA, et al. (2002) Grassland responses to global environmental changes suppressed by elevated CO$_2$. Science 298: 1987–1990. doi:10.1126/science.1075312.

82. Ollinger SV, Aber JD, Reich PB (1997) Simulating ozone effects on forest productivity: interactions among leaf-, canopy-, and stand-level processes. Ecological Applications 7: 1237–1251. doi:10.1890/1051-0761(1997)007[1237:SOEOFP]2.0.CO;2.

83. Morales P, Hickler T, Rowell DP, Smith B, Sykes MT (2007) Changes in European ecosystem productivity and carbon balance driven by regional climate model output. Global Change Biology 13: 108–122. doi:10.1111/j.1365-2486.2006.01289.x.

84. Long S (1991) Modification of the response of photosynthetic productivity to rising temperature by atmospheric CO$_2$ concentrations: has its importance been underestimated? Plant, Cell & Environment 14: 729–739. doi:10.1111/j.1365-3040.1991.tb01439.x.

85. Hickler T, Smith B, Prentice IC, Mjöfors K, Miller P, et al. (2008) CO$_2$ fertilization in temperate FACE experiments not representative of boreal and tropical forests. Global Change Biology 14: 1531–1542. doi:10.1111/j.1365-2486.2008.01598.x.

86. Kirschbaum M (1994) The sensitivity of C3 photosynthesis to increasing CO$_2$ concentration: a theoretical analysis of its dependence on temperature and background CO$_2$ concentration. Plant, Cell & Environment 17: 747–754. doi:10.1111/j.1365-3040.1994.tb00167.x.

87. Joshi M, Shine K, Ponater M, Stuber N, Sausen R, et al. (2003) A comparison of climate response to different radiative forcings in three general circulation models: towards an improved metric of climate change. Climate Dynamics 20: 843–854. doi:10.1007/s00382-003-0305-9.

88. Ainsworth EA, Long SP (2005) What have we learned from 15 years of free-air CO$_2$ enrichment (FACE)? A meta-analytic review of the responses of photosynthesis, canopy properties and plant production to rising CO$_2$. New Phytologist 165: 351–372. doi:10.1111/j.1469-8137.2004.01224.x.

89. Long SP, Ainsworth EA, Leakey AD, Nösberger J, Ort DR (2006) Food for thought: lower-than-expected crop yield stimulation with rising CO$_2$ concentrations. Science 312: 1918–1921. doi:10.1126/science.1114722.

90. Evans J, Poorter H (2001) Photosynthetic acclimation of plants to growth irradiance: the relative importance of specific leaf area and nitrogen partitioning in maximizing carbon gain. Plant, Cell & Environment 24: 755–767. doi:10.1046/j.1365-3040.2001.00724.x.

91. Walther G-R, Post E, Convey P, Menzel A, Parmesan C, et al. (2002) Ecological responses to recent climate change. Nature 416: 389–395. doi:10.1038/416389a.

92. Hui D, Luo Y, Katul G (2003) Partitioning interannual variability in net ecosystem exchange between climatic variability and functional change. Tree physiology 23: 433–442. doi:10.1093/treephys/23.7.433.

93. Dangal SR, Felzer BS, Hurteau MD (2014) Effects of agriculture and timber harvest on carbon sequestration in the eastern US forests. Journal of Geophysical Research: Biogeosciences 119: 35–54. doi:10.1002/2013JG002409.

13

In Situ CO_2 Efflux from Leaf Litter Layer Showed Large Temporal Variation Induced by Rapid Wetting and Drying Cycle

Mioko Ataka[1]*, **Yuji Kominami**[2], **Kenichi Yoshimura**[2], **Takafumi Miyama**[2], **Mayuko Jomura**[3], **Makoto Tani**[1]

1 Laboratory of Forest Hydrology, Division of Environmental Science and Technology, Graduate School of Agriculture, Kyoto University, Kyoto, Japan, **2** Kansai Research Center, Forestry and Forest Products Research Institute (FFPRI), Kyoto, Japan, **3** College of Bioresource Sciences, Nihon University, Fujisawa, Kanagawa, Japan

Abstract

We performed continuous and manual in situ measurements of CO_2 efflux from the leaf litter layer (R_{LL}) and water content of the leaf litter layer (LWC) in conjunction with measurements of soil respiration (R_S) and soil water content (SWC) in a temperate forest; our objectives were to evaluate the response of R_{LL} to rainfall events and to assess temporal variation in its contribution to R_S. We measured R_{LL} in a treatment area from which all potential sources of CO_2 except for the leaf litter layer were removed. Capacitance sensors were used to measure LWC. R_{LL} increased immediately after wetting of the leaf litter layer; peak R_{LL} values were observed during or one day after rainfall events and were up to 8.6-fold larger than R_{LL} prior to rainfall. R_{LL} declined to pre-wetting levels within 2–4 day after rainfall events and corresponded to decreasing LWC, indicating that annual R_{LL} is strongly influenced by precipitation. Temporal variation in the observed contribution of R_{LL} to R_S varied from nearly zero to 51%. Continuous in situ measurements of LWC and CO_2 efflux from leaf litter only, combined with measurements of R_S, can provide robust data to clarify the response of R_{LL} to rainfall events and its contribution to total R_S.

Editor: Ben Bond-Lamberty, DOE Pacific Northwest National Laboratory, United States of America

Funding: Funding was provided by the Japan Society for the Promotion of Science (JSPS; grant number 25–2482 (http://www.jsps.go.jp/english/index.html)) Grant-in-Aid for Scientific Research (B) (20380182 (http://www.jsps.go.jp/english/index.html)). The funders had no role in study design, data collection and analysis, decision to publish, or preparation of the manuscript.

Competing Interests: The authors have declared that no competing interests exist.

* Email: teshimamioko@yahoo.co.jp

Introduction

Efflux of CO_2 from the soil surface (soil respiration; R_S), which is the sum of respiration by autotrophs and heterotrophs, is an important component of total CO_2 efflux from forest ecosystems [1–3]. The R_S: total ecosystem respirations varied from 58% to 76% in a mixed coniferous-deciduous forest [4], depending on interannual and seasonal changes in autotrophic and heterotrophic respiration; variability in R_S can affect the forest carbon balance on daily and seasonal time scales. To explain the cause of variability in R_S, many studies have attempted to separate differing sources of Rs and to examine factors controlling CO_2 efflux rate from each source [5–7]. Especially in forest ecosystems, heterotrophic respiration consists of CO_2 efflux from various sources (e.g., leaf and root litter, woody debris, soil organic matter) and their rates are controlled by their specific environmental condition such as water content (WC) and temperature [8], physical properties of the substrate (e.g., density and structure) [9,10], and chemical properties (e.g., labile and recalcitrant carbon) [11,12]. Moreover, CO_2 efflux from the various heterotrophic sources responds differently to these controlling factors, which illustrates the complexity of R_S. In recent decades, a variety of methods for separating components of heterotrophic respiration

and for determining their contribution to total R_S have been developed [9,13].

Among heterotrophic sources of CO_2, the leaf litter layer (L-layer) is a significant reservoir of degradable carbon and a large potential source of CO_2 efflux from forest soils [14]. In temperate forests, the contribution of CO_2 efflux from the L-layer (leaf litter respiration; R_{LL}) to R_S is reported to range from 23% to 48% [13,15,16]. The L-layer is in direct contact with rainfall, solar radiation, and wind, and environmental conditions (e.g., WC and temperature) can change more dynamically in the L-layer than in lower soil layers. Rapid and transient temporal variation in WC of the L-layer has been observed, especially in warm climates [16,17]. Heterotrophic respiration responds rapidly to changes in moisture status [17,18]; therefore, rapid and transient wetting and drying cycles would produce large temporal variations in R_{LL}. This would significantly affect variation in R_S [17,19], suggesting that R_{LL} is an important controller of temporal (daily and seasonal) patterns in the carbon balance in warm regions [19,20].

Several methods for measuring R_{LL} and for calculating its contribution to R_S have been explored. Cisneros-Dozal et al. [21] used an isotope mass balance method and reported that the contribution of R_{LL} to R_S increased from 5% to 37% in response to water addition after transient drought. Deforest et al. [15]

determined that the annual contribution of R_{LL} to R_S was 48% $\pm 12\%$ by measuring R_S with and without the L-layer, and the ratio was consistent over a range of environmental conditions. However, there is little information about temporal variation in R_{LL} in relation to rainfall events because of the difficulty of continuous and direct measurement of R_{LL} in situ.

To continuously measure CO_2 efflux from the L-layer only, in parallel with measurement of R_S, we developed an approach for measuring R_{LL} using an automated chamber method in a treatment area from which all CO_2 sources except for the L-layer were removed. In parallel with R_{LL} and R_S measurements, we continuously measured water content of the L-layer (LWC) and soil water content (SWC). LWC was measured using a method developed by Ataka et al. [22], in which intact leaf litter was attached to surrounding capacitance sensors. Sensors were also placed on top of the L-layer and at the boundary between the L- and mineral layers. From these continuous in situ measurements, we investigated the response of R_{LL} to rainfall events by comparing R_{LL} with R_S, and examined temporal variation in the contribution of R_{LL} to R_S in a warm temperate forest in Japan.

Materials and Methods

Ethics statement

The study site (Yamashiro Experimental Forest) is maintained by the Forestry and Forest Products Research Institute. All necessary permits were obtained for the field study, and the study did not involve endangered or protected species.

Study site

Our observations of R_{LL} and R_S were conducted at the Yamashiro Experimental Forest in southern Kyoto Prefecture, Japan (34°47'N, 135°50'E). The study site is a 1.7-ha watershed characterized by an annual mean air temperature of 15.5°C (maximum, 34.8°C; minimum, −3.9°C) and annual precipitation of 1449 mm [2]. The rainy season generally occurs from early June to mid-July. Daily rates of evaporation from the forest floor are 0.4–0.8 mm day^{-1} for 1–2 days after precipitation, declining thereafter to 0.2–0.3 mm day^{-1} [23]. The soils are Regosols with sandy loam or loamy sand texture and contain fine gravel (53% by mass) composed of residual quartz crystals from granite parent material [24]. These are immature soils in which the thickness of the A horizon is 2–3 cm. Deciduous broad-leaved, evergreen broad-leaved, and coniferous tree species account for 66%, 28%, and 6% of the living tree biomass, respectively [25]. The forest is dominated by Quercus serrata Thunb., which accounts for approximately 33% of the biomass. The L-layer (approximately 3–4 cm thick) consists mainly of fresh Q. serrata litter. There is no substantial organic horizon below the L-layer.

Automated chamber method for measuring leaf litter respiration and soil respiration

We measured R_{LL} and R_S using an automated dynamic chamber system with an infrared gas analyzer (IRGA, GMP343; Vaisala Group, Vantaa, Finland) (Fig. 1A). The system consisted of two automated circular chambers for R_{LL} and R_S measurement, four solenoid valves, a pump, mass flow meter, and IRGA. The chambers (surface area 320 cm^2) were made from PVC collars with clear acrylic lids that can be opened and closed automatically using an air cylinder. Air was supplied to the cylinder from a compressor. To ensure a seal between the chamber and the closed lid, a soft rubber gasket was attached to the top edge of the chamber. Opening and closing of the chamber lid and solenoid

valves of each chamber were regulated synchronously by a control unit (ZEN, OMRON, Kyoto, Japan).

The duration of measurement of CO_2 concentration inside each chamber was 6 min and was performed twice per hour. The CO_2 concentration in each chamber was recorded at 1-s intervals using a data logger (GL220, Graphtec, Kanagawa, Japan). We calculated R_{LL} and R_S from the increase in CO_2 concentration (ΔC_{CO2}) using linear regression. Data from the first 2 min were discarded to avoid effects of closing the chamber. R_{LL} and R_S were calculated using the following equation:

$$R = \frac{\Delta C_{CO_2}}{10^6} \times \frac{V}{V_{air}} \frac{273.2}{273.2+T} \times M_{CO_2} \times \frac{1}{A}, \quad (1)$$

where R is respiration (mg CO_2 m^{-2} s^{-1}), ΔC_{CO2} is the change in CO_2 concentration per unit time (CO_2 ppm s^{-1}), V is the volume of the system (L), V_{air} is the standard gas volume (22.41 L mol^{-1}), T is temperature inside the chamber (°C), M_{CO2} is the molecular weight of CO_2 (44.01 g mol^{-1}), and A is the soil surface area covered by the chamber (m^2).

To continuously measure CO_2 efflux from the L-layer only, we developed an approach for measuring R_{LL} by using an automated chamber method in a treatment area in which all potential CO_2 sources (e.g., organic soil and fine roots) except for the L-layer were replaced with combusted granite soil (Fig. 1B). To prepare the treatment area (1 m^2), we removed surface soil (approximately 5 cm). An acrylic board was placed on the bottom and sides of the treatment area to prevent penetration of roots; a drain tube was located at the bottom of the board to prevent the treatment area from flooding with rainwater. The treatment area was then filled with granite soil combusted in a muffle furnace (500°C for 1 day). For R_{LL} measurement, we placed a PVC collar (320-cm^2 surface area) and acrylic board below the collar. The board was set at a slight incline to drain rainwater from the collar. We added 15 g of newly fallen leaf litter, which represents the average litterfall mass per unit ground surface area at this site, to the collar. We added the leaf litter to each chamber on January 2012. To acquire data on the temporal variation in R_{LL} of fresh leaf litter, we replaced the litter with newly fallen leaf litter in January 2013. The collar for measurement of R_S was placed near the treatment area for R_{LL} measurement and the L-layer inside the collar was removed and leaf litter was supplied similarly as for measurement of R_{LL}. To prevent incorporation of newly fallen litter, we placed a mesh sheet (1×1 mm mesh) on the L-layer inside the chamber, and fallen litter was removed weekly. CO_2 efflux from combusted granite soil was measured 6 months from the start of the R_{LL} measurements. The mean CO_2 flux rate ($\pm$ standard deviation) was 0.00063 $\pm$ 0.00068 mg CO_2 m^{-2} s^{-1} ($n = 16$) when SWC ranged from 0.05 to 0.3 m^3 m^{-3} at temperatures of 24°C. Thus, we assumed that CO_2 efflux from the combusted granite soil was negligible throughout the measurement period.

For continuous in situ measurement of LWC, we used capacitance sensors as described by Ataka et al. [22]. The measurements were performed on the top surface of the L-layer and at the boundary between the L-layer and mineral soil (Fig. 1B), to capture the large vertical distribution of WC within the L-layer. We estimated average LWC from the output voltage (V) of the two sensors using the conversion equation LWC = 12.73 V−3.42 presented by Ataka et al. [22]. LWC at the forest floor shows spatial variability associated with tree canopy conditions. Thus, to reflect the LWC of the L-layer by direct measurement, two capacitance sensors were placed on the L-layer inside the chamber. To check the validity of continuous LWC monitoring, we compared the sensor values with LWC measured

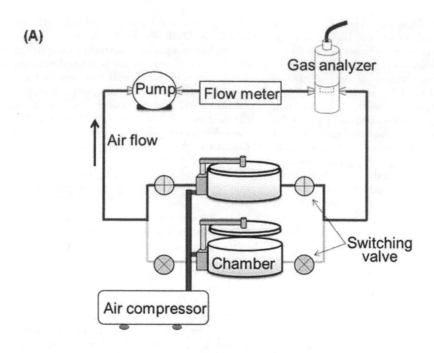

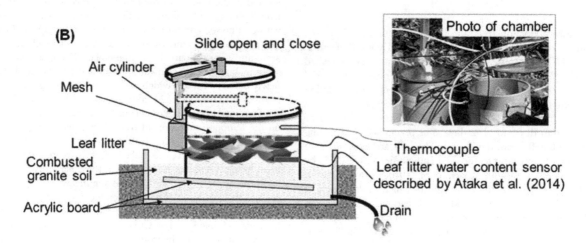

Figure 1. Schematic of the automated chamber system and the experimental design for measurement of CO_2 efflux from the leaf litter layer. A. Schematic of the automated dynamic-closed chamber system for measuring leaf litter respiration and soil respiration. **B.** The experimental design for continuous measurement of CO_2 efflux from the leaf litter layer only using automated chamber system.

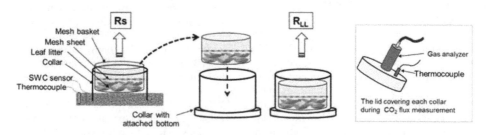

Figure 2. Schematic of the manual chamber system and the experimental design for measurement of CO_2 efflux from the leaf litter layer (R_{LL}) and soil (R_S).

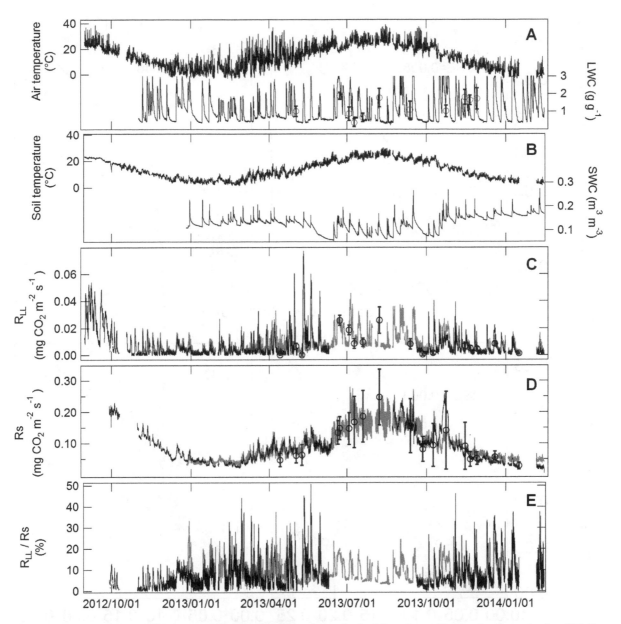

Figure 3. Seasonal variation in environmental factors, CO_2 efflux from the leaf litter layer (R_{LL}), and soil respiration (R_S). Data were measured every 30 min between September 2012 and January 2014. **A.** Bold and fine lines show air temperature and water content of the leaf litter layer (LWC), respectively. **B.** Bold and fine lines show soil temperature and soil water content (SWC), respectively. **C.** Black and grey lines show observed and estimated R_{LL}, respectively. **D.** Black and grey lines show observed and estimated R_S, respectively. **E.** Black and grey lines show the ratio of observed and estimated R_{LL} to R_S, respectively. Circles and bars show mean values and standard deviation of manual measurements. Estimated R_{LL} and R_S were calculated from regression equations using temperature (T) and water content (WC): $R_{LL}=0.29e^{0.059T}[WC/(95.04+WC)]$ and $R_S=0.031e^{0.10T}[WC/(0.032+WC)]$.

Table 1. Q_{10} of leaf litter respiration (R_{LL}) and soil respiration (R_s) for different water contents of the leaf litter layer (LWC) and soil (SWC).

	R_{LL}			R_s		
	LWC≤1	1<LWC≤2	2<LWC	SWC≤0.1	0.1<SWC≤0.15	0.15<SWC
Q_{10}	1.54	1.88	2.07	1.97	2.12	2.73
a	0.0019	0.0044	0.0064	0.027	0.032	0.025
b	0.043	0.063	0.073	0.068	0.075	0.10

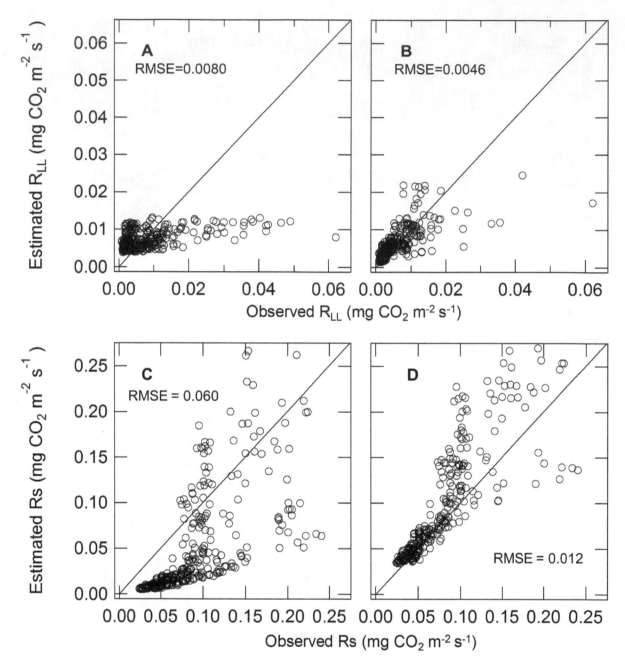

Figure 4. Relationship between observed and estimated CO$_2$ efflux rate from leaf litter respiration (R_{LL}) and soil respiration (R_S). R_{LL} (A, B) and Rs (C, D) show daily mean values. Estimated respiration rates were calculated using a function of temperature (A, C) from Eq. (5,6) and a function of temperature and water content (B, D) from Eq. (7,8) in the Results. Lines represent the 1:1 ratio. RMSE: root mean square error.

manually as described in the following section. In parallel with LWC measurement, soil temperature (copper-constantan thermocouple) and soil volumetric water content (ECH$_2$O EC-5 sensors; Decagon Devices, Pullman, WA, USA) were measured at 5-cm depth near each chamber. The output voltage of all environmental data was recorded every 1 min with a data logger (Datamark LS-3000 PtV; Hakusan, Japan) and average values were computed every 30 min. The environmental data, R_{LL}, and R_S were measured continuously between September 2012 and January 2014. Malfunction of IRGA resulted in a lack of data for R_{LL} and R_S for 31% of the measurements.

Manual chamber method for measuring leaf litter respiration and soil respiration

To determine the validity of R_{LL} and R_S measured using the automated chamber method, respiration was measured using the manual chamber method. We assumed that manual chamber method allow to measure under conditions that were closer to natural than the automated chamber method. We measured R_{LL} and R_S manually using a static chamber system at midday on 18 days between April 2013 and January 2014. Twelve PVC collars (320 cm^2 surface area) were placed in a 2×4 m area in January 2013. The edges of the collars were inserted approximately 1.5 cm into the soil. To measure R_{LL}, mesh baskets (1×1 mm mesh, the

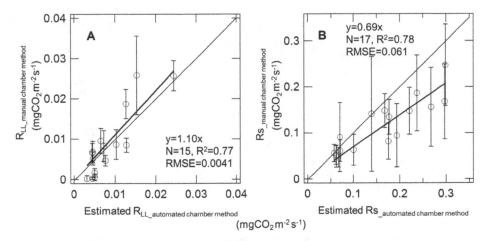

Figure 5. Relationship between respirations measured using a manual chamber method and estimated from automated chamber data. Respiration rate measured with the manual chamber method ($R_{\_manual\ chamber\ method}$) show mean value obtained from measurement of 12 collars. Bars show standard deviation. Respiration estimated from automated chamber data (estimated $R_{\_automated\ chamber\ method}$) shows daily mean respiration. The estimated R was calculated using a function based on temperature and water content (Eq. 8, 9).

same diameter as the PVC collars; 20 cm) were set into each collar and 15 g (dry weight) of newly fallen leaf litter was placed on the L-layer inside each basket (Fig. 2). To prevent supply of newly fallen litter, we placed a mesh sheet (1×1 mm mesh) on the L-layer inside the chamber, and fallen litter was removed weekly.

For measurement of R_S, the collars were completely covered with lids to which an IRGA and copper-constantan thermocouple were attached. Soil temperature and SWC (5 cm depth) were measured close to the collars when R_S was measured. After completing the measurements of R_S, the mesh baskets were carefully removed from the collars and placed in PVC chambers (20 cm diameter, 7 cm high; Fig. 2). We measured R_{LL} using the same methods as used for R_S measurement. The temperature and CO_2 concentrations in the chamber were recorded at 1-s intervals using a data logger (GL220). Linearity of the CO_2 flux was checked on the data logger monitor at each measurement. The measurement period for each chamber was 10 min and CO_2 data for the middle 5-min intervals were used to determine R_{LL} according to Eq. (1), excluding data from the first 3 min.

For measurement of LWC in the mesh baskets, four or five leaves were removed from each basket and immediately placed in sealed plastic bags. Fresh weight of the leaf litter was measured in the laboratory within 24 h of sampling. Leaf litter samples were oven dried at 65°C for 48 h, and water content (WC; g g^{-1}) was calculated using Eq. 2 as follows:

$$WC = \frac{(FW - DW)}{DW}, \qquad (2)$$

where FW is the fresh mass of the sample (g), and DW is the dry mass of the sample (g). Samples were returned to each mesh basket within 1 week after sampling.

Leaf litter respiration and soil respiration rates as a function of environmental factors

Respiration models are fundamentally described by nonlinear functions. We used the following function to investigate the response of respiration to temperature:

$$R = a \exp(bT), \qquad (3)$$

where T is temperature (leaf litter temperature for R_{LL} measurement or soil temperature for R_S measurement) and a and b are constants. Leaf litter temperature was assumed to be same as air temprature. b is related to the Q_{10} parameter ($Q_{10} = e^{10b}$). To determine the effects of temperature and water content on R_{LL} and R_S, we used a function that was previously applied to estimate soil respiration by Subke and Schlesinger [26]:

$$R = a \exp(bT) \left(\frac{WC}{c + WC}\right), \qquad (4)$$

where a, b, and c are constants. LWC or SWC was used as WC in this equation. These nonlinear regressions were performed using a modified Levenberg–Marquardt method with Igor Pro 6.0 software (WaveMetrics, Lake Oswego, OR, USA). The estimated respiration values presented in this manuscript were calculated using Eq. 4.

Short-term changes in R_{LL} and LWC on wetting and drying cycle

To evaluate short-term changes in R_{LL} and LWC after rainfall events, we chose eight typical periods that included one wetting and drying cycle and had consecutive no rainfall days for at least 3 days. We used daily mean R_{LL} and LWC before the day on which precipitation occurred as the pre-wetting condition, and these values after precipitation as the post-wetting condition. Daily mean R_{LL} was calculated from R_{LL} values observed using the automated chamber method.

Effect of wetting and drying cycle of the L-layer on R_{LL} and Rs on the annual time scale

To investigate the effects of wetting and drying of the L-layer on R_{LL} on the annual time scale, we separated the estimated daily mean R_{LL} in 2013 into 'Dry' and 'Wet' periods based on daily mean LWC as a threshold value. The threshold LWC value that separated 'Dry' and 'Wet' periods for R_{LL} was estimated by the abovementioned short-term analyses. Daily mean R_{LL} was calculated from the estimated R_{LL} values because there were gaps in the continuous R_{LL} data observed using the automated

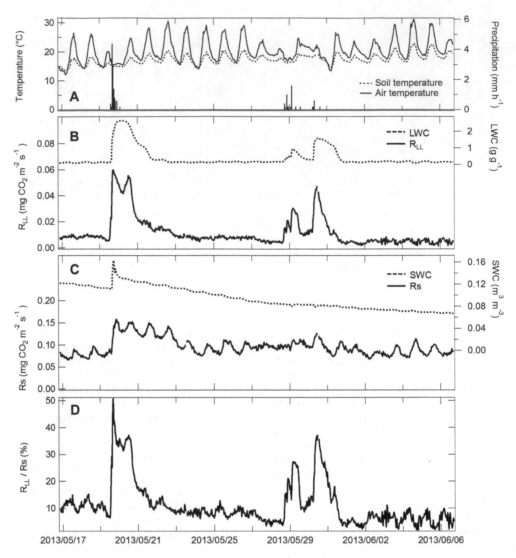

Figure 6. Temporal variation in environmental factors, CO_2 efflux from the leaf litter layer (R_{LL}), soil respiration (R_S), and the ratio of R_{LL} to R_S. Data was measured at one collar every 30 min between May 17 and June 6, 2013. **A.** Soil and air temperature. Spikes on the x-axis indicate precipitation events (mm h^{-1}). **B.** R_{LL} and water content of the leaf litter layer (LWC). **C.** R_S and soil water content (SWC). **D.** The ratio of R_{LL} to R_S (%).

chambers. We estimated the contribution of R_{LL} accumulated during the wet and dry period to total R_S.

Results

Seasonal variation in R_{LL} and R_S

The magnitude of the peak in the observed R_{LL} pulse was higher in summer than in winter (Fig. 3C). R_{LL} values were low when LWC was low (Fig. 3A, C). R_S changed substantially according to temperature (Fig. 3B, D), with higher values in summer than in winter. The relationships between respiration and temperature were described by the following functions:

$$R_{LL}(mg\ CO_2\ m^{-2}\ s^{-1}) = 0.0038\exp(0.065 \times T_{LL}), \quad (5)$$

$$Rs\ (mg\ CO_2\ m^{-2}\ s^{-1}) = 0.0031\exp(0.19 \times Ts), \quad (6)$$

where T_{LL} is leaf litter temperature and Ts is soil temperature (°C).

To evaluate effect of WC on the temperature sensitivity of respiration, the measured respiration data was separated into three groups based on WC (Table 1). More than 14% of total respiration data was included in each WC group. R_{LL} showed low values when WC values were low in spite of high temperature. Consequently, calculated Q_{10} values for not only R_{LL} but also R_S decreased with decreasing WC. The relationships between respiration and temperature and WC were described by the following functions:

$$R_{LL}(mg\ CO_2\ m^{-2}\ s^{-1}) =$$
$$0.29\exp(0.059 \times T_{LL})\left(\frac{LWC}{95.04 + LWC}\right), \quad (7)$$

$$Rs\ (mg\ CO_2\ m^{-2}\ s^{-1}) =$$
$$0.031\exp(0.10 \times Ts)\left(\frac{SWC}{0.032 + SWC}\right), \quad (8)$$

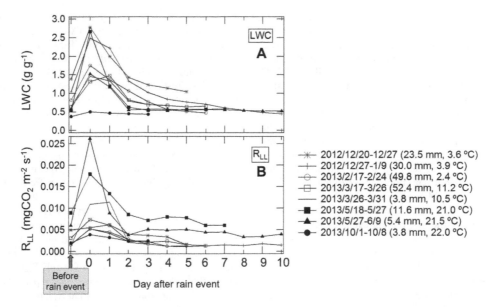

Figure 7. Temporal variation in water content of the leaf litter layer (LWC) and CO$_2$ efflux from the leaf litter layer (R_{LL}) after rainfall events. LWC (A) and R_{LL} (B) show the daily mean values. The rainfall intensity of each precipitation event was 23.5 mm in 2 days (2012/12/20–12/27, mean air temperature; 3.6°C); 30.0 mm in 3 days (2012/12/27–1/9, 3.9°C); 49.8 mm in 2 days (2013/2/17–2/24, 2.4°C); 52.4 mm in 3 days (2013/3/17–3/26, 11.2°C); 3.8 mm in 2 days (2013/3/26–3/31, 10.5°C); 11.6 mm in 2 days (2013/5/18–5/27, 21.0°C); 5.4 mm in 3 days (2013/5/27–6/9, 21.5°C); and 3.8 mm in 4 days (2013/10/1–10/8, 22.0°C).

where LWC (g g^{-1}) and SWC (m^3 m^{-3}) are water content of leaf litter and soil, respectively. The RMSE between observed and estimated daily mean respiration based on temperature (R_{LL}, 0.0080 mg CO$_2$ m^{-2} s^{-1}; R_S, 0.060 mg CO$_2$ m^{-2} s^{-1}) was larger than that based on temperature and WC (R_{LL}, 0.0046 mg CO$_2$ m^{-2} s^{-1}; R_S, 0.012 mg CO$_2$ m^{-2} s^{-1}) (Fig. 4). Estimated respiration was calculated using the equation based on temperature and WC because of the lower RMSE. Throughout the measurement period, the contribution of observed R_{LL} to variation in R_S changed from nearly zero to 51% following a rainfall event (Fig. 3E).

To consider the validity of R_{LL} and R_S estimated from continuous measurement, we compared these values with respiration rates measured using the manual chamber method (Fig. 5). Estimated respiration was very similar to that observed using manual measurements. The RMSE between estimated and observed respiration were 0.0041 and 0.061 mg CO$_2$ m^{-2} s^{-1} for R_{LL} and R_S, respectively.

Temporal changes in R_{LL} and R_S on the short-term scale

To show clear temporal variation in R_{LL} and R_S, the period between May 17 and June 6, 2013 (Fig. 6) was chosen because this

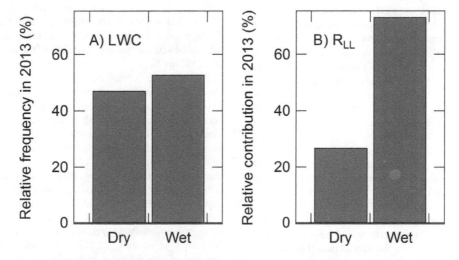

Figure 8. Histograms of the relative frequency of "Dry" and "Wet" periods in relation to water content of the leaf litter layer (LWC), and the relative contribution of estimated leaf litter respiration (R_{LL}) in 2013. The daily mean LWC (A) and R_{LL} (B) were used to present histograms. Estimated respiration rates were calculated using a function based on temperature (T) and water content (WC). $R_{LL}=0.29e^{0.059T}$[WC/(95.04+WC)]. The daily mean LWC and R_{LL} were defined as Dry or Wet based on LWC. Days in which daily mean LWC <0.75 g g^{-1} were defined as Dry periods, while days in which daily mean LWC ≥0.75 g g^{-1} were defined as Wet periods.

period included two characteristic rainfall events. The rainfall intensity was 11.6 mm over 13 h during the first event and 5.4 mm over 46 h during the second event. LWC and SWC increased from 0.11 to 2.64 g g^{-1} and from 0.11 to 0.16 m^3 m^{-3}, respectively, following the first rainfall event (Fig. 6B, C). LWC increased from 0.16 to 1.58 g g^{-1} but SWC did not increase after the second rainfall event.

Temporal variation in R_{LL} measured using the automated chamber system changed according to wetting and drying of the L-layer (Fig. 6B), reaching a maximum of 0.060 and 0.047 mg CO_2 m^{-2} s^{-1} during first and second rainfall events, respectively. R_S increased following the increase in SWC and subsequently decreased gradually with diurnal variation according to temperature (Fig. 6C). Between May 17 and June 6, 2013, the contribution of R_{LL} to R_S increased from 6.5% to 51%, with a peak value of 51% during the first rainfall event and 37% during the second rainfall event (Fig. 6D).

Both R_{LL} and LWC reached a peak during or one day after rainfall events (Fig. 7). The peak of R_{LL} and LWC varied from 0.0020 to 0.026 mg CO_2 m^{-2} s^{-1} and from 0.50 to 2.66 g g^{-1}, respectively. Peak value of each rainfall event highly depended on air temperature. High peaks of R_{LL} were observed in the warm season (0.017 mg CO_2 m^{-2} s^{-1}; 2013/5/18–5/27, 0.026 mg CO_2 m^{-2} s^{-1}; 2013/5/27–6/9 in Fig. 7). Also, the peak value was related to LWC: low peak of R_{LL} was observed when LWC was low (0.004 mg CO_2 m^{-2} s^{-1}; 2013/10/1–10/8 in Fig. 7). The relationship between LWC and amout of precipitation was not clear. In the cold season, peak values of R_{LL} were relatively low (e.g., 0.005 mg CO_2 m^{-2} s^{-1}; 2013/2/17–2/24, 0.006 mg CO_2 m^{-2} s^{-1}; 2012/12/20–12/27 in Fig. 7) even when the L-layer was wet enough (LWC more than 1.5 g g^{-1}). The peak values of R_{LL} were 1.2- to 8.6-fold higher than the R_{LL} values before rainfall events, and R_{LL} fell to pre-wetting levels within 2–4 days after rainfall events and peak LWC values were 1.3- to five-fold higher than LWC before rainfall, and LWC also dropped to pre-wetting levels within 2–4 days after rainfall events. We defined R_{LL} from the period just after rainfall events through 2–4 days later as the "R_{LL} pulse".

Effects of wetting and drying of the L-layer on R_{LL} and R_S on the annual time scale

Estimated daily mean R_{LL} in 2013 was separated into 'Dry' and 'Wet' periods based on daily mean LWC. Days for which mean LWC was <0.75 g g^{-1} were categorized as Dry, while days for which mean LWC ≥0.75 g g^{-1} were categorized as Wet. The threshold value (0.75 g g^{-1}) was obtained from mean LWC 3 days after a rainfall event (Fig. 7A). The relative frequency of Dry and Wet periods in 2013 were 47.2% and 52.8%, respectively, while the relative contributions of daily mean R_{LL} during the Dry and Wet periods in 2013 were 26.9% and 73.2%, respectively (Fig. 8). Annual R_{LL} and R_S in 2013 were estimated to be 0.69 and 7.94 t C ha^{-1} y^{-1}, respectively. The RMSE between continuous respiration measured and estimated based on temperature and WC was 0.011 and 0.029 t C ha^{-1} y^{-1}, respectively.

The contribution of annual R_{LL} to R_S was 8.6%. The relative frequency of LWC was similar during Dry and Wet periods, while the contribution of R_{LL} during the Wet period was approximately three-fold higher than that during the Dry period (Fig. 8).

Discussion

As seen in Fig. 6, R_{LL} immediately increased with wetting of the L-layer and decreased to pre-wetting levels within 2–4 days after rainfall events, which was consistent with observations made in previous studies [17,19]. R_{LL} showed no diurnal variation despite a diurnal temperate range >10°C. Consequently, the Q_{10} of R_{LL} increased with increasing LWC (Table. 1). The variation in Q_{10} would be directly related to water stress experienced by microorganism. This indicated that LWC can reach to adequate low value, suspected as water stress for microorganism, within several days after rainfall. On the one hand, R_S increased during rainfall and subsequently decreased, showing diurnal variation. The Q_{10} of R_S also increased with increasing SWC. Dannoura et al. [27] reported that root respiration showed little change with variation in SWC compared with changes in R_S. Therefore, the increased Q_{10} of R_S with increasing SWC might be highly affected by not only R_{LL} but also by respiration from other heterotrophic sources.

Although the relative frequency of LWC was similar during Dry and Wet periods, the contribution of annual R_{LL} during the Wet period was approximately three-fold higher than that during the Dry period (Fig. 8), indicating strong effect of rainfall on R_{LL}. Although the R_{LL} pulse can last for only 3–4 days after a rainfall event, this pulse would determine a large part of annual R_{LL}. This suggests that the magnitude of total R_{LL} may be influenced by the frequency of rainfall events, especially in summertime, rather than the intensity of rainfall. Still, the cumulative R_{LL} in the Dry period contributed 26.9% of annual R_{LL} in 2013, even though instantaneous R_{LL} was very low. There may be large vertical variability in WC and R_{LL} within the L-layer, indicating that higher WC and R_{LL} occur in lower parts of the L-layer during the drying process because the upper L-layer dries more rapidly [28]. In that case, although the mean WC of the L-layer was very low, local wetting in lower sections would produce small CO_2 fluxes. Despite low instantaneous R_{LL}, the accumulation of R_{LL} over a long time period (approximately 6 mo) resulted in a substantial contribution (27%) of Dry-period respiration to annual R_S.

Raindrops first reach the L-layer and then percolate to the soil layers below. Small amounts of precipitation caused no change in SWC or R_S, but R_{LL} increased rapidly with increasing LWC (Fig. 6). In semi-arid and arid ecosystems, wetting of the L-layer and surface soil by small fog-drop pulses during the dry season can contribute up to 35% of R_S [29]. Although such small water inputs (e.g., brief rain showers and fog), which mainly affect the surface of the forest floor, can be significant drivers of temporal variation in R_S, the soil water content sensors (generally inserted at depths > 5 cm) could not capture these inputs. Continuous measurement of LWC allowed for realistic modeling of the effects of rapid changes in LWC on R_{LL}.

Although the annual contribution of R_{LL} to R_S was relatively small (8.6%), this contribution showed large temporal variation according to rainfall, ranging from nearly zero to 51%. Several other studies have described similar results [17,21]. For example, Borken et al. [17] reported that peaks in R_{LL} during addition of water ranged from 0.031 to 0.071 mg CO_2 m^{-2} s^{-1} in vitro, which represented 11–26% of maximum in situ R_S in the Harvard forest, although R_{LL} before addition of water was nearly zero. These findings indicate that R_{LL} is a significant component of rapid and transient temporal variation in R_S in relation to rainfall events. Although numerous studies have examined CO_2 efflux from mineral soils in relation to the intensity, duration, and frequency of rainfall [30,31], few studies have focused on R_{LL} because of the difficulty in measuring this dynamic. Here, R_{LL} pulses were observed only during and several days after rainfall events. Thus, periodic sampling (e.g., twice per week) might be insufficient to capture the contribution of the R_{LL} pulse to R_S. Moreover, manual flux measurements are usually not performed during precipitation events because of difficulties that can occur with electronic instruments and sampling methods. In our view, conducting in situ measurements of CO_2 efflux from the L-layer only over short time intervals (e.g., up to 1 h) produces robust data for understanding the response of R_{LL} to rainfall events and its contribution to R_S.

The contribution of R_{LL} to annual R_S was 8.6% in our site. In an oak forest, the contribution of R_{LL} to R_S was 23%, according to model simulation based on temperature and LWC by Hanson et al. [13]. Ngao et al. [32] reported a lower contribution (8%) in a beech forest, estimated using an isotope mass balance approach, which was close to the value observed at our site (8.6%). However, simple quantitative comparisons between studies are difficult because of the use of different methods. In addition, some technical problems remain at our site. First, we performed R_{LL} measurements in the treatment area in which the mineral soil below the L-layer was replaced with combusted granite soil. This treatment may have affected the microbial community and environmental conditions in the L-layer. Secondly, each continuous measurement of R_{LL} and R_S was performed with single chambers, so spatial heterogeneity in R_{LL} and R_S were not considered. Automated chamber methods allowed high-interval measurements of temporal variation in respiration but had poorer spatial distribution compared with the manual chamber method. The balance of trade-offs between automated and manual chamber method is subject to the relative importance of characterizing temporal and spatial variability of individual CO_2 sources. The number of chambers used can enhance the accuracy of measured mean values. Loescher et al. [33] reported that the number of chambers needs to be >100 to adequately represent spatial variability. However, this is not a feasible experimental design because of practical limitations to sampling efforts. To improve estimation of R_{LL} and R_S at the forest stand level, and to better understand the soil carbon budget, a comprehensive comparison of the diverse C pools and fluxes in forest soils is required.

Conclusions

In our study, the rapid and transient variation in R_{LL} induced by rainfall; the peak R_{LL} was observed during or one day after rainfall, and R_{LL} subsequently decreased to pre-wetting levels

within 2–4 days after rainfall events, following the decrease in LWC. On the one hand, CO_2 efflux from coarse woody debris found in our site decreased during rainfall events, and subsequently, a gradual increase in CO_2 efflux continued for at least 14 days until next rainfall [34]. Therefore, coarse woody debris was a CO_2 efflux source over longer time scales, while R_{LL} approached nearly zero within a few days after rainfall events, even at high temperatures. Such specific temporal CO_2 efflux patterns for each heterotrophic source when subjected to wetting and drying cycles would be a result of substrate properties (e.g., specific surface area). In our view, continuous and direct measurements of CO_2 efflux and environmental conditions characterized by substrate properties of individual CO_2 sources could improve understanding of the processes that regulate variation in heterotrophic respiration and R_S and enable progress beyond empirical models that are primarily based on simple temperature and SWC relationships.

Moreover, the magnitude of heterotrophic respiration under wetting and drying cycles is strongly related to microbial physiology and community composition. For example, Schnurer et al. [35] showed that longer-duration wetting could promote microbial biomass, causing an increase in basal respiration. Fierer et al. [36] showed the influence of drying and rewetting frequency on microbial (fungi and bacteria) community composition. To improve understanding of heterotrophic respiration associated with response and adaptation of microorganisms under climatic changes, collected continuous in situ data for CO_2 efflux and environmental conditions (e.g., temperature and WC) of individual CO_2 sources should be combined with analyses of microbial physiology and community composition.

Acknowledgments

We greatly thank Dr. Yoshiko Kosugi and the staff of the Forest Hydrology Laboratory of Kyoto University for assistance in the field and for helpful advice.

Author Contributions

Conceived and designed the experiments: MA YK TM. Performed the experiments: MA KY MT. Analyzed the data: MA YK MT. Contributed reagents/materials/analysis tools: MA MJ. Contributed to the writing of the manuscript: MA.

References

1. Curtis PS, Hanson PJ, Bolstad P, Barford C, Randolph JC, et al. (2002) Biometric and eddy-covariance based estimates of annual carbon storage in five eastern North American deciduous forests. Agricultural and Forest Meteorology 113: 3–19.

2. Kominami Y, Jomura M, Dannoura M, Goto Y, Tamai K, et al. (2008) Biometric and eddy-covariance-based estimates of carbon balance for a warm-temperate mixed forest in Japan. Agricultural and Forest Meteorology 148: 723–737.

3. Keith H, Leuning R, Jacobsen KL, Cleugh HA, van Gorsel E, et al. (2009) Multiple measurements constrain estimates of net carbon exchange by a Eucalyptus forest. Agricultural and Forest Meteorology 149: 535–558.

4. Yuste JC, Nagy M, Janssens IA, Carrara A, Ceulemans R (2005) Soil respiration in a mixed temperate forest and its contribution to total ecosystem respiration. Tree Physiology 25: 609–619.

5. Hanson PJ, Edwards NT, Garten CT, Anderson JA (2000) Separating root and soil microbial contributions to soil respiration: A review of methods and observations. Biogeochemistry 48: 115–146.

6. Kuzyakov Y (2006) Sources of CO_2 efflux from soil and review of partitioning methods. Soil Biology & Biochemistry 38: 425–448.

7. Moyano FE, Kutsch WL, Rebmann C (2006) Soil respiration fluxes in relation to photosynthetic activity in broad-leaf and needle-leaf forest stands. Agricultural and Forest Meteorology 148: 135–143.

8. Suseela V, Conant RT, Wallenstein MD, Dukes JS (2012) Effects of soil moisture on the temperature sensitivity of heterotrophic respiration vary seasonally in an old-field climate change experiment. Global Change Biology 18: 336–348.

9. Jomura M, Kominami Y, Tamai K, Miyama T, Goto Y, et al. (2007) The carbon budget of coarse woody debris in a temperate broad-leaved secondary forest in Japan. Tellus B 59: 211–222.

10. Matsumoto A, Kominami Y, Ishii H (2010) Field measurement of heterotrophic respiration of root litter using a small chamber system. Journal of Forest Research 92(5): 269–272.

11. Tewary CK, Pandey U, Singh JS (1982) Soil and litter respiration rates in different microhabitats of a mixed oak-conifer forest and their control by edaphic conditions and substrate quality. Plant and Soil 65: 233–238.

12. Kirschbaum MUF (2013) Seasonal variations in the availability of labile substrate confound the temperature dependence of organic matter decomposition. Soil Biology & Biochemistry 57: 568–576.

13. Hanson PJ, O'Neill EG, Chambers MLS, Riggs JS, Joslin JD, et al. (2003) Soil respiration and litter decomposition. In: North America Temperate Deciduous Forest Responses to Changing Precipitation Regimes (eds Hanson PJ and Wullschleger SD). Springer, New York.

14. Andersson M, Kjoller A, Struwe S (2004) Microbial enzyme activities in leaf litter, humus and mineral soil layers of European forests. Soil Biology & Biochemistry 36: 1527–1537.

15. DeForest JL, Chen J, McNulty SG (2009) Leaf litter is an important mediator of soil respiration in an oak-dominated forest. International Journal of Biometeorology 53(2): 1432–1254.

16. Wilson TB, Kochendorfer J, Meyers TP, Heuer M, Sloop K, et al. (2014) Soil respiration in a mixed temperate forest and its contribution to total ecosystem respiration. Agricultural and Forest Meteorology 192–193: 42–50.

17. Borken W, Davidsona EA, Savagea K, Gaudinskib J, Trumborec SE (2003) Drying and wetting effects on carbon dioxide release from organic horizons. Soil Science Society of American Journal 67: 1888–1896.

18. Orchard VA, Cook FJ (1983) Relationship between soil respiration and soil moisture. Soil Biology & Biochemistry 15(4): 447–453.

19. Lee X, Wu HJ, Sigler J, Oishi C, Siccama T (2004) Rapid and transient response of soil respiration to rain. Global Change Biology 10: 1017–1026.

20. Goulden ML, Miller SD, da Rocha HR, Menton MC, de Freitas HC, et al. (2004) Diel and seasonal patterns of tropical forest CO_2 exchange. Ecological Application 14: 42–54.

21. Cisneros Dozal LM, Trumbore S, Hanson PJ (2007) Effect of moisture on leaf litter decomposition and its contribution to soil respiration in a temperate forest. Journal of Geophysical Research 112: 148–227.

22. Ataka M, Kominami, Miyama T, Yoshimura K, Jomura M, et al. (2014) Using capacitance sensors for the continuous measurement of the water content in the litter layer of forest soil. Applied and Environmental Soil Science 2014:

23. Tamai K, Hattori S (1994) Modeling of evaporation from forest floor in a deciduous broad-leaved forest and its application to basin. Journal of the Japanese Forestry Society 76: 233–241 (in Japanese with English summary).

24. Kaneko S, Akieda N, Naito F, Tamai K, Hirano Y (2007) Nitrogen budget of a rehabilitated forest on a degraded granitic hill. Journal of Forest Research 12: 38–44.

25. Goto Y, kominami Y, Miyama T, Tamai K, Kanazawa Y (2003) Aboveground biomass and net primary production of a broad-leaved secondary forest in the southern part of Kyoto prefecture, central Japan. Bulletin of FFPRI 387: 115–147 (in Japanese with English summary).

26. Subke JA, Reichstein M, Tenhunen JD (2003) Explaining temporal variation in soil CO_2 efflux in a mature spruce forest in Southern Germany. Soil Biology & Biochemistry 35: 1467–1483.

27. Dannoura M, Kominami Y, Tamai K, Jomura M, Miyama T, et al. (2006) Development of an automatic chamber system for long-term measurements of CO_2 flux from roots. Tellus 58B: 502–512.

28. Ataka M, Kominami Y, Jomura M, Yoshimura K, Uematsu C (2014) CO_2 efflux from leaf litter focused on spatial and temporal heterogeneity of moisture. Journal of Forest Research 19: 295–300.

29. Carbone MS, Still CJ, Ambrose AR, Dawson TE, Williams AP, et al. (2011) Seasonal and episodic moisture controls on plant and microbial contributions to soil respiration. Oecologia 167: 265–278.

30. Borken W, Matzner E (2009) Reappraisal of drying and wetting effects on C and N mineralization and fluxes in soils. Global Change Biology 15: 808–824.

31. Birch HF (1958) The effect of soil drying on humus decomposition and nitrogen availability. Plant and Soil 10: 9–31.

32. Ngao J (2005) Estimating the contribution of leaf litter decomposition to soil CO_2 efflux in a beech forest using 13C-depleted litter. Global Change Biology 11(10): 1768–1776.

33. Loescher HW, Law BE, Mahrt L, Hollinger DY, Campbell J, et al. (2006) Uncertainties in, and interpretation of, carbon flux estimates using the eddy covariance technique. Journal of Geophysical Research 111: D21S90.

34. Jomura M, Kominami Y, Kanazawa Y (2005) Long-term measurements of the CO_2 flux from coarse woody debris using an automated chamber system. Journal of the Japanese Forest Society 87(2): 138–144 (in Japanese with English summary).

35. Schnurer J, Clarholm M, Bostrom S, Rosswall T (1986) Effects of moisture on soil microorganisms and nematodes: A field experiment. Microbial Ecology 12(2): 217–230.

36. Fierer N, Schimel JP, Holden PA. (2003) Influence of drying-rewetting frequency on soil bacterial community structure. Microbial Ecology 45(1): 63–71.

14

Accounting for Biomass Carbon Stock Change Due to Wildfire in Temperate Forest Landscapes in Australia

Heather Keith[1]*, David B. Lindenmayer[1], Brendan G. Mackey[2], David Blair[1], Lauren Carter[1], Lachlan McBurney[1], Sachiko Okada[1], Tomoko Konishi-Nagano[3]

1 The Fenner School of Environment and Society, Australian National University, Building 48, Canberra, ACT, Australia, 2 Griffith Climate Change Response Program, Griffith University, Queensland, Australia, 3 Fujitsu Laboratories Ltd., Kawasaki, Japan

Abstract

Carbon stock change due to forest management and disturbance must be accounted for in UNFCCC national inventory reports and for signatories to the Kyoto Protocol. Impacts of disturbance on greenhouse gas (GHG) inventories are important for many countries with large forest estates prone to wildfires. Our objective was to measure changes in carbon stocks due to short-term combustion and to simulate longer-term carbon stock dynamics resulting from redistribution among biomass components following wildfire. We studied the impacts of a wildfire in 2009 that burnt temperate forest of tall, wet eucalypts in south-eastern Australia. Biomass combusted ranged from 40 to 58 tC ha^{-1}, which represented 6–7% and 9–14% in low- and high-severity fire, respectively, of the pre-fire total biomass carbon stock. Pre-fire total stock ranged from 400 to 1040 tC ha^{-1} depending on forest age and disturbance history. An estimated 3.9 TgC was emitted from the 2009 fire within the forest region, representing 8.5% of total biomass carbon stock across the landscape. Carbon losses from combustion were large over hours to days during the wildfire, but from an ecosystem dynamics perspective, the proportion of total carbon stock combusted was relatively small. Furthermore, more than half the stock losses from combustion were derived from biomass components with short lifetimes. Most biomass remained on-site, although redistributed from living to dead components. Decomposition of these components and new regeneration constituted the greatest changes in carbon stocks over ensuing decades. A critical issue for carbon accounting policy arises because the timeframes of ecological processes of carbon stock change are longer than the periods for reporting GHG inventories for national emissions reductions targets. Carbon accounts should be comprehensive of all stock changes, but reporting against targets should be based on human-induced changes in carbon stocks to incentivise mitigation activities.

Editor: Benjamin Poulter, Montana State University, United States of America

Funding: The funder provided support in the form of salaries for the authors [HK, SO], but did not have any additional role in the study design, data collection and analysis, decision to publish, or preparation of the manuscript. The specific roles of these authors are articulated in the 'author contributions' section. The author affiliated with the funding organization contributed to the conceptual framework of the study.

Competing Interests: The authors declare that one of the authors [TKN] is affiliated with one of the funding organizations of this research study, Fujitsu Laboratories Ltd.

* Email: Heather.Keith@anu.edu.au

Introduction

Stabilising the carbon stock in the atmosphere to prevent dangerous anthropogenic interference with the climate system is the objective of the United Nations Framework Convention on Climate Change (UNFCCC) [1]. Forest ecosystems naturally exchange carbon dioxide (CO_2) with the atmosphere, but losses of carbon stocks from forests due to land management activities is a significant source of greenhouse gas (GHG) emissions [2]. Losses of carbon as emissions of CO_2 from forest ecosystems to the atmosphere occur due to biological decomposition processes, natural disturbance events like fire, as well as human activities. Removals of CO_2 from the atmosphere to forest ecosystem carbon stocks occur by the uptake of carbon in growing vegetation.

Under the UNFCCC and Kyoto Protocol, each country constructs GHG inventories and reports on their net annual emissions, that is, flows of CO_2 and other GHGs, due to specific activities and sectors including Land Use, Land Use Change and Forestry. Flow-based inventories in the land sector, however, obscure important differences between ecosystems types and between the impacts of human activities and natural disturbances. A more comprehensive, stock-based approach to accounting can address these problems and provide information that complements flow-based accounts (Ajani et al. 2013). The flow-based accounts are currently used for national reporting against emission reduction targets over short time periods, such as annual or 5 to 8 year commitment periods, as determined by political negotiations. However, this short time period exposes nations to high variability in emissions due to natural disturbance events, especially wildfires, which are outside human control, yet create large GHG fluctuations. The resulting variability in national net emissions due to natural disturbance can be much greater than the changes due to land use impacts, thus confounding attempts to provide incentives for reducing emissions in the land sector through improved forest management.

Losses of carbon due to wildfire are a significant component of global annual GHG balances and so it is important that they are included in national inventories. Seasonal losses of carbon from wildfires are detected as high concentrations of trace gases in the troposphere [3,4]. Estimates of gross losses of carbon from forest fires globally range from 2.0–2.5 Pg C yr^{-1} [5,6] to 3.8–4.3 Pg C yr^{-1} [7] as annual averages. As a comparison of the magnitude of this flux, the global net losses are 1.1 PgC yr^{-1} from land-use change and 7.9 PgC yr^{-1} from fossil fuel emissions [8]. For nations with large areas of forest that are subject to periodic wildfires, such as Australia, Canada, USA, Russia and South Africa [9,10,11], the treatment of wildfires in carbon accounting for forest management is an important consideration. The rules and methodologies are complex, often controversial, and raise many issues in international accounting systems.

Carbon stock changes due to wildfires occur at different temporal scales that should be included in comprehensive stock-based accounts. Large losses occur over periods of hours to days or weeks due to combustion and are highly variable each year due to occurrence of fire. Uptake through forest regrowth and losses through decomposition of dead biomass occur over decades to centuries. These changes in carbon stocks are balanced in comprehensive accounts, and in the accumulated concentration of CO_2 in the atmosphere, when assessed over a sufficiently long time period, assuming the fire regime is stationary. The following information is required to enable inclusion of wildfire disturbance in comprehensive stock-based national carbon accounts: carbon stock loss due to combustion, redistribution of carbon stocks between living and dead biomass components; subsequent rates of decomposition; and carbon uptake by regenerating vegetation. These estimates of carbon stock change need to be calculated on a landscape-wide basis and integrated over appropriate time periods [12] with data from a range of forest types. However, methodologies for measuring carbon stock change due to wildfire are problematic for a number of reasons, including: (i) wildfires typically extend over large areas, within which fire severity varies spatially; (ii) initial carbon stocks vary across the landscape in response to differences in forest type, age, disturbance history and environmental factors; (iii) biomass components are combusted with differing efficiencies; (iv) fire occurrence is stochastic [13,14]; and (v) the dynamics of carbon stocks occur over long time periods in response to wildfire.

Of these methodological issues, the parameter with the greatest uncertainty is the amount of carbon combusted, calculated as either an amount or a proportion of the fuel load or total stock, and then upscaled from individual components to sites and landscapes [5,15]. Combustion efficiency varies due to characteristics of individual fuel materials, such as logs of varying size, wood density, state of decay and moisture content [16], and characteristics of the fire, such as intensity and residence time, which determine the oxidation conditions of flaming or smouldering combustion [7,17,18]. At the landscape scale, the proportion combusted depends on the areas burnt within the fire boundary, including small scale heterogeneity in fire occurrence and severity. The high degree of uncertainty created by the limited existing data is demonstrated when aggregated spatially in regional and global inventories. For example, estimates of carbon stock losses from wildfires in Australia derived from global analyses differ by up to six-fold [6,19,20,21,22].

The objectives of this study were to investigate the magnitude and timeframe of carbon stock changes due to wildfire and to evaluate the results in terms of implications for national carbon accounts.

(i) We assessed the magnitude of carbon stock losses due to combustion in a wildfire in terms of the amount and proportion of the total forest ecosystem carbon stock. We then considered the significance of these carbon losses compared with potential impacts from human activities.

(ii) We considered the impact of fire on the ecosystem carbon stock in terms of the timeframe of stock changes. The mean residence times of carbon stocks change in response to redistribution of carbon among biomass components postfire and their subsequent relative rates of losses and gains. We investigated the dynamics of carbon stock change within the current disturbance regime and not predicted future regimes or climate.

(iii) We evaluated our results about the magnitude and timeframe of carbon stock changes in terms of their significance for the rules and methodologies governing national carbon accounting for forest management. Now is an important time to contribute quantitative information to the development of accounting guidelines for post-2020 agreements which include forest management and disturbance events under the Durban Platform for Enhanced Action [23].

To address these issues about accounting for carbon stock changes due to wildfire, we examined changes in carbon stocks in a temperate eucalypt forest in south-east Australia that was subjected to a wildfire in 2009. We selected a temperate forest because they contain high carbon stocks and contribute 34% to the global forest carbon sink; a proportion that has been increasing over the last two decades [24]. Changes in carbon stocks due to combustion during the fire were assessed empirically and we also investigated the long-term dynamics of biomass carbon pools postfire using a simulation model, both at the site level and up-scaled across the landscape.

Methods

Description of study region

Access to field sites was granted by Parks Victoria and the Department of Environment and Primary Industries Victoria. Our study region (2,326 km^2) was the tall, wet sclerophyll montane ash forests (*Eucalyptus regnans, E. delegatensis* and *E. nitens)* in the Central Highlands of Victoria (Figure 1). These temperate forests are among the most carbon-dense in the world and protection of these carbon stocks offers the potential for mitigation [25]. These forests are subject to a disturbance regime of infrequent, high intensity wildfires [26,27].

Wildfire disturbance regime

The known historical fire frequency has been one in the 1700s, three in the 1800s, nine in the 1900s, and three in the 2000s [27,28], and their spatial extent was mapped across the study region (Figure 1). Occurrence of these fires does not necessarily mean that each conflagration occurred at the same locations or at the same intensity. The mean interval for fires that kill ash trees at a site has been estimated to be 75 to 150 years [29,30]. However, an estimate that approximately half the trees survive within the boundary of a wildfire [30] means that only every second fire in any one location is of sufficient intensity to kill ash trees. Hence, the mean interval of all fires is estimated to be 37 to 75 years. A maximum fire interval of 350–400 years was estimated based on maximum age of the montane eucalypt species [31].

Unlike most eucalypt species, *E. regnans* and *E. delegatensis* are obligate seeders and usually killed if their canopies are predominantly scorched by fire [32,33,34]. However, wildfires are highly

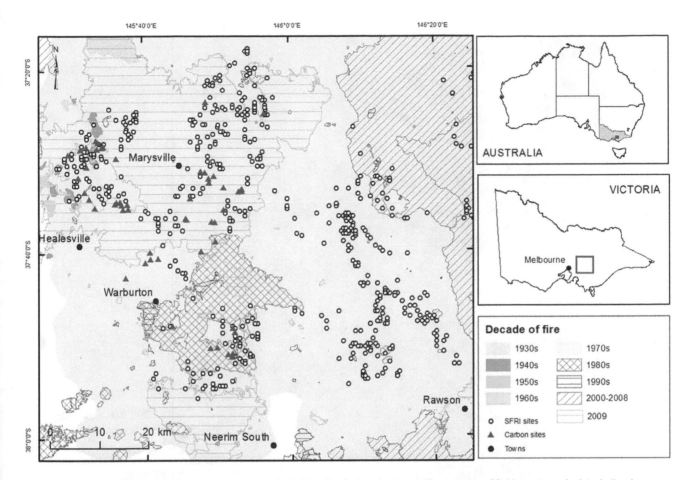

Figure 1. Location of the study region in the Central Highlands of Victoria, Australia. Location of field sites is marked, including long-term ecological monitoring sites (n = 54) where carbon stocks in biomass components were measured, and inventory sites (n = 876) where carbon stocks were estimated and used for upscaling. The spatial extent of wildfires shown represents the outer boundary of the burnt area.

variable in extent and severity, which is reflected in current stand age structures of a mosaic of even-aged and multi-aged forests [26,35,36]. Even-aged stands result from widespread mortality and regeneration following high-severity wildfire. Multi-aged stands result from partial tree death and subsequent regeneration alongside surviving trees [29,37,38]. Hence, the variability of historical fires is important for determining current stand structures. For example, our entire study area is within the boundary of the 1939 fire. However, evidence that individual trees and forest stands survived is seen by the current age structure of patches of old-growth forest where many living trees have fire scars from the 1939 fire event [39].

A wildfire started on 7th February 2009 that burnt over 450,000 ha in Victoria under extreme hot, dry weather conditions (46.4 °C in the nearby city of Melbourne), with winds up to 100 km hr^{-1} [40,41]. Within our study region of montane ash forest, 485 km^2 burnt at low-severity and 283 km^2 burnt at high-severity (Figure 2). Fire severity was mapped by the Victorian Department of Environment and Primary Industries using air photo interpretation and categorised by degree of crown scorch and subsequent tree survival or mortality.

Field sites

We selected 54 sites within the montane ash forest region of the Central Highlands of Victoria (Figure 1) from an existing network of 170 long-term ecological monitoring sites [27] to provide sites

with a range of conditions of forest age and severity of the 2009 wildfire. The site categories of forest age were 1983 and 1939 regrowth after wildfire followed by salvage logging or clearcut logging, and old growth. Salvage logging was common practice after wildfire to remove the timber resource [42,43,44,45]. Selection of these regrowth sites provided reasonably even-aged stands of similar structure, although some residual older trees occur [46]. Old growth forest consisted of dominant trees greater than 120 years old but mostly 250 years old that had regenerated following a fire in approximately 1750 [39]. The site categories of severity of the 2009 wildfire were unburnt, low-severity and high-severity burns. Our distinction between low- and high-severity fire was based on site observations of full scorch of the canopy and mortality of *E. regnans* and *E. delegatensis* trees. The three forest age and three fire severity categories produced a matrix of nine categories, each with six replicate sites.

While the 54 sites covered a range of forest stand structures and disturbance histories, the assumption that the sites had comparable pre-fire biomass carbon stocks, disturbance history, and local biophysical conditions did not necessarily hold. Their carbon stocks likely varied because of the natural heterogeneity of the landscape and the fact that fires are less likely to burn in wet, highly productive mature and old growth forest areas [47,48,49]. Furthermore, the monitoring protocols used prior to the 2009 fire at the sites were aimed at biodiversity assessments, and did not include all the measurements required to quantify ecosystem

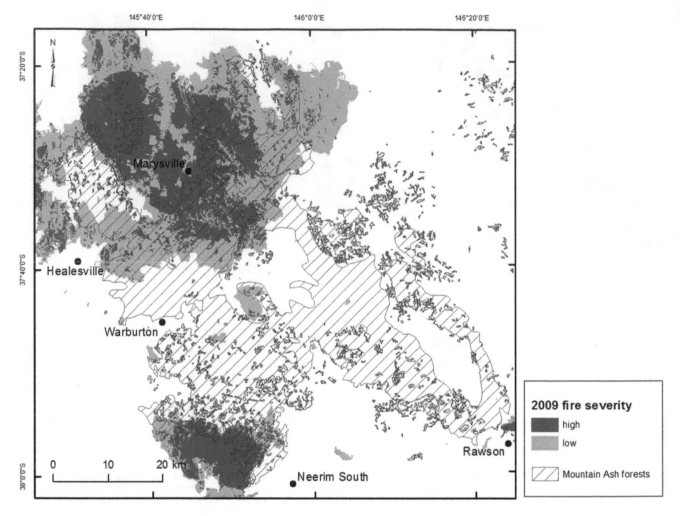

Figure 2. Spatial extent of the 2009 wildfire within the study region. The fire was categorised as low-severity where most trees survived, and high-severity where most trees were killed.

carbon stocks. Therefore, we undertook additional field survey to ensure there was a consistent set of measurements across sites.

Measurement of biomass carbon stocks and stock loss due to fire

We measured carbon stocks in all aboveground biomass components in the forest (living vegetation, dead standing trees, coarse woody debris (CWD or dead and downed woody debris), and litter) at the selected 54 sites in 2010. We defined biomass as all intact organic components of the ecosystem, both living and dead, above- and below-ground, but excluding soil organic matter. Biomass components were classified for ease of measurement and to track movement of carbon between ecosystem components. Classification was based on the status of living or dead, size and vertical location. We used different sampling strategies for each biomass component within 1 ha sites to maximise accuracy and efficiency of measurements of different sizes, densities and distributions of components (Table 1 and Figure 3).

All woody stems, both living and dead, greater than 2 m height and less than 100 cm diameter (diameter at breast height, DBH) were assessed in height and diameter categories for each species within three 10 m×10 m plots (0.03 ha). Trees greater than 100 cm DBH were measured in two perpendicular and intersect-

ing 100 m transects by 30 m width (0.51 ha). In large trees that form buttresses with fluted stems [50], there is a cross-sectional area deficit that was accounted for by converting to a 'functional' diameter [51]. The allometric equation for *E. regnans* derived by Sillett et al. [51] was used to calculate stem and branch volume, and multiplied by wood density and carbon concentration to derive aboveground biomass carbon content. Average stem wood density was 0.520 g cm^{-3} [52,53,54,55] and branch wood density was 0.677 g cm^{-3} [51]. A carbon concentration of 0.5 gC g^{-1} was used for all biomass components [56,57]. Internal decay or hollows in stems were accounted for in the calculation of biomass derived from stem volume [58,59]. Equations to predict occurrence and volume of decay related to tree size were derived from a subset of inventory data (734 trees) [60] where dimensions of defect in sawlogs had been measured. Allometric equations for other vegetation components included *Acacia* spp as mid-storey trees [61], a general rainforest equation for other mid- and under-storey species [62], and treeferns were calculated as a cylindrical volume and measured wood density. Root mass in *E. regnans* forests has not been measured, hence an average root: shoot ratio for eucalypt forests of 0.25 [63] was used to convert aboveground biomass to total tree biomass for living and dead trees.

Table 1. Methods for measuring carbon stocks in biomass components within 1 ha sites post-fire at burnt and unburnt sites.

Biomass component	Sampling strategy	Measurement
Trees (living and dead) eucalypt and rainforest species		
<100 cm diameter	Three 10 m×10 m plots (0.03 ha)	Estimated height and diameter in size categories
>100 cm diameter	Two perpendicular intersecting 100 m×30 m transects (0.51 ha)	Measured DBH and height.
Mid- and Under-storey (living and dead) non-eucalypt species		
<20 m height	Three 10 m×10 m plots (0.03 ha)	Estimated height and diameter in size categories
Coarse woody debris		
<60 cm diameter	Line intersect method along 6×10 m transects	Log diameter, decay class, hollows, charcoal and bulk density [64,65]
>60 cm diameter	2×100 m transects	
Litter layer		
<2.5 cm diameter	15 points along transects per site. 30 quadrats randomly located at sites to cover range in litter depths	Measured litter depth.
		Measured litter dry mass.

Coarse woody debris consisted of all woody material ≥25 mm diameter on the ground, and was measured using the line intersect method [64,65]. Logs less than 60 cm diameter were measured along 6×10 m transects within the site and logs greater than 60 cm diameter were measured along 2×100 m perpendicular transects. Each piece of CWD was assessed for hollows and degree of decay in three categories. Bulk density was measured for each category of decay to determine biomass. The litter layer consisted of organic material less than 25 mm diameter including leaves, twigs, insect detritus, animal scats, and comminuted material that was recognisable as organic material. Litter depth was measured at 15 points per site on a compressed litter pack to standardise the quantity of loose material. Depth was converted to biomass using a relationship derived from 30 quadrats sampled for dry weight of litter. Our results for carbon stocks refer to the total biomass of living and dead, above- and below-ground components.

We estimated biomass carbon stock loss by combustion in the 2009 wildfire from the following lines of evidence: (i) field observations and photographs taken immediately after the fire (within 2 months) [66], (ii) measurements of some components of the forest that were measured pre-fire compared with post-fire, and (iii) measurements of biomass components post-fire taken at burnt and unburnt sites. Biomass components combusted in the fire were measured specifically; including, hollow trees, decorticating and rough bark, canopy, shrub biomass, CWD and litter (Table 2 and Figure 4).

Hollow trees, both living and dead, are highly susceptible to combustion because fire is funnelled through the pipe [67]. Data for individual tree heights and diameters from 2005 and 2011 [67] were compared to assess the change pre- and post-2009 fire. The difference in carbon stock between years represented a combination of natural mortality, decay and collapse of trees plus combustion at the burnt sites. Carbon stock loss was calculated as a proportion of the initial stock and the difference between burnt and unburnt used to distinguish the effect of natural decay and combustion.

Biomass of decorticating or ribbon bark hanging from the upper stem and branches was related to tree DBH by a non-destructive sampling method using estimated standard units [68]. All decorticating bark was combusted in the fire. Biomass of rough or fibrous bark on the lower stem was estimated from measure-

ments of bark thickness and bulk density related to tree DBH. Combustion of rough bark was determined from the difference between burnt and unburnt sites. Both decorticating and rough bark mass were calculated for all the inventory trees at the 54 sites.

Canopy leaf biomass was estimated to be approximately 7 tC ha^{-1}, derived from annual litterfall of 4.7 tC ha^{-1} yr^{-1} in ash forests [69,70] and longevity in the canopy of about 18 months [50,71]. Twig (diameter <4 mm) biomass is a similar amount to the leaves [72]. The proportion of the canopy combusted was estimated from photographs after the fire of the leaves remaining in the canopy and the scorched leaves that had fallen to the ground.

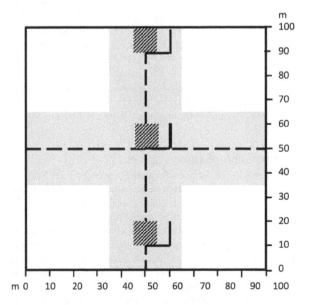

Figure 3. Design of 1 ha experimental sites. Sites of 100 m×100 m included central and perpendicular transects (dashed lines), 10 m×10 m plots (hatched), 10 m transects (black lines), and area sampled for large trees 2×30 m×100 m=0.51 ha) (grey shaded area). Different biomass components were sampled in different parts of the site.

Table 2. Methods for measuring changes in carbon stocks of biomass components after wildfire, within 1 ha sites.

Biomass component	Sampling time	Sampling strategy	Measurement	Proportion of biomass combusted	
				Low-severity fire	High-severity fire
Hollow trees	Measured in 2005 and 2011	All trees in 1 ha	Measured DBH and height.	0.12 to 0.29	0.16 to 0.56
Decorticating bark	Post-fire, burnt and unburnt sites	Representative size range of trees.	Measured relationship between tree DBH and bark mass.	1.0	1.0
Rough bark	Post-fire, burnt and unburnt sites	Representative size range of trees.	Measured relationship between tree DBH and bark thickness plus bulk density.	0.3 (for trees >200 cm)	0.6 (for trees >200 cm)
Canopy leaves	Post-fire, burnt and unburnt sites	Observations and photos	Estimated biomass from litterfall and leaf longevity	0.5	1.0 (mostly)
Canopy twigs	Post-fire, burnt and unburnt sites	Observations and photos	Relationship between leaf and twig mass.	0.5	1.0 (mostly)
Shrubs	Post-fire, burnt sites	Representative range of shrub sizes and species.	Minimum size of tips remaining on stems after fire. Calculated biomass in each stem diameter class in plots.	0.25 to 0.5 (from branch diameter <5 cm)	0.3 to 0.75 (from branch diameter <5 cm)
Coarse woody debris	Post-fire burnt sites	Line intersect transects	Proportions of logs charred. Lines of charcoal remaining post-fire.	Highly variable dependent on log size	Highly variable dependent on log size
Litter	Photos soon after the fire	Average from all low- and high-severity burnt sites.	Proportion of litter layer combusted.	1.0	1.0

Combustion of shrub biomass in woody stems, branches and twigs was assessed using measurements of: (i) minimum sizes of tips remaining on stems after the fire, (ii) the proportion of biomass lost from a shrub when different tip sizes were removed, and (iii) the amount of biomass in each stem diameter size class of shrubs from the inventory data. A range of species and sizes of understorey shrubs was sampled.

Combustion of CWD was determined on each piece along the measurement transects by the proportion of the piece charred, or by the number of lines of charcoal remaining on the ground. The initial carbon stock of the logs that were completely converted to charcoal was assumed to be represented by the mean size of logs at the site. CWD was too variable between burnt and unburnt sites within stand age categories to allow a comparison to be indicative of the amount combusted as the mean of a category. The proportion of the litter layer combusted was estimated from photographs of quadrats taken soon after the fire in low- and high-severity burnt areas and compared with litter biomass in unburnt sites.

Spatial estimation of carbon stocks

Spatial up-scaling of the site data to the landscape level followed the statistical modelling approach of Keith et al. [73]. Calibration data for the model used the pre-2009 carbon stocks from the 54 sites described above, as well as an additional 876 sites with inventory data [60]. The statistical model regressed site carbon stocks against the following explanatory variables at a 250 m grid resolution: climate variables of precipitation, water availability index, temperature (minimum, mean, maximum) and radiation; topographic variables of topographic position, elevation, aspect and slope; substrate variables of soil parent material, lithology and soil organic carbon content; gross primary productivity (GPP, derived from remote sensing); forest type; forest management area; and disturbance history. Methods used to derive spatial estimates of the environmental and GPP variables are documented in Mackey et al. and Berry et al. [26,74,75]. Disturbance history included wildfires and logging under different silvicultural systems,

both of which have complex effects on forest age structure. The statistical regression model was used to generate spatial estimates of carbon stocks across the study region.

Impacts of the wildfire on ecosystem carbon stocks were quantified by determining first, the loss in carbon stock of each biomass component combusted, and second, the redistribution of carbon among biomass components within the ecosystem. Spatial estimation of the carbon stock post-2009 fire was based on measured reductions in the stock according to the fire severity category in the burnt areas and forest age.

Modelling change in carbon stocks

Changes in carbon stocks were simulated for each forest age category after high-severity fire at the average historical return time of 112 years. Initial conditions for the model used site data of the current carbon stock in each age category and the proportion of the stock combusted. After the fire, carbon stocks were redistributed among living and dead components according to the proportions measured in each age category. For the second fire cycle, the initial condition of total biomass at the time of the fire was the stock simulated for year 112, and the proportions of the stock combusted and redistributed were taken as the average of the site data from the 1939 regrowth and old growth forests. Changes in carbon stocks over time were simulated using functions to describe the processes of regeneration, mortality, collapse of dead trees, and decomposition of CWD derived from existing data and observations for montane ash forest (Table 3).

The rate of regeneration was based on carbon stocks at sites with different times since stand-replacing disturbance, derived from the current study and inventory data [60]. Mortality due to self-thinning of the regeneration was estimated from changes in stem density with age, based on data from the literature [76,77]. Mortality of living trees that survived the fire was based on measured rates of mortality over 28 years [67] and the average longevity of *E. regnans* of at least 250 years [78], but up to 500 years [79]. The rate of collapse of dead standing trees was described by a logistic function, with a slow initial rate that

Total Carbon Stocks (tC ha⁻¹)

	Living Biomass	Litter	Course Woody Debris	Dead Standing Trees
Pre-fire	822	8	51	41
Post-fire	795.3	0	50.6	19.3

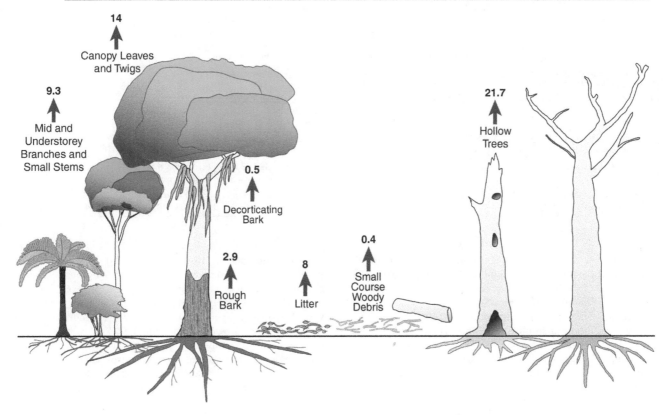

Figure 4. Schematic of carbon stocks in the forest ecosystem and stock changes resulting from fire.

increased to a maximum and then a slow final rate when few trees remain. This pattern was based on observations and evidence that dead trees remain standing for approximately 10 to 75 years [66,80]. The biomass in dead trees was then transferred to the CWD pool. The rate of decomposition of CWD and dead root biomass used in the model was 0.07 yr⁻¹, which had been derived from empirical data for *E. regnans* [55]. This rate was intermediate among published rate constants [80,81,82], and produced amounts of CWD that were consistent with observed amounts in montane ash forest.

Results

Current carbon stock

Biomass density varied substantially among sites within the montane ash forest due to the disturbance history and subsequent age distribution of the forest. Biomass density (mean ±standard error, n = 6 sites) in unburnt stands was 405±33 tC ha⁻¹ in 1983 regrowth, 603±74 tC ha⁻¹ in 1939 regrowth, and 1039±44 tC ha⁻¹ in old growth sites. The structure of these forest stands is illustrated in terms of the distribution of their carbon stock by tree sizes (Figure 5). These tree sizes demonstrate that the regrowth stands were not necessarily even-aged and included some larger residual trees which contained high, but variable, levels of biomass.

CWD biomass was highly variable among sites in the age/fire categories, ranging from 15 to 186 tC ha⁻¹, with a mean of 67 tC ha⁻¹. There was no significant difference in CWD biomass between forest age categories, reflecting the varied disturbance history of these forests where CWD remained after logging and wildfire, and its biomass was not related to the age of the stands. Litter biomass in the unburnt sites ranged from 6 tC ha⁻¹ in the 1983 regrowth to 9 tC ha⁻¹ in mature and old growth stands. This litter biomass corresponded to average site depth of litter when compressed of 2 cm and 3 cm, respectively, with the majority of measured points less than 5 cm and only a few points with 5–10 cm depth.

Biomass components combusted

In areas burnt by low-severity fire, most of the over-storey trees survived as well as variable proportions of mid- and under-storey vegetation with green, scorched or combusted leaves. In areas burnt by high-severity fire, most of the leaves, twigs and decorticating bark in all canopy strata were combusted, although scorched leaves remained in patches. Some montane ash forests included a few *E. nitens* trees that regenerated by epicormic growth. Woody stems of most shrubs and mid-storey trees remained, but some were entirely combusted. The average branch size combusted from shrubs was 4 mm (maximum 19 mm) in low-severity fire, and 8 mm (maximum 42 mm) in high-severity fire.

Table 3. Input data and equations to predict biomass carbon stock (B in tC ha⁻¹) as functions of time (t) since disturbance in the model of change in carbon stocks after a high-severity wildfire in old growth montane ash forest.

Process	Function	Data	Source
Regeneration (= gain in carbon)	$B_{regen}(t) = 1200 \times (1 - \exp(-0.0045 \times t))^{0.7}$	Inventory data from 99 sites	46,60
	$Log_{10}(SD_t) = 1.28 + 3.16 \times 0.913^t + 1.9 \times 0.99^t$ $Log_{10}(SD_t) = 11.61 - 1.624 \log_{10}(t)$ (average from the two functions used)	Double exponential function fitted to chronosequence site data for *E.regnans*. Function derived from site data	76,77
Mortality of regeneration (= input to dead trees)	$B_{dead}(t) = [(SD(t-1) - SD(t))/SD(t)] \times B_{regen}(t)$	Empirically derived from stem density changes	Current
Branch fall (= input to CWD)	$B_{branch}(t) = B_{regen}(t) \times 0.005$	Rate constant derived empirically to produce CWD biomass in the range observed (3–255 tC ha⁻¹).	96
Mortality of surviving trees (= input to dead trees)	$B_{dead}(t) = 135 \exp(-0.015 \times t)$	Rate of mortality estimated from site data.	67
Collapse of dead trees (= input to CWD)	$B_{CWD\_in}(t) = 687/(1 + \exp(0.1 \times t - 5)$	Logistic function derived to fit observations that dead trees remain standing for 10 to 75 years.	66
Decomposition of CWD (= loss of carbon)	$B_{CWD\_loss}(t) = 124 \exp(-0.07 \times t)$	Rate constant. Modelled range in CWD consistent with site data (mean of 51 tC ha⁻¹ and max. of 255 tC ha⁻¹)	55,96 Current

The carbon stock losses from each biomass component and the proportion of the total biomass combusted differed with age class of the forest (Table 4). The distribution of readily combusted components within the ecosystem influenced the proportion of the stock emitted during the fire. Much of the observed change due to fire was combustion of shrubs in the understorey. However, this represented a small proportion of the total biomass in the ecosystem. Combustion of 7.8–16.2 tC ha⁻¹ in the small branches on shrubs of all stem diameter categories in the understorey (Table 4) represented 1.6–4% of the total living biomass (above-

and below-ground) in low-severity fire, and 1.9–6.7% in high-severity fire (range from young regrowth to old growth stands).

Some CWD was combusted during the fire, but both the absolute amount and proportion of initial biomass were highly variable spatially. Higher amounts of CWD were combusted in the 1983 and 1939 regrowth stands (range 1.37–2.41 tC ha⁻¹) than in the old growth stands (0.16–0.43 tC ha⁻¹) (Table 4). Large amounts of CWD occur in these regrowth stands due to self-thinning and residual material from previous logging and fire events. The high rates of combustion in regrowth stands occur

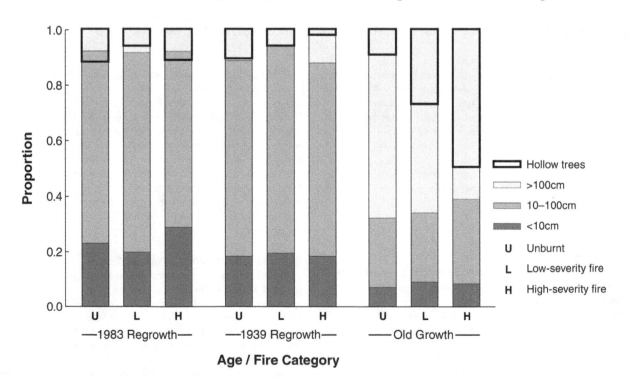

Figure 5. Distribution of carbon stocks in trees according to their size. Carbon stock is shown as the proportion in stem size classes of DHB and trees with hollows, averaged for the sites in each age/fire category (n = 6 sites per category).

Table 4. Summary of biomass components combusted by low- and high-severity fires.

	Mass of carbon[1] combusted (tC ha^{-1})					
	1983 regrowth		1939 regrowth		Old growth	
Low-severity fire	mean	SE	mean	SE	mean	SE
Litter	6	1.0	9	1.0	9	1.0
Canopy leaves	3.5		3.5		3.5	
Small stems	1.5	0.6	0.6	0.4	0.2	0.2
Small branches	11.9	1.2	11.8	2.7	7.6	3.4
Decorticating bark	0.3	0.02	0.3	0.06	0.5	0.07
Rough bark	0.6	0.2	0.6	0.1	1.3	0.3
CWD	1.4	0.6	1.8	0.7	0.2	0.2
Dead standing trees	3.8	0.9	9	1.7	17.6	5.2
Total biomass[2] combusted[2]	29	4.5	36.6	6.7	39.9	10.4
Total carbon stock[3]	433	35	475	91	564	88
Proportion combusted[4]	0.063		0.072		0.066	
High-severity fire						
Litter	6	1.0	9	1.0	9	1.0
Canopy leaves	7		7		7	
Canopy twigs	7		7		7	
Small stems	2.2	0.5	0.8	0.3	0.3	0.2
Small branches	14	3.0	15	4.0	9	2.0
Decorticating bark	0.3	0.05	0.4	0.1	0.5	0.1
Rough bark	0.8	0.1	0.7	0.2	2.9	0.4
CWD	2.4	1.7	1.7	0.5	0.4	0.3
Dead standing trees	7	1.9	14.8	1.2	21.7	6.2
Total biomass[2] combusted[2]	46.7	8.3	56.4	7.9	57.8	10.2
Total carbon stock[3]	28	46	459	102	589	63
Proportion combusted[4]	0.140		0.109		0.089	

[1]Mass of carbon (mean ±standard error, n=6 sites).
[2]Total biomass combusted is the sum of all components combusted.
[3]Total carbon stock (above- and below-ground) pre-fire is current carbon stock before the fire.
[4]Proportion combusted is the total biomass combusted divided by the pre-fire carbon stock.

because large proportions of the CWD are small logs and drier logs (decay class 1), compared with the old growth stands. Hollow or damaged trees, either living or dead, were a major source of carbon stock loss (3.8–21.7 tC ha^{-1}) (Table 4). Where these trees had hollows in the centre of the stem, flames funnel up like a chimney and ignite exposed dry wood. In regrowth stands, hollow trees comprised 2–12% of the total tree biomass, and contained a carbon stock ranging from 7–50 tC ha^{-1}. In old-growth stands, trees with hollows generally consisted of the older cohort of trees but also included damaged younger trees. They represented a highly variable proportion of the total biomass, from 9–49%, and contained a carbon stock of 45–241 tC ha^{-1}.

The average carbon stock loss from biomass components was 40 tC ha^{-1} and 58 tC ha^{-1} in low- and high-severity fires, respectively, and this represented an estimated maximum amount (Table 4). As a proportion of the total biomass carbon stock (above- and below-ground), this loss represented 6–7% in low-severity fires and 9–14% in high-severity fires.

Distribution of carbon among components due to fire

The largest impact of fire on the ecosystem carbon stock was a shift in distribution among biomass components. The proportion of total biomass that was in living vegetation in unburnt sites was 72%, 71% and 89%, respectively in the 1983, 1939 regrowth and old growth sites. The shift from living to dead biomass was 3%, 6% and 19% due to low-severity fire, and 85%, 97% and 84% due to high-severity fire, respectively in the age categories (Figure 6). CWD increased by only a small proportion in the burnt sites; some logs were combusted and a similar amount was transferred into the CWD component.

Carbon stock dynamics

Combustion of biomass during the fire resulted in an initial loss of some carbon, but most of the biomass remained at the site. Our simulation model predicted that carbon stocks would increase for several decades after the fire due to the combination of regeneration of living biomass and persistence of dead biomass components (Figure 7). Mortality of trees during the fire produced large amounts of dead biomass, but the trees only collapsed slowly and then contributed to the CWD pool, from where carbon stock loss by decomposition occurred over many decades. The lowest carbon stock in an old growth forest occurred nearly a century after a single fire when most of the dead biomass produced by the fire had decomposed. The maximum fluctuation in carbon stock

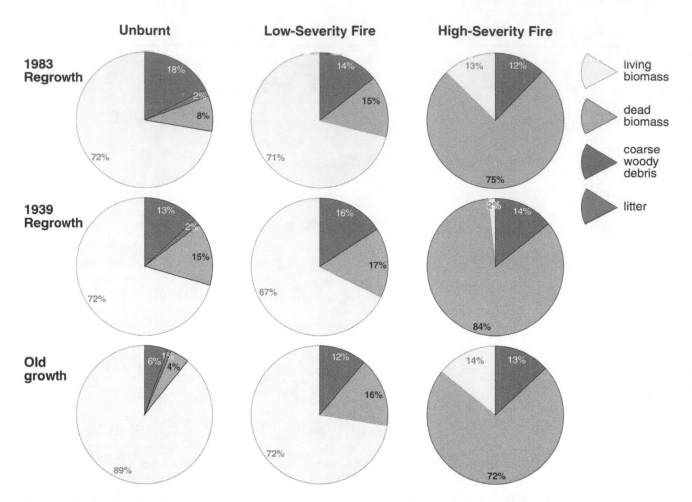

Figure 6. Distribution of carbon stocks among components in relation to stand age and fire severity. Distribution is shown as the proportion of carbon in each biomass component, including living vegetation, dead standing trees, litter and coarse woody debris.

over time as a result of fire was estimated to be 20% from the scenario of a single high-severity fire in old growth forest where there was no trend in total biomass due to changes in forest age (Figure 7a).

The carbon stock dynamics in the forests of initially different ages reflect the differences in the total biomass stock, its distribution among living and dead trees and CWD, the proportion combusted, and the proportions remaining in each of the components after the fire. The old growth forest had the largest increase in total biomass with the combined regeneration plus remaining dead trees, but subsequently the largest decline in total biomass as the dead trees collapsed and decomposed (Figure 7b). In the regrowth forests from 1939 and 1983, there was less dead biomass after the fire and the regeneration resulted in increasing total biomass carbon stock over most of the century following the fire (Figure 7c and d).

Under a regime of repeated fires on a 112-year return time, trends occurred in the total biomass carbon stock because forest age changed from that of the initial forest. The old growth forest decreased in carbon stock because of the more frequent disturbance, whereas the 1983 and 1939 regrowth forests increased in carbon stock because they remained undisturbed for a longer period than at their initial age. These scenarios of carbon stock dynamics represent an average forest stand of a given initial age subject to the average wildfire regime. A range of forest age classes occur across the

landscape, including forests much older than 112 years. Forests that remain undisturbed for several hundred years have the capacity to continue accumulating carbon stocks.

Spatial estimation of carbon stocks post-2009 wildfire

The carbon stock loss post-2009 fire was estimated spatially from the distribution of biomass carbon density in the unburnt montane ash forest and the proportion of biomass combusted in low- and high-severity fire for each forest age category (Figure 8). The total carbon stock in the montane ash forest region (2,326 km^2) was 112.8 TgC pre-2009 and an estimated 3.9 TgC was emitted from biomass combustion during the 2009 wildfire (burnt area of 768 km^2 or 33% of the area of montane ash forest). The carbon emitted represented 8.5% of the total biomass carbon stock in the area of montane ash forest that was burnt.

Discussion

Although the total amount of carbon stock loss across the landscape burnt during the wildfire was large as an annual flow in the national inventory, we found that the proportion of total ecosystem carbon stock combusted was relatively small, both at site and landscape scales. The major impact of wildfire was the redistribution of carbon among biomass components, from living to dead biomass and CWD, and their longer-term rates of change due

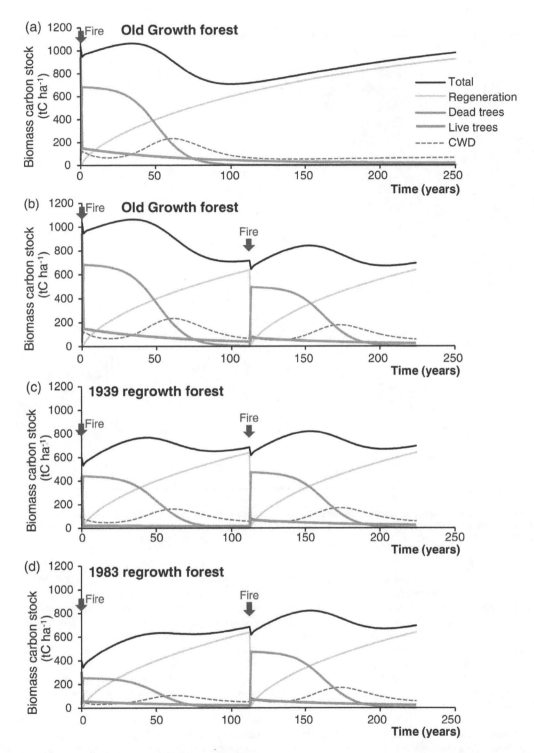

Figure 7. Simulated carbon stock change in biomass components of forests of different ages after high-severity wildfire. Initial fire occurred in year 1 and the change over time in each biomass component, plus the total biomass, is simulated; (a) old growth forest initially that is burnt in year 1 and recovery is simulated over 250 years assuming no other disturbance; (b) old growth forest initially that is burnt in years 1 and 113, according to the average fire frequency, and recovery simulated over two fire cycles; (c) 1939 regrowth forest initially that is simulated over two fire cycles; (d) 1983 regrowth forest initially that is simulated over two fire cycles.

to decomposition and regeneration. Assessing the carbon stock loss by combustion in relation to the total stock, rather than fuel load, meant that changes in stock could be tracked over time and compared between different human activities and disturbance types.

Amount and proportion of forest ecosystem carbon stock combusted

We found that the proportion of total ecosystem biomass carbon stock combusted during a wildfire in the montane ash forest was

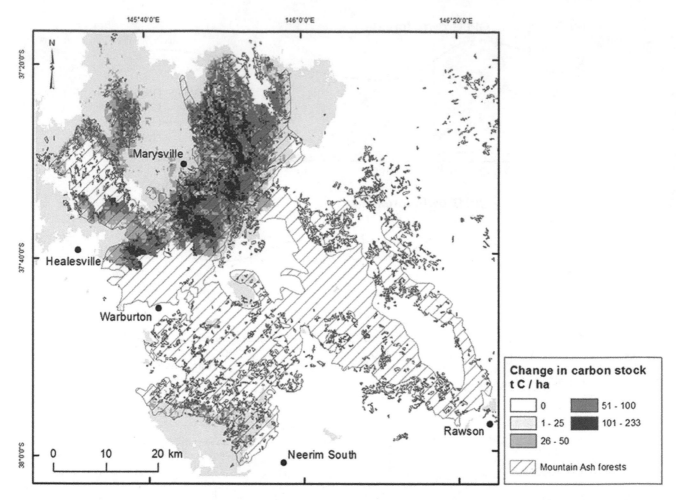

Figure 8. Spatially modelled change in carbon density (tC ha⁻¹) across the montane ash forest region. Change in carbon stock was calculated as the difference between pre- and post-2009 wildfire, including total biomass of living and dead trees, above- and below-ground, shrubs, litter and coarse woody debris.

6–14% at the site scale depending on fire severity and forest age (27 years to old growth), with an average of 8.5% across the landscape that was burnt. This is a relatively small proportion compared with impacts of human disturbance events, such as logging that removes biomass off-site [46,83,84]. Our results are similar to studies in other forest ecosystems (boreal, temperate mixed conifer, savannah woodlands, and other temperate wet sclerophyll) that estimated the relationship between carbon stock loss with total stock, but are lower than estimates of proportion of biomass combusted related to fine and coarse fuel loads (data and references in Table S1).

The estimated loss of only a small proportion of ecosystem carbon stock combusted in a fire concords with simulation results of net losses of carbon by 4 to 8% over 180 years under a fire regime scenario of increased frequency and intensity in south-east Australia [85]. Under this scenario, the loss was due to increased mortality leading to a younger age distribution of trees. Similarly, a sensitivity analysis of the carbon balance model for Canadian forests showed that a three-fold increase in area burned decreased the total forest carbon stock by 6% over 100 years [86]. These predicted losses of carbon stocks are relatively small over a century or two compared with the impact of many human activities.

Differences in the proportion of fine fuel combusted in different ecosystems are insignificant in terms of the total ecosystem carbon

stock, representing 0.5 to 1% or 0.5 to 3 tC ha⁻¹. The proportion of coarse fuel combusted varies greatly depending on the size of material, which is influenced not only by forest age but also by disturbance history. Although proportions of biomass combusted were lower in montane ash forest than in other forest types, the amounts combusted in these fuel size classes were greater (Table S1). Amounts of carbon emitted during combustion in different ecosystems depend on vegetation type, past disturbances resulting in dead biomass, and fire weather conditions [87].

We conclude from our methodology that estimating carbon stock loss as a proportion of the total ecosystem carbon stock (based on tracking each biomass component) is a useful metric to compare with other impacts on ecosystem carbon stocks and other sectors of the national GHG inventory. This metric is thus more appropriate for national accounts than the current default reporting of emissions as the product of a combustion factor and fuel load in the IPCC guidelines and Australian national inventory report [6,15,82,88,89]. In our method, all biomass components are assessed without making assumptions about which components constitute fuel.

Our values of carbon stock loss in a wildfire may be small overestimates because some of the biomass combusted is not volatilised and therefore not emitted to the atmosphere. Under conditions of reduced oxygen, partial combustion forms char that remains on the ground, or particulate carbon that is transported in smoke but settles to the ground elsewhere. The proportion of char produced in wildfires has been estimated to be 1–10% of the biomass combusted [90]. Both these forms of carbon are highly resistant to decay and form a long-term stable component of soil carbon stocks [90].

Changes in forest biomass carbon dynamics

The major impact of wildfire in the montane ash forest was the redistribution of carbon among biomass components, from living to dead, and subsequent long-term stock losses from decomposition and gains by regeneration. The relative rates of these processes and subsequent mean residence times of the biomass components determine the long-term dynamics of carbon stock in the ecosystem. These long-term dynamics resulted in changes of up to 20% about the mean stock, whereas direct losses from combustion represented 6–14% of the total biomass stock.

Mean residence time of the biomass components combusted is critical for determining the impact of wildfire on longer-term carbon dynamics. More than half the stock losses from combustion were derived from biomass components with short lifetimes; in the canopy, bark, litter layer and shrubs. These components are described as fine fuel, and this amount remains reasonably constant after about 10 years post-fire in a range of forest types [91]. For example, fine litter in montane ash forest has a mean residence time of 1–4 years [50,70], and 85–95% of the litter mass has accumulated within a decade after fire [91]. Recovery of the carbon stock within a decade in the montane ash forest is within the range of up to 10–20 years for recovery of rates of carbon uptake estimated from eddy covariance measurements of net ecosystem carbon exchange at chronosequence sites across a range of forest types in North America [92].

The components combusted with longer mean residence times, such as CWD and large hollow trees, have variable stocks and proportions combusted depending on forest type, conditions for decomposition, logging and fire history. This coarse material contributed 18 to 45% of the carbon combusted in the montane ash forest, with an increasing proportion in older stands. The residence time of this coarse material is in the order of decades to centuries [67,93].

Rates of carbon accumulation in regenerating stands are difficult to estimate due to limited site data of known forest age, particularly in older aged stands. We compared carbon uptake in regenerated living biomass at a given age estimated from our site data, for example, 3.8 tC ha^{-1} yr^{-1} at 80 years, with the net ecosystem exchange (NEE) estimated from eddy covariance measurements. An average NEE over 10 years of measurements was 5.8 tC ha^{-1} yr^{-1} and the maximum was 9.3 tC ha^{-1} yr^{-1} in an 80 year old stand of *Eucalyptus delegatensis* or Alpine Ash [94]. This average NEE is comparable to the 6 tC ha^{-1} yr^{-1} in a temperate forest in Oregon, which is the highest value reported for forests of a similar age in North America [92]. Carbon accumulated in living biomass does not include uptake of carbon that is then transferred to dead biomass pools and soil organic matter.

Our simulated carbon stocks for each biomass component were within the range observed in the field. Actual magnitudes of stocks at different ages were difficult to verify because of the confounding disturbances of wildfire and salvage logging that have occurred in these forests. Among our sites, there were no younger stands that had been burnt but not logged and such stands would be rare in the landscape. Our simulated turnover times appeared to be slower than in some other ecosystems, such as boreal forests [95]. However, observations of large logs remaining after 60–90 years in the montane ash forest [50,96] suggested that even slower rates may be characteristic of large biomass components. Uncertainty about the magnitude of the changes and their timing in the long-term dynamics of carbon occurred mainly because of limited information about rates of tree fall and decomposition.

Inclusion of wildfire in national carbon accounts

Changes in carbon stocks due to natural disturbances, such as wildfire, must be included in national carbon accounts. The current UNFCCC carbon accounting guidelines for forest management (Article 3 paragraph 4 clause 33) address carbon stock change due to natural disturbances under the '*force majeure*' provision (Decision 2/CMP.7, CoP 17 Durban Climate Conference) [97]. Natural disturbances, such as wildfire, outbreaks of pests and diseases, extreme weather events and geological disturbances, are considered as extraordinary events that are outside human control. An average net stock change due to natural disturbances is calculated as a background level within the forest management reference level. These accounts for forest management and natural disturbances must be included in the 2015 inventory reports from signatory nations to the Kyoto Protocol (38 of the 43 Annex 1 nations), which are assessed for compliance against emissions reduction targets [98].

The current negotiations under the Durban Platform for Enhanced Action [23] represent a critical time to provide quantitative data on carbon stock changes due to disturbance and to interpret the implications of the inventory for maximising outcomes of mitigation activities. Nations can develop and choose methodologies, within the UNFCCC guidelines, that are appropriate for their accounts [99]. The default method is calculated as an average of net stock change from natural disturbances from 1990 to 2009, plus a margin of twice the standard deviation. Annual stock changes that are above this background level are excluded from the accounts. When this exclusion occurs, the area of land affected is removed from the inventory so that subsequent uptake by the regenerating vegetation is not counted, until pre-disturbance carbon stocks are restored [99]. The UNFCCC guidelines provide some flexibility, but the methodology should adhere to the objectives of the *force majeure* provision of minimising annual variability in the reported net stock changes due to natural disturbance events. The rationale is that this variability should not negate reductions in emissions due to human mitigation activities over the timeframe of the reporting [97].

Our results about the magnitude and timeframes of carbon stock changes due to wildfires suggest that carbon accounts need to be comprehensive and include all stocks and flows. Such accounts enable differentiation of stock changes due to natural disturbances and human activities. There are four reasons why constructing carbon accounts which differentiate carbon stock changes are beneficial.

First, our results show that in terms of national annual net change in carbon stocks, the magnitude of carbon stock loss due to combustion in a wildfire is sufficient to mask reduced emissions due to mitigation activities. Our estimated carbon stock loss was 3.9 TgC from a single (albeit large) fire in 2009 within the montane ash forest. The annual average for all fires in Australia is estimated to be 8.7 TgC yr^{-1} [100]. These single and average wildfire emissions represent about 27% and 59%, respectively, of Australia's annual emissions reduction target of 14.7 TgC yr^{-1} [100]. Similar magnitudes of emissions from individual wildfires have been quantified elsewhere, for example 2.9 TgC from the 2003 Californian fires [101]. Hence, differentiation of carbon stock changes due to natural disturbances and human activities is important when national GHG inventory reports are used for assessing progress in meeting emission reduction targets over short time periods.

Second, we demonstrated that the changes in carbon stocks in response to a disturbance event occur over the following decades. Hence, derivation of a background level of net carbon stocks over a two decade period (1990–2009) is not adequate to account for the variations in carbon dynamics within a disturbed area. However, the carbon stock at the landscape scale represents the sum of forest stands that have been burnt at different times and intensities, thus producing a mosaic of stand age classes, amounts of debris, and stages of recovery from disturbance. Under a stable disturbance regime, the carbon stock at the landscape scale remains stable when accounted for over sufficiently long time periods. Hence, carbon stock losses from combustion of biomass components do not increase the accumulated atmospheric CO_2 concentration under a stable fire regime. Carbon stock dynamics may alter at the landscape level if fire regimes change due to climate change, or due to the effects of growth from a forest age class distribution affected by historical activities [9,85,86, 102,103,104].

Third, the increases and decreases in carbon stock post-fire resulting from differential rates of regeneration, mortality and decomposition mean that excluding burnt areas from the national inventory until carbon stocks are restored to pre-fire levels, is not an adequate method for excluding the effect of natural disturbances on net changes in carbon stocks. In fact, the increase in carbon stocks for some decades post-fire would produce artificial credits, but later decreases due to decomposition would produce artificial debits in the national inventory, which are not related to human activities. Additionally, the high levels of uncertainty in these long-term dynamics could produce errors in the national inventory.

Fourth, an averaging period of net stock changes from a disturbance regime over two decades, or within the time period of adequate records, does not encompass the temporal scale of natural variability in wildfire regimes in many temperate forest ecosystems [105], such as montane ash forest with a return interval of many decades [30,31]. The frequency and intensity of wildfire is influenced by interactions of large-scale features of atmospheric circulation, such as the El Niña-Southern Oscillation, Indian Ocean Dipole, Southern Annular Mode and the Sub-Tropical Ridge, which drive multi-decadal climate variability [13,106,107]. If a background level is to be used in an accounting system, then the preferred method would be to re-calculate the actual incidence of disturbance events at the end of the reporting period [108,109,110].

Methodological and data uncertainties

Our results suggest that data from ecological monitoring sites can be used *a posteriori* to estimate carbon stock losses from wildfire, when supplemented with additional field data and ancillary data sources. Our approach used the basic framework of estimating carbon stock changes due to wildfire from the area burned, fuel load and combustion efficiency [87,90] However, our sampling also incorporated spatial variability in fire severity and initial carbon stocks at the landscape level. This provided a more representative range in carbon stock losses at the regional scale. Covering this range is important even though a balanced experimental design is difficult to obtain in studies of stochastic disturbance events where field sites may not be comparable because of uncertainties in disturbance histories and stand ages. Previous studies in Australia have used either experimental fires imposed on a planned design of sites and pre-fire measurements, where the fires were either lower severity than wildfire or

regeneration burns of logging slash [89,111,112,113], or post-fire comparisons of burnt and unburnt sites albeit with limited replicate sites [41,114].

Improved assessment of fire severity as a continuous variable at the landscape scale is important to allow more accurate estimates of carbon stock changes spatially [115]. New techniques in remote sensing are providing quantitative information about fire radiative power that can be related to factors that control fire severity [116,117,118]. However, adequate field calibration of the remotely sensed metric currently limits their application [116].

The results of up-scaling the site data to estimate carbon stocks and stock losses across our study region revealed that the effect of disturbance history on tree age structure and CWD was the major source of uncertainty in spatial estimates of biomass carbon stocks. Other studies have drawn similar conclusions that the determination of forest stand age is the main factor contributing to uncertainty in assessment of regional carbon stocks [5].

Our results suggest that current national accounting that uses default values (Tier 2 methodology) for fuel combustion efficiency and fuel load can be improved by using site data from different ecosystems. Data from existing state forest inventories need to be augmented by comprehensive sampling of forest age structure, including old growth forests, and all components of the carbon stock, including dead wood. The methods used here are generic and would constitute Tier 3 methodology that is appropriate for national and project-based accounting. The sampling methods were non-destructive and feasible to apply to a large number of sites.

Conclusions

Carbon stocks in forest ecosystems change in response to wildfire in the short-term by combustion and in the long-term by redistribution among biomass components and their subsequent differential rates of decomposition and regeneration. It is important that carbon accounts include these processes because all stock changes, activities and land areas contribute to quantifying the global carbon cycle. However, the timeframes of the ecological processes that determine carbon stock changes are longer than the time periods for reporting of national inventories for compliance with emissions reductions targets. This problem of incongruous timeframes means that reporting against targets should be based on net changes in carbon stocks from human activities to incentivise mitigation activities. If the carbon accounts differentiate stock changes between the sources from natural and human-induced disturbances, then the net stock change from human activities can be extracted from the accounts for use in reporting. This distinction is currently made between the comprehensive UNFCCC accounts and the reports against targets for the Kyoto Protocol, except that the guidelines for the Protocol include net stock changes from natural disturbances that can be discounted above an averaged background level. However, our results indicate that the spatial and temporal variability in carbon dynamics is too high and the uncertainty of estimates too great to enable a realistic averaged background level that would prevent distortions of the inventory by stock changes that were outside human control.

Acknowledgments

We wish to thank Clive Hilliker for expert assistance with the figures and Steve Meacher for assistance with field measurements of shrub biomass. Helpful discussions about fire ecology with Malcolm Gill, Geoff Cary and Michael Doherty are gratefully acknowledged. The inventory plot data and spatial data for the Central Highlands were obtained from the Victorian Department of Sustainability and Environment under data access licence agreement DALA_2009_082. We thank anonymous reviewers for comments that have improved the manuscript.

Author Contributions

Conceived and designed the experiments: HK DL BM. Performed the experiments: HK DB LM SO. Analyzed the data: HK LC. Contributed to the writing of the manuscript: HK DL BM. Developed the conceptual framework and global context: HK DL BM TKN.

References

1. United Nations Framework Convention on Climate Change 1992. Available: http://unfccc.int/resource/docs/convkp/conveng.pdf. Accessed 2013 Oct 9.
2. Cubasch U, Wuebbles D, Chen D, Facchini MC, Frame D, et al. (2013) Introduction. In: Stocker TF, Qin D, Plattner G-K, Tignor M, Allen SK, et al., editors. Climate Change 2013: The Physical Science Basis. Contribution of Working Group I to the Fifth Assessment Report of the Intergovernmental Panel on Climate Change. Cambridge University Press, Cambridge, United Kingdom and New York, NY, USA. pp. 119–158.
3. Watson CE, Fishman J, Reichle HG (1990) The significance of biomass burning as a source of carbon monoxide and ozone in the southern hemisphere tropics: a satellite analysis. J Geophys Res 95: 16444–16450.
4. Cahoon DR, Stocks BJ, Levine JS, Cofer WR, O'Neill KP (1992) Seasonal distribution of African savanna fires. Nature 359: 812–815.
5. Andreae MO, Merlet P (2001) Emission of trace gases and aerosols from biomass burning. Global Biogeochem Cycles 15(4): 955–966.
6. van der Werf GR, Randerson JT, Giglio L, Collatz GJ, Kasibhatla PS, et al. (2006) Interannual variability in global biomass burning emissions from 1997 to 2004. Atmosph Chem Phys 6: 3423–3441.
7. Kasischke ES, Bruhwiler LP (2003) Emissions of carbon dioxide, carbon monoxide, and methane from boreal forest fires in 1998. J Geophys Res 108: D1,8146.
8. Le Quéré C, Raupach MR, Canadell JG, Marland G, Bopp L, et al. (2009) Trends in the sources and sinks of carbon dioxide. Nature Geosci doi: 10.1038/ngeo689.
9. Stocks BJ, Mason JA, Todd JB, Bosch EM, Wotton BM, et al. (2003) Large forest fires in Canada, 1959–1997. J Geophys Res 108:(D1), 8149.
10. Kurz WA, Stinson G, Rampley GJ, Dymond CC, Neilson ET (2008) Risk of natural disturbances makes future contribution of Canada's forests to the global carbon cycle highly uncertain. Proc Natl Acad Sci USA 105: 15551–1555.
11. Gill AM, Stephens SL, Cary GJ (2013) The worldwide "wildfire" problem. Ecol Appl 23(2): 438–454.
12. Hurst DF, Griffith DWT, Cook GD (1994) Trace gas emissions from biomass burning in tropical Australian savannas. J Geophys Res 99(D8): 16,441–6,456.
13. van der Werf GR, Randerson JT, Collatz GJ (2004) Continental-scale partitioning of fire emissions during the 1997 to 2001 El Niño/La Niña period. Science 303: 73–76.
14. Gill AM and Allan G (2008) Large fires, fire effects and the fire-regime concept. International J Wildland Fire 17: 688–695.
15. de Groot WJ, Pritchard JM, Lynham TJ (2009) Forest floor fuel consumption and carbon emissions in Canadian boreal forest fires. Can J For Res 39: 367–382.
16. Hollis JJ, Matthews S, Ottmar RD, Prichard SJ, Slijepcevic A, et al. (2010) Testing woody fuel consumption models for application in Australian southern eucalypt forest fires. For Ecol Manag 260: 948–964.
17. Scholes RJ, Kendall J, Justice CO (1996a) The quantity of biomass burned in southern Africa. J Geophys Res 101, D19: 23667–23676.
18. Scholes RJ, Ward DE, Justice CO (1996b) Emissions of trace gases and aerosol particles due to vegetation burning in southern hemisphere Africa. J Geophys Res 101: D19: 23677–23682.
19. Horowitz LW, Walters S, Mauzerall DL, Emmons LK, Rasch PJ, et al. (2003) A global simulation of tropospheric ozone and related tracers: Description and evaluation of MOZART, version 2. J Geophys Res 108: D24,4784.
20. Shirai T, Blake DR, Meinardi S, Rowland FS, Russell-Smith J, et al. (2003) Emission estimates of selected volatile organic compounds from tropical savannah burning in northern Australia. J Geophys Res 108: D3, 8406.
21. Hoelzemann J, Schulz M, Brasseur GP, Granier C (2004) Global wildland fire emission model (GWEM): Evaluating the use of global area burnt satellite data. J Geophys Res 109: D14S04.
22. Ito A, Penner JE (2004) Global estimates of biomass burning emissions based on satellite imagery for the year 2000. J Geophys Res Atm 109: D14S05.
23. UNFCCC (2012) Ad Hoc Working Group on the Durban Platform for Enhanced Action. Available: http://unfccc.int/bodies/body/6645.php Accessed 8 August 2013.
24. Pan Y, Birdsey RA, Fang J, Houghton R, Kauppi PE, et al. (2011) A large and persistent carbon sink in the world's forests, 1990–2007. Science 333: 988–993.
25. Keith H, Mackey BG, Lindenmayer DB (2009) Re-evaluation of forest biomass carbon stocks and lessons from the world's most carbon dense forests. Proc Natl Acad Sci USA 106(28): 11635–11640.
26. Mackey BG, Lindenmayer DB, Gill AM, McCarthy AM, Lindesay JA (2002) Wildfire, Fire and Future Climate: A forest ecosystem analysis. CSIRO Publ., Melbourne.
27. Lindenmayer DB (2009) Forest Pattern and Ecological Processes: A Synthesis of 25 Years of Research. CSIRO Publishing, Collingwood, Australia.
28. Department of Sustainability and Environment (1996) Study of old growth forest in Victoria's Central Highlands. Available: http://www.dse.vic.gov.au/forests/publications/research-reports/study-of-old-growth-forests. Accessed 2013 Aug 8.
29. McCarthy MA, Lindenmayer DB (1998) Multi-aged mountain ash forest, wildlife conservation and timber harvesting. For Ecol Manag 104: 43–56.
30. McCarthy MA, Gill AM, Lindenmayer DB (1999) Fire regimes in mountain ash forest: evidence from forest age structure, extinction models and wildlife habitat. For Ecol Manag 124: 193–203.
31. McCarthy MA, Gill AM, Bradstock RA (2001) Theoretical fire interval distributions. Int J Wildland Fire 10: 73–77.
32. Ashton DH (1975a) The root and shoot development of Eucalyptus regnans F. Muell. Aust J Bot 23: 867–887.
33. Flinn D, Squire R, Waring K (2007) Victoria. In: Raison RJ, Squire RO, editors. Forest Management in Australia: Implications for Carbon Budgets. National Carbon Accounting System Technical Report No. 32, Australian Greenhouse Office, Canberra. Ch 3, 103–146.
34. Tng DYP, Williamson GJ, Jordan GJ, Bowman DMJS (2012) Giant eucalypts – globally unique fire – adapted rainforest trees? New Phytol doi: 10.1111/j.1469–8137.2012.04359.x
35. Simkin R, Baker PJ (2008) Disturbance history and stand dynamics in tall open forest and riparian rainforest in the Central Highlands of Victoria. Austral Ecol 33: 747–760.
36. Turner PAM, Balmer J, Kirkpatrick JP (2009) Stand-replacing wildfires? The incidence of multi-cohort and single-cohort Eucalyptus regnans and E. obliqua forests in southern Tasmania. For Ecol Manag 258: 366–373.
37. Banks JCG (1993) Tree ring analysis of two Mountain Ash trees from the Watts and O'Shannassy catchments, Central Highlands, Victoria. Report to the Central Highlands Old Growth Forest Project, Department of Conservation and Natural Resources, Melbourne.
38. Vivian LM, Cary GJ, Bradstock RA, Gill AM (2008) Influence of fire severity on the regeneration, recruitment and distribution of eucalypts in the Cotter River Catchment ACT. Austral Ecol 33: 55–67.
39. Lindenmayer DB, Cunningham RB, Donnelly CF, Franklin JF (2000) Structural features of old growth Australian montane ash forests. For Ecol Manag 134: 189–204.
40. Victorian Bushfires Royal Commission 2009, Australia. Available: http://www.royalcommission.vic.gov.au/Documents/HR-Chapters/. Accessed 2013 Aug 8.
41. Cruz MG, Sullivan AL, Gould JS, Sims NC, Bannister AJ, et al. (2012) Anatomy of a catastrophic wildfire: The Black Saturday Kilmore East fire in Victoria, Australia. For Ecol Manag 284: 269–285.
42. Noble WS (1977) Ordeal by Fire: The Week a State Burned Up. Hawthorn Press, Melbourne, Australia.
43. Ough K (2001) Regeneration of wet forest flora a decade after clear-felling or wildfire – is there a difference? Aust J Bot 49: 645–664.
44. Lindenmayer DB, Ough K (2006) Salvage harvesting in the montane ash forests of the Central Highlands and its potential impacts on biodiversity. Conserv Biol 20: 1005–1015.
45. Lindenmayer DB, Burton PJ, Franklin JF (2008) Salvage Logging and its Ecological Consequences. CSIRO Publishing, Collingwood, Australia.
46. Keith H, Lindenmayer DB, Mackey B, Blair D, Carter L, et al. (2014) Managing temperate forests for carbon storage: impacts of logging versus forest protection on carbon stocks. Ecosphere 5(6): 75 doi 10.1890
47. Jackson WD (1968) Fire, air, water and earth – an elemental ecology of Tasmania. Proc Ecol Soc Aust 3: 9–16.
48. Lindenmayer DB, Hobbs RJ, Likens GE, Krebs CJ, Banks SC (2011) Newly discovered landscape traps produce regime shifts in wet forests. Proc Natl Acad Sci USA doi 10.1073
49. Taylor C, McCarthy MA, Lindenmayer DB (2014) Non-linear effects of stand age on fire severity. Conserv Letters doi: 10.1111/conl.12122
50. Ashton DH (1975b) Studies in litter in Eucalyptus regnans forests. Aust J Bot 23: 413–433.

51. Sillett SC, van Pelt R, Koch GW, Ambrose AR, Carroll AL, et al. (2010) Increasing wood production through old age in tall trees. For Ecol Manag 259: 976–994.

52. Illic J, Boland D, McDonald M, Downes G, Blakemore P (2000) Woody density phase 1– state of knowledge. National Carbon Accounting System Technical Report No. 18, Australian Greenhouse Office, Canberra.

53. Bootle KR (2005) Wood in Australia: types, properties and uses. McGraw-Hill, Sydney.

54. Chafe SC (1985) The distribution and interrelationship of collapse, volumetric shrinkage, moisture content and density in trees of Eucalyptus regnans F. Muell. Wood Sci Tech 19: 329–345.

55. Mackensen J, Bauhus J (1999) The decay of coarse woody debris. National Carbon Accounting System, Technical Report No. 6, Australian Greenhouse Office, Canberra.

56. Gifford RM (2000) Carbon contents of above-ground tissues of forest and woodland trees. National Carbon Accounting System Technical Report No. 22, Australian Greenhouse Office, Canberra.

57. Keith H, Leuning R, Jacobsen KL, Cleugh HA, van Gorsel E, et al. (2009a) Multiple measurements constrain estimates of net carbon exchange by a Eucalyptus forest. Agric For Meteorol 149: 535–558.

58. Mackowski CW (1987) Wildlife hollows and timber management in Blackbutt forest. Masters' Thesis, University of New England, Armidale, NSW.

59. Gibbons P, Lindenmayer DB (2002) Tree hollows and wildlife conservation in Australia. CSIRO Publications, Collingwood.

60. Department of Sustainability and Environment (2007) Victoria's Statewide Forest Resource Inventory – Central, Central Gippsland and Dandenong Forest Management Areas, Natural Resources Report Series 07–1, Victorian Government.

61. Feller MC (1980) Biomass and nutrient distribution in two eucalypt forest ecosystems. Aust J Ecol 5: 309–333.

62. Keith H, Barrett D, Keenan R (2000) Review of allometric relationships for estimating woody biomass for New South Wales, the Australian Capital Territory, Victoria, Tasmania and South Australia. National Carbon Accounting System Technical Report No. 5b, Australian Greenhouse Office, Canberra.

63. Snowdon P, Eamus D, Gibbons P, Khanna P, Keith H, et al. (2000) Synthesis of allometrics, review of root biomass and design of future woody biomass sampling strategies. National Carbon Accounting System Technical report No. 17, Australian Greenhouse Office, Canberra.

64. Van Wagner CE (1968) The line intersect method in forest fuel sampling. For Sci 14: 20–26.

65. McKenzie N, Ryan P, Fogarty P, Wood J (2001) Sampling, measurement and analytical protocols for carbon estimation in soil, litter ad coarse woody debris. National Carbon Accounting System Technical Report No. 14, Australian Greenhouse Office, Canberra.

66. Lindenmayer DB, Blair D, McBurney L, Banks S (2010) Forest Phoenix: How a Great Forest Recovers after Wildfire. CSIRO Publishing, Melbourne. 128 pp.

67. Lindenmayer DB, Blanchard W, McBurney L, Blair D, Banks S, et al. (2012) Interacting factors driving a major loss of large trees with cavities in a forest ecosystem. PLoS 1: 7(10), e41864.

68. Andrew MH, Noble IR, Lange RT (1979) A non-destructive method for estimating the weight of forage on shrubs. Aust Range J 1(3): 225–231.

69. Ashton D H (1975a) Studies of litter in Eucalyptus regnans forests. Aust J Bot 23: 413–433.

70. Polglase PJ, Attiwill PM (1992) Nitrogen and phosphorus cycling in relation to stand age of Eucalyptus regnans F. Muell. I. Return from plant to soil in litterfall. Pl Soil 142: 157–166.

71. Jacobs M (1955) Growth Habits of the Eucalypts. Forestry and Timber Bureau, Canberra.

72. Keith H (1991) Effects of fire and fertilization on nitrogen cycling and tree growth in a subalpine eucalypt forest. PhD Thesis, Australian National University, Canberra. 369 pp.

73. Keith H, Mackey BG, Berry S, Lindenmayer D, Gibbons P (2010) Estimating carbon carrying capacity in natural forest ecosystems across heterogeneous landscapes: addressing sources of error. Glob Change Biol 16: 2971–2989.

74. Mackey B, Berry S, Brown T (2008) Reconciling approaches to biogeographic regionalization: a systematic and generic framework examined with a case study of the Australian continent. J Biogeog 35: 213–229.

75. Berry S, Mackey B, Brown T (2007) Potential applications of remotely sensed vegetation greenness to habitat analysis and the conservation of dispersive fauna. Pac Conserv Biol 13(2): 120–127.

76. Ashton DH (1976) The development of even-aged stand of Eucalyptus regnans F. Muell. in Central Victoria. Aust J Bot 24: 397–414.

77. Watson FGR, Vertessy RA (1996) Estimating leaf area index from stem diameter measurements in Mountain Ash forest. Cooperative Research Centre for Catchment Hydrology, Report 96/7.

78. Attiwill PM (1994) Ecological disturbance and the conservative management of eucalypt forests in Australia. For Ecol Manag 63: 301–346.

79. Wood SW, Hua Q, Allen KJ, Bowman DMJS (2010) Age and growth of a fire prone Tasmanian temperate old-growth forest stand dominated by Eucalyptus regnans, the world's tallest angiosperm. For Ecol Manag 260, 438–447.

80. Grove SJ, Stamm L, Barry C (2009) Log decomposition rates in Tasmanian Eucalyptus obliqua determined using an indirect chronosequence approach. For Ecol Manag 258: 389–397.

81. IPCC (2006) Guidelines for National Greenhouse Gas Inventories. Eggleston S, Buendia L, Miwa K, Ngara T, Tanabe K, editors. Institute for Global Environmental Strategies, Hayama, Japan. Available: http://www.ipcc-nggip.iges.or.jp/public/2006gl/index.html. Accessed 2013 Oct 30.

82. DCCEE (2012) Department of Climate Change and Energy Efficiency, Australian National Greenhouse Gas Accounts, National Inventory Report 2010, Australian Government Submission to the United National Framework Convention on Climate Change, April 2012.

83. Harmon ME, Ferrell WK, Franklin JF (1990) Effects on carbon storage of conversion of old growth forests to young forests. Science 247, 699–702.

84. Gough CM, Vogel CS, Harrold KH, George K, Curtis PS (2007) The legacy of harvest and fire on ecosystem carbon storage in a north temperate forest. Glob Change Biol 13, 1935–1949.

85. King KJ, deLigt RM, Cary GJ (2011) Fire and carbon dynamics under climate change in south-eastern Australia: insights from FullCAM and FIRESCAPE modelling. Int J Wildland Fire 20: 563–577.

86. Kurz WA, Apps MJ (1999) A 70-year retrospective analysis of carbon fluxes in the Canadian forest sector. Ecol Appl 9(2): 526–547.

87. French NHF, de Groot WJ, Jenkins LK, Rogers BM, Alvarado E, et al. (2011) Model comparisons for estimating carbon emissions from North American wildland fire. J Geophys Res 116: G00K05.

88. IPCC (2003) Good practice guidance for land use, land use change and forestry. Penman J et al., editors. National Greenhouse Gas Inventories Programme, Institute for Global Environmental Strategies, Kanagawa, Japan. Available: http://www.ipcc-nggip.iges.or.jp/public/gpglulucf/gpglulucf-files. Accessed 2013 Oct 30.

89. Gould JS, Cheney NP, Hutchings PT, Cheney S (1996) Final Report on Prediction of Bushfire Spread for Australian Co-ordination Committee International Decade of Natural disaster Reduction. Project 4/95. CSIRO Forestry and Forest Products.

90. Seiler W, Crutzen PJ (1980) Estimates of gross and net fluxes of carbon between the biosphere and the atmosphere from biomass burning. Clim Change 2: 207–247.

91. Gould JS, Cheney NP (2007) Fire management in Australian forests. In: Forest management in Australia: implications for carbon budgets. Raison RJ and Squire RO, editors. 341–371. National Carbon Accounting System Technical Report No.32.

92. Amiro BD, Barr AG, Barr JG, Black TA, Bracho R, et al. (2010) Ecosystem carbon dioxide fluxes after disturbance in forests of North America. J Geophys Res 115: G00K02.

93. Lindenmayer DB, Wood JT (2010) Long-term patterns in the decay, collapse and abundance of trees with hollows in the mountain ash (Eucalyptus regnans) forests of Victoria, southeastern Australia. Can J For Res 40: 48–54.

94. van Gorsel E, Berni JAJ, Briggs P, Cabello-Leblic A, Chasmer L, et al. (2013) Primary and secondary effects of climate variability on net ecosystem carbon exchange in an evergreen Eucalyptus forest. Agric For Meteorol 182–183: 248–256.

95. Manies KL, Harden JW, Bond-Lamberty BP, O'Neill KP, (2005) Woody debris along an upland chronosequence in boreal Manitoba and its impact on long-term carbon storage. Can J For Res 35: 472–482.

96. Lindenmayer DB, Incoll RD, Cunningham RB, Donnelly CF (1999) Attributes of logs on the floor of Australian Mountain Ash (Eucalyptus regnans) forests of different ages. For Ecol Manag 123: 195–203.

97. UNFCCC (2011) 17th Conference of the Parties to the UNFCCC in Durban, CMP.7 Land Use, Land Use Change and Forestry. Available: http://unfccc.int/files/meetings/durban_nov_2011/decisions/application/pdf/awgkp_lulucf.pdf. Accessed 2013 Aug 8.

98. UNFCCC (2013b) National Inventory Submissions. Available: http://unfccc.int/national_reports/annex_i_ghg_inventories/national_inventories_submissions/. Accessed 2013 Aug 8.

99. UNFCCC (2013a) Revised Supplementary Methods and Good Practice guidance arising from the Kyoto Protocol. Good Practice Guidance. Available: http://www.ipcc-nggip.iges.or.jp/home/docs/kpsg/KP_Supplement_precopyedit.pdf. Accessed 2013 Aug 8.

100. Australian Treasury (2011) Strong growth, low pollution: modelling a carbon price. Australian Government, Canberra.

101. Mühle J, Lueker TJ, Su Y, Miller BR, Prather KA, et al. (2007) Trace gas and particulate emissions from the 2003 southern Californian wildfires. J Geophys Res 112, D03307.

102. Williams A, Karoly D, Tapper N (2001) The sensitivity of Australian fire danger to climate change. Clim Change 49: 171–191.

103. Pitman A, Narisma G, McAneney J (2007) The impact of climate change on the risk of forest and grassland fires in Australia. Clim Change 84: 383–401.

104. Clarke H, Smith P, Pitman A (2011) Regional signatures of future fire weather over eastern Australia from global climate models. Int J Wildland Fire 20: 550–562.

105. Macintosh A (2012a) Are forest management reference levels incompatible with robust climate outcomes? A case study on Australia. Carbon Manag 2(6): 691–707.

106. Esplin B, Gill AM, Enright N (2003) Report of the Inquiry into the 2002–2003 Victorian Bushfires. Office of the Emergency Services Commissioner, State

Government of Victoria. Available: http://www.oesc.vic.gov.au/home/reviews+and+inquiries/report+of+the+inquiry+into+the+2002-2003+Victorian+bushfires. Accessed 2013 Oct 15.

107. Murphy B, Timbal B (2008) A review of recent climate variability and climate change in southeastern Australia. Int J Clim 28: 859–879.

108. Canadell J, Kirschbaum M, Kurz W, Sanz M, Schlamadinger B, et al. (2007) Factoring out natural and indirect human effects on terrestrial carbon sources and sinks. Environ Sci Policy 10, 370–384.

109. Schlamadinger B, Johns T, Ciccarese L, Braun M, Sato A, et al. (2007) Options for including land use in a climate agreement post-2012: improving the Kyoto Protocol approach. Environ Sci Policy 10, 295–305.

110. Macintosh A (2011) Potential carbon credits from reducing native forest harvesting in Australia. ANU Centre for Climate Law and Policy, Working Paper Series 2011/1.

111. Marsden-Smedley JB, Slijepcevic A (2001) Fuel characteristics and low intensity burning in *Eucalyptus obliqua* wet forest at the Warra LTER site. Tasforests 13(2): 261–277.

112. Slijepcevic A (2001) Loss of carbon during controlled regeneration burns in *Eucalyptus obliqua* forest. Tasforests 13(2): 281–290.

113. Tolhurst KG, Anderson WR, Gould JS (2006) Woody fuel consumption experiments in an undisturbed forest. In: Conference Proceedings of the V International Conference on Forest Fire Research. Viegas DX, editor.

114. Hollis JJ, Anderson WR, McCaw WL, Cruz MG, Burrows ND, et al. (2011) The effect of fireline intensity on woody fuel consumption in southern Australian eucalypt forest fires. Aust For 74: 81–96.

115. Turetsky MR, Kane ES, Harden JW, Ottmar RD, Manies KL, et al. (2011) Recent acceleration of biomass burning and carbon losses in Alaskan forests and peatlands. Nature Geosciences 4: 27–31.

116. Campbell J, Donato D, Azuma D, Law B (2007) Pyrogenis carbon emissions from a large wildfire in Oregon, United States. J Geophys Res 122: G04014.

117. Barrett K, Kasischke ES, McGuire AD, Turetsky MR, Kane ES (2010) Modelling fire severity in black spruce stands in the Alaskan boreal forest using spectral and non-spectral geospatial data. Remote Sens Environ 114: 1494–1503.

118. Barrett K, Kasischke ES (2013) Controls on variation in MODIS fire radiative power in Alaskan boreal forests: Implications for fire severity conditions. Remote Sens Environ 130: 171–181.

15

Impacts of Diffuse Radiation on Light Use Efficiency across Terrestrial Ecosystems Based on Eddy Covariance Observation in China

Kun Huang[1,2], Shaoqiang Wang[1]*, Lei Zhou[1], Huimin Wang[1], Junhui Zhang[3], Junhua Yan[4], Liang Zhao[5], Yanfen Wang[2], Peili Shi[1]

1 Key Laboratory of Ecosystem Network Observation and Modeling, Institute of Geographic Sciences and Natural Resources Research, Chinese Academy of Sciences, Beijing 100101, China, 2 University of Chinese Academy of Sciences, Beijing 100049, China, 3 Institute of Applied Ecology, Chinese Academy of Sciences, Shenyang 110016, China, 4 South China Botanical Garden, Chinese Academy of Sciences, Guangzhou 510650, China, 5 Northwest Plateau Institute of Biology, Chinese Academy of Sciences, Xining 810001, China

Abstract

Ecosystem light use efficiency (LUE) is a key factor of production models for gross primary production (GPP) predictions. Previous studies revealed that ecosystem LUE could be significantly enhanced by an increase on diffuse radiation. Under large spatial heterogeneity and increasing annual diffuse radiation in China, eddy covariance flux data at 6 sites across different ecosystems from 2003 to 2007 were used to investigate the impacts of diffuse radiation indicated by the cloudiness index (CI) on ecosystem LUE in grassland and forest ecosystems. Our results showed that the ecosystem LUE at the six sites was significantly correlated with the cloudiness variation ($0.24 \leq R^2 \leq 0.85$), especially at the Changbaishan temperate forest ecosystem ($R^2 = 0.85$). Meanwhile, the CI values appeared more frequently between 0.8 and 1.0 in two subtropical forest ecosystems (Qianyanzhou and Dinghushan) and were much larger than those in temperate ecosystems. Besides, cloudiness thresholds which were favorable for enhancing ecosystem carbon sequestration existed at the three forest sites, respectively. Our research confirmed that the ecosystem LUE at the six sites in China was positively responsive to the diffuse radiation, and the cloudiness index could be used as an environmental regulator for LUE modeling in regional GPP prediction.

Editor: Dafeng Hui, Tennessee State University, United States of America

Funding: This research was jointly supported by the CAS for Strategic Priority Research Program (Grant No. XDA05050602), National Basic Research Program of China (973Program) (Grant No. 2010CB833503) and the Key Project in the National Science & Technology Pillar Program of China (Grant No. 2013BAC03B00). The funders had no role in study design, data collection and analysis, decision to publish, or preparation of the manuscript.

Competing Interests: The authors have declared that no competing interests exist.

* Email: sqwang@igsnrr.ac.cn

Introduction

Terrestrial ecosystems play an increasingly important role in global carbon cycle under climate change [1]. Light use efficiency (LUE) was first presented in the context of agricultural ecosystem focusing on the linear relationship between yield and solar irradiance, and gross primary production (GPP) was defined as the overall photosynthetically fixation of carbon per unit space and time [2]. The fact that GPP represents the critical flux component driving the terrestrial ecosystem carbon cycle implies that subtle fluctuations in GPP have substantial implications for future climate warming scenarios [3,4]. With the quantification terrestrial ecosystem GPP for regions, continents, or the globe, we can gain insight into the feedbacks between the terrestrial biosphere and the atmosphere under global change and climate policy-making facilitation [5,6]. Still, GPP predictions at regional scale to global scale are a major challenge due to the spatial heterogeneity [7,8]. Moreover, with the great carbon sequestration potential of the terrestrial ecosystem of China in global carbon budget [9], large

uncertainties exist in terrestrial ecosystem GPP simulation in China.

A number of modeling approaches have been developed for regional/global GPP estimations, including ecological process-based models and light use efficiency models driven by remote sensing data [10]. Among all the models, LUE models encompassing the LUE algorithm proposed by [2] may have the highest potential to identify the spatio-temporal dynamics of regional GPP due to the simplicity of concept and availability of remote sensing data [11]. With this method, GPP was defined as product of photosynthetically active radiation (PAR) absorbed by the vegetation canopy and a conversion factor, LUE [2,12]. Various LUE models have been developed for this purpose, including MODIS GPP algorithm [13], Vegetation Photosynthesis Model (VPM) [14], EC-LUE model [15], Vegetation Index (VI) model [16], C-Fix model [17], Temperature and Greenness Rectangle (TGR) model [18], Temperature and Greenness (TG) model [19] and so on. In order to acquire GPP estimations of high accuracy, the biophysical controls on the ecosystem LUE are significantly important to be fully understood [8,20]. Recent studies indicated that GPP and LUE were affected by both the quantity and composition of the incoming solar radiation [10,21–23]. With a given value of total incoming radiation, LUE of the entire canopy will increase with the increasing fraction of diffuse radiation (FDR) [23–25]. Under cloudy or aerosol-laden skies, incoming radiation was more diffuse and more uniformly distributed in the canopy with a smaller fraction of the canopy that was light saturated [10]. Consequently, canopy photosynthesis was inclined to be more light-use efficient under diffuse sunlight than under direct sunlight condition [10,21,23,24,26,27]. Evidences showed that global secondary organic aerosol in the atmosphere will increase by 36% in 2100 [28]. The aerosol influenced the cloud formation, which was the main contributor to the increment on FDR in the atmosphere [29–31]. Furthermore, an increasing trend of annual diffuse radiation in China has been proved to be 7.03 $MJ.m^{-2}.yr^{-1}$ per decade from 1981 to 2010 [32]. However, few studies on ecosystem GPP predictions took into account effects of the FDR variations of the incoming radiation on LUE based on the LUE models.

Up to now, the eddy covariance (EC) technique provides an alternative way to measure NEE continuously that can be used for GPP calculation by subtracting the modeled ecosystem respiration components [5,33–35]. Multi-sites and continuous eddy covariance (EC) flux and meteorological observation from the China-FLUX network provided a valuable tool for GPP and LUE calculation across ecosystems in China [34]. Therefore, in order to reveal the biophysical controls on measured ecosystem LUE for better regional GPP predictions in terrestrial ecosystems of China which is of high spatial heterogeneity, the impact of diffuse radiation resulting from cloud condition on LUE is of growing concern to be characterized by a uniform proxy. Despite the study that effect of cloudiness change on ecosystem LUE and water use efficiency was detected by the clearness index [36], it was difficult to incorporate the clearness index into LUE model for regional GPP estimates due to the specification of the highest interval of solar elevation angle in each grid. Here we employed an cloudiness index algorithm based on simple inputs [13,22], flux and metrological measurements from six sites of ChinaFLUX encompassing three forest ecosystems and three grassland ecosystems, to address the impact of diffuse radiation on light use efficiency (defined as GPP/PAR) [21,23,26]. The objectives of this study are to: (1) illustrate the seasonal dynamics of the cloudiness index and light use efficiency at different sites; (2) address the influence of fraction of diffuse radiation on ecosystem light use efficiency; (3) identify whether the cloudiness index thresholds favorable for enhancing ecosystem carbon sequestration exist or not.

Materials and Methods

Sites descriptions and measurements

In this study, flux observations were implemented at three forest ecosystems and three grassland ecosystems attached to the Chinese Terrestrial Ecosystem Flux Observational Network (ChinaFLUX). The three forest sites were comprised of the Changbaishan temperate mixed forest (CBS), Qianyanzhou subtropical evergreen needle leaf planted forest (QYZ), and Dinghushan subtropical evergreen broad-leaved forest (DHS). Subject to moosoon-influenced, temperate continental climate, CBS was located in the Jilin province of China, in which growing season ranged from May to September [37]. The QYZ site was located in the subtropical continental monsoon region, in which the mean annual air temperature was 17.9°C [38,39]. Located in the Guangdong province with a subtropical monsoon humid climate, DHS had a wet season from April to September and dry season

from November to March [39]. The three grassland ecosystems were the Inner Mongolia semi-arid L. chinensis steppe (NMG) which is C3 grassland, Haibei alpine frigid P. fruticosa shrub(HB), and Damxung (DX) alpine meadow-steppe ecosystem with short sparse vegetation(about 10 cm). NMG was located in the Xilin River Basin, Inner Mongolia Autonomous Region of China with a temperate semiarid continental climate. Its growing season lasted from late April to early October [40]. HB was located in the northeast of the Qinhai-Tibet Plateau with a plateau continental climate, which was characterized by lengthy cold winters and very short warm summers. Being situated in a frigid highland, HB receives strong solar radiation, with a mean annual global radiation of up to 6000–7000 $MJ.m^{-2}$ [36,41]. The DX site was located in the Lhasa City, Tibet, categorized as plateau monsoon climate. Its growing season duration was from May to September. The PAR was usually high, similar to that in alpine meadow area located in eastern Tibetan Plateau and higher than other grassland ecosystems [42]. The locations of six sites were shown in Figure1, and the detailed information of the six sites was provided in Table 1.

Routine meteorological variables were measured simultaneously with the eddy fluxes at each site. Air humidity and air temperature were measured with shielded and aspirated probes (HMP45C, Vaisala, Helsinki, Finland) at different sites. Global radiation and net radiation were recorded with radiometers (CM11 and CNR-1, Kipp & Zonen, Delft, the Netherlands). Photosynthetically active radiation (PAR) above the canopy was measured with a quantum sensor (LI-190Sb, LiCor Inc., USA). All meteorological observations were recorded at 30-min intervals with dataloggers (Model CR10X & CR23X, Campbell Scientific Inc.) [37,39,41–44].

In the study, we only used data measured during the periods of relatively stable leaf area index (LAI) each year from 2003 to 2007 in order to eliminate the potential effect of changing LAI [36]. The LAI of temperate ecosystems (CBS, NMG, HB and DX) remain stable in the mid-growing season. DHS and QYZ were evergreen forest ecosystems, and their LAI did not vary much with season. Therefore, data of mid growing season (June-August) from all the six flux sites were used to analyze the impact of diffuse radiation on ecosystem light use efficiency and photosynthesis.

Ethics statement

Three forest ecosystems (CBS, QYZ and DHS) and three grassland ecosystems (NMG, HB and DX) attached to China-FLUX were maintained by different institutions of Chinese Academy of Sciences (CAS), respectively. The CBS site was maintained by the Institute of Applied Ecology, CAS; the QYZ site and DX site was maintained by the Institute of Geographic Sciences and Natural Resources Research, CAS; the DHS site was maintained by the South China Botanical Garden, CAS; the NMG site was maintained by University of CAS, and the HB site was maintained by Northwest Plateau Institute of Biology, CAS. All necessary permits were obtained for the described field study. The field study did not involve endangered or protected species. Data will be made available upon request.

Eddy flux data

Carbon flux data (GPP and NEP) observed at 6 typical sites from 2003 to 2007 across China were applied to in this study (Figure1). The raw 30-min flux data procedure included: (1) 3D coordinate rotation was applied to force the average vertical wind speed to zero and to align the horizontal wind to mean wind direction, (2) flux data was corrected according the variation of air density caused by transfer of heat and water vapor [45], (3)the storage below EC height was corrected for forest sites [46], and (4)

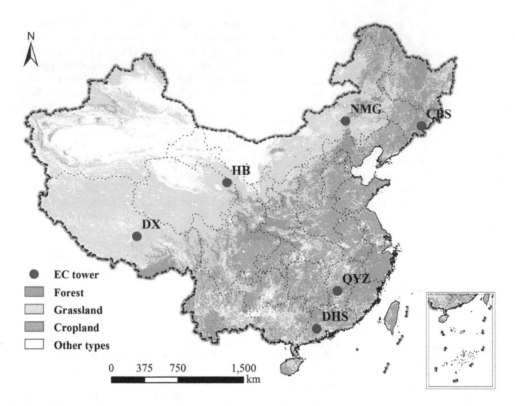

Figure 1. Distribution of the 6 eddy covariance flux sites in China in this study. The background was the MODIS land cover map.

the outlier data were filtered and data gaps were filled by using the look-up table method and mean diurnal variation(MDV) [37,39,47]. In the end, continuous 30 min flux data was performed.

The flux of net ecosystem CO_2 exchange (NEE, mg CO_2 $m^{-2} s^{-1}$) between the ecosystem and the atmosphere was calculated with equation (1), the net ecosystem productivity (NEP) was assigned to $-$NEE. Negative NEE values denote carbon uptake, while positive values denote carbon source.

$$NEP = -(\overline{w'\rho'c(zr)} + \int_0^{zr} \frac{\delta\overline{\rho c}}{\delta t}dz), \qquad (1)$$

where the first term on right-hand side is the eddy flux for carbon dioxide or water vapor below the height of observation (z_r), and all advective terms in the mass conservation equation were ignored.

Daily GPP data are partitioned from NEP data measured every 30-min using the eddy covariance technique. GPP was derived from the measured NEP, which was processed using the same method as [36]. Gross primary production (GPP) was calculated employing the following equation:

$$GPP = Re + NEP \qquad (2)$$

Table 1. Site descriptions.

Site (ab.)[a]	Changbaishan (CBS)	Qianyanzhou (QYZ)	Dinghushan (DHS)	Haibei (HB)	Inner Mongolia (NMG)	Damxung (DX)
Location	42°24′N	26°45′N	23°10′N	37°40′N	43°32′N	30°51′N
	128°06′E	115°04′E	112°32′E	101°20′E	116°40′E	91°05′E
Elevation(m)	738	102	300	3293	1189	4333
LAI(m^2m^{-2})	6.1	5.6	4.0	2.8	1.5	1.88
Annual mean precipitation(mm)	600–900	1489	1956	580	350–450	480
Annual mean temperature(°C)	3.6	18.6	21	−1.7	−0.4	1.3
Vegetation type	Mixed forest	Evergreen needle leaf forest	Evergreen broadleaf forest	Alpine frigid shrub	Temperate steppe	Alpine steppe-meadow

[a]Abbreviation for sites.

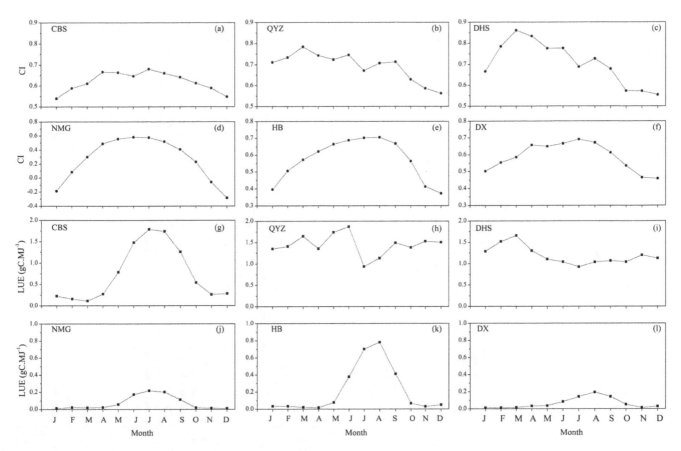

Figure 2. The seasonal variations of monthly mean cloudiness index (*CI*), monthly mean light use efficiency (LUE) from 2003 to 2007 at the five sites.

NEP was obtained directly from the eddy covariance measurement. Ecosystem respiration (Re) of the seven sites was estimated using the Lloyd-Taylor equation (1994) [41,48]. The nighttime NEP data under turbulent conditions were used to establish Re-temperature response relationship Eq.(3):

$$Re = R_{ref} e^{E_0(1/(T_{ref}-T_0)-1/(T-T_0))}$$ (3)

where R_{ref} represents the ecosystem respiration rate at reference temperature (Tref, 10°C); E_0 is the parameter that determines the temperature sensitivity of ecosystem respiration, and T_0 is a constant and set as −46.02°C; T is the air temperature or soil temperature(°C). Eq. (3) was also used to estimate daytime Re.

Calculation of light use efficiency

In this study, LUE (gC.MJ^{-1}) was defined as the ratio of daily GPP (gC.m^{-2}.d^{-1}) to incident PAR (MJ^{-1}.m^{-2}.d^{-1}, using 217 kJ mol^{-1} photons),

$$LUE = \frac{GPP}{PAR}$$ (4)

where PAR was directly measured by the in situ meteorological equipment simultaneous with the flux tower observation.

Cloudiness index

A cloudiness index implemented in CFLUX model was used in our model, since an increase on light use efficiency under overcast conditions at both hourly and daily time steps has been proved in previous studies [22,49]. The cloudiness index was calculated as [22]:

$$CI = 1 - \downarrow PAR / \downarrow PAR_{po}$$ (5)

where CI is the cloudiness index, $\downarrow PAR$ is incident PAR(MJd^{-1}) from daily observation input, $\downarrow PAR_{po}$ is potential incident PAR as a derivation of the algorithm of [50]. With the simple inputs of digital elevation model (DEM) data and readily available parameters, the $\downarrow PAR_{po}$ can be calculated as the global solar radiation at daily time scale in each grid. The spatial resolution of the DEM data was 500 m×500 m, provided by Institute of Geographical Sciences and Natural Resources Research, Chinese Academy of Sciences. More details of the algorithm can be found in the previous literature [50].

The clear sky LUE (LUE$_{cs}$) was specified for each site based on observations of LUE at eddy covariance flux towers. The clear sky LUE was based on the value when $\downarrow PAR / \downarrow PAR_{po}$ (decreasing cloud cover) approximated 1.0 by a function of LUE under low stress conditions plotted against $\downarrow PAR / \downarrow PAR_{po}$ [51].

Statistical analysis

The relationships between different variables were fitted with linear and non-linear equations. All analyses were conducted using the origin package v.8.0 (OriginLab Corporation, Northampton, MA, USA). Statistically significant differences were set with P< 0.05 (α = 0.05) unless otherwise stated.

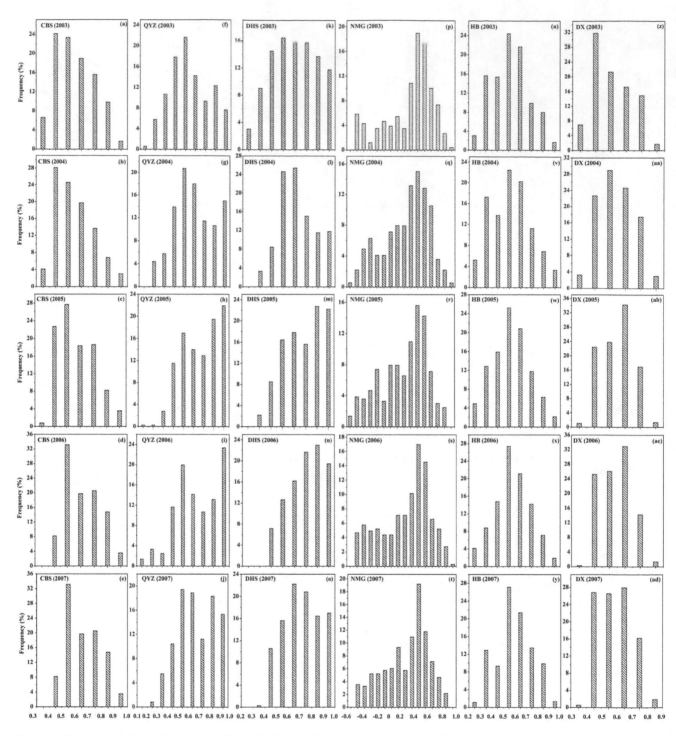

Figure 3. Histograms of the cloudiness index (*CI*) value at the six sites during the mid-growing seasons from 2003 to 2007.

Table 2. Light use efficiency under clear sky in different ecosystems.

Sites	CBS	QYZ	DHS	NMG	HB	DX
$LUE_{cs}(gC.MJ^{-1})$	0.29	0.425	0.569	0.003	0.013	0.009

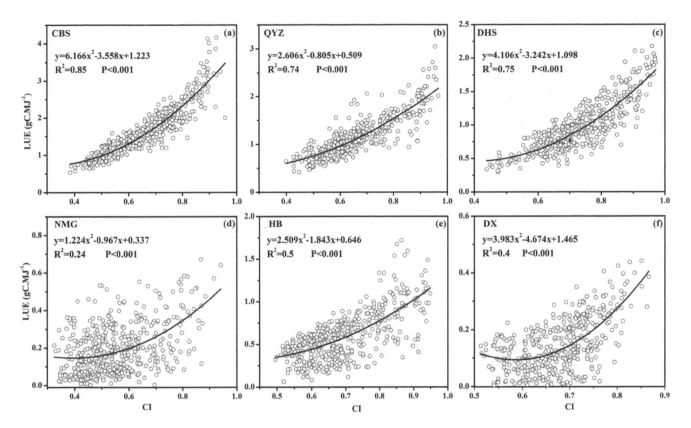

Figure 4. Relationships between LUE and the cloudiness index (CI, positive values) during the mid-growing season from 2003 to 2007 at the six sites.

Results and Discussion

Seasonal variation of cloudiness index and light use efficiency across ecosystems

Figure 2 showed the seasonal variations of the cloudiness index and light use efficiency of the six sites from 2003 to 2006. Mostly, cloudiness index (CI) was greater at QYZ and DHS than the other temperate ecosystems (CBS, NMG, HB and DX). The CI values of subtropical ecosystems (QYZ and DHS) reached the maximum in March, and were higher during the mid-growing season than the two ends of the year (Figure 2b, c). At the temperate ecosystems, the CI values peaked during the mid-growing season (Figure 2a, d, e and f), while the CI values of the subtropical ecosystems failed to show substantial variations with the seasonal changes. This indicated that sky conditions of two subtropical ecosystem sites were cloudier than those of four temperate ecosystem sites, and cloudy days were more during the mid-growing seasons at the temperate sites. It was also noted that negative CI values were found at NMG site, which was in consistency with meteorological observation that the NMG site received stronger solar radiation during the non-growing season. Meanwhile, the forest ecosystems LUE were significantly higher than grassland ecosystem LUE (Figure 2g, h, i, j, k and l). The LUE at subtropical forest sites (QYZ and DHS) failed to show significantly seasonality, while LUE of the temperate ecosystems (CBS, NMG, HB and DX) peaked during mid-growing season. Furthermore, the ecosystem LUE at QYZ site reached its turning point in July during mid-growing season, presented by a sharp fall resulting from the epidemic summer drought [38]. Among grassland sites, the LUE at HB site exhibited apparently higher values than the other two grassland sites and reached its maximal value in August, whereas the ecosystem LUE at NMG and DX site peaked in July and August, respectively (Figure 2j, k and l).

Frequency distribution of cloudiness index value across ecosystems

Apart from the seasonal dynamics of sky conditions (Figure 3a–f), the temporal patterns of cloudiness at the six sites in the mid-growing seasons were showed by the frequency distribution of CI values (Figure 3). Despite inter-annual variations resulting from climatic variability, common characteristics of the cloudiness pattern were found to be among the six sites. The CI values at CBS site occupied the largest frequency around 0.4 in 2003 and 2004(Figure 3a, b), while the CI value frequency took the most part around 0.5 from 2005 to 2007(Figure 3c, d and e). The peaks of CI value frequency at QYZ site located around 0.5 (Figure 3f, g and j) and 0.9 (Figure 3h, i). The CI value frequency at the DHS peaked between 0.5 and 0.7, except for 2005 and 2006, in which the largest frequency occurred around 0.8 in the mid-growing seasons (Figure 3m, n). As to the NMG site, the largest CI value frequency occurred between 0.4 and 0.5. Meanwhile, the CI frequency peaked around 0.5 at the HB site (Figure3u–z). The CI frequency between 0.5 and 0.7 occupied the largest proportion at DX site, except that peaked around 0.4 in 2003(Figure 3z). Overall, the CI frequencies occurred between 0.8 and 1.0 in the subtropical forest sites (QYZ and DHS) were much larger than what in the temperate ecosystems (CBS, NMG, HB and DX), which was verified by the report that spatial patterns of annual diffuse radiation in China showed strong regional heterogeneity, lower in the north but higher in the south [32].

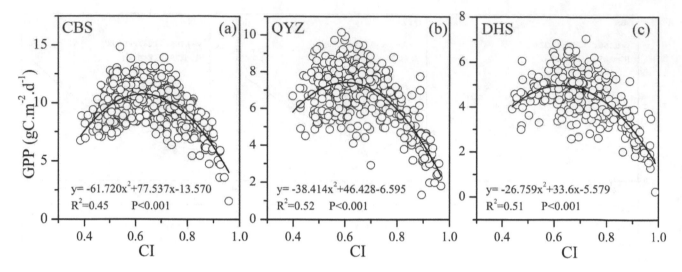

Figure 5. Relationships between ecosystem daily GPP and *CI* during the mid-growing season at three forest ecosystems.

Clear sky light use efficiency in the six ecosystems

The ecosystem LUE was plotted against the ratio of PAR to potential PAR (decreasing cloud cover), and the clear sky LUE (LUE_{cs}) was acquired when PAR/PAR_{po} was around 1.0 [51]. LUE_{cs} values were greater at forest sites than those at grassland sites (Table 2). Among the forest sites, the ecosystem LUE_{cs} was greatest at DHS site, intermediated at QYZ site and lowest at CBS site. The LUE_{cs} of forest ecosystem decreased with the degree of latitude. For grassland sites, the LUE_{cs} peaked at the HB site, followed by the DX site and NMG site, respectively. This was verified by the fact that measured ecosystem LUE was higher than the other two grassland ecosystem sites (Figure 2).

Impacts of diffuse radiation on ecosystem light use efficiency

Figure 4 exhibited the interactive responses of ecosystem LUE to the variation in the diffuse radiation fraction of incoming solar radiation (indicated by the cloudiness index) in different ecosystems. At all sites, significantly quadratic regression relationships were found between the ecosystem LUE and *CI* during the mid-growing season. Once the value of *CI* exceeded a certain one, determined by the minimal value (zero) of the first derived function of each quadratic regression function, the ecosystem LUE increased with *CI* dramatically. LUE of forest ecosystem showed more significantly positive relationship with *CI* ($R^2 \geq 0.74$), compared with three grassland ecosystem sites ($R^2 \leq 0.5$). Also, differences in enhancement on ecosystem LUE induced by the variation of diffuse PAR existed within the ecosystem type across sites. For the forest sites, the ecosystem LUE at CBS site demonstrated stronger increasing trend than the other two subtropical forest sites with the largest correlation coefficient

($R^2 = 0.85$) and quadratic term coefficient (6.166)(Figure 4a, 4b and 4c). The ecosystem LUE at NMG site exhibited least increasing potential ($R^2 = 0.24$, quadratic term coefficient $= 1.224$) with the variation of *CI* among the three grassland ecosystem sites. The expectation that canopy LUE could be enhanced by the diffuse components of solar radiation compared to direct radiation has been reported in previous studies [21,26,52]. Under cloudy skies, incoming radiation was more diffuse and more uniformly distributed in the canopy with a smaller fraction of the canopy that was light saturated [10]. Consequently, canopy photosynthesis was inclined to be more light-use efficient under diffuse sunlight than under direct sunlight conditions [10,21,23,24,26,27]. In addition, differences in canopy structure density across different ecosystem types was presented to contribute to the increasing rate differences of LUE to fraction of diffuse radiation [36], due to its effective penetration to the lower depths of canopy [32,53].

Cloudiness threshold for enhancing forest ecosystem carbon sequestration

Similar significances of quadratic regression relationships between daily GPP and *CI* during mid-growing season were confirmed at three forest sites (Figure 5). The quadratic regression relationships implied that the ecosystem GPP would peak at a certain value of *CI*, and then decreased with increasingly values of *CI*. Specifically, it was noted that the impact of cloudiness on ecosystem carbon exchange process was also dependent on local thermal, moisture and light conditions [36]. At the beginning stage, the forest ecosystems GPP were in positive association with the *CI*. Instead, the forest ecosystems GPP were gradually restrained by the cloudiness when the value of *CI* exceeded a

Table 3. Cloudiness thresholds for enhancing forest ecosystem LUE and GPP.

Cloudiness index	CBS	QYZ	DHS
Lower bounds	0.288	0.154	0.394
Upper bounds	0.628	0.604	0.629

certain one where the symmetric axis of the parabolic curve regression functions located. This phenomenon could partly be ascribed to the decreasing PAR absorbed by the vegetation canopy, based on the radiation conversion efficiency concept of Monteith (1972) [2]. Consequently, the cloudiness thresholds (Table 3) were calculated by a range that began from the value where the symmetric axis of the parabolic curve regression functions (response of LUE to *CI*) located (Figure 4a, 4b and 4c), and stopped at the point where symmetric axis of the parabolic curve regression functions (response of GPP to *CI*) located (Figure 5). However, the optimal cloudiness index threshold was not available for the three grassland sites because of the poor quadratic relationship between GPP and *CI* (P>0.05). The difference in responses of GPP to the variation of diffuse PAR received by the ecosystem between forest sites and grassland sites was likely to result from the difference in canopy structure [36]. The LAI of forest ecosystem at CBS, QYZ and DHS were higher than those of grassland ecosystem (Table 1). Previous studies reported that LUE and GPP of an ecosystem with low LAI, such as grassland and shrubs, did not increase on cloudy days [54,55]. This inconsistency was partly attributed to the differences of climate conditions of the studied ecosystems, including light, water and thermal conditions [36].

Conclusions

Eddy covariance flux observations from six sites encompassing two ecosystem types and the cloudiness index were used to detect the response of LUE and GPP to diffuse radiation during mid-growing season. Results indicated that (1) cloudiness index (*CI*) was mostly greater at two subtropical forest ecosystem sites (QYZ and DHS) than the other temperate ecosystem sites (CBS, NMG, HB and DX), and LUE in the temperate ecosystem peaked during mid-growing season;(2) LUE under clear sky were greater at forest sites than at grassland sites, and the LUE under clear sky of forest ecosystem decreased with the degree of latitude; (3) significantly quadratic regression relationships were found between the ecosystem LUE and *CI* during the mid-growing season at all sites;(4) cloudiness thresholds favorable for enhancing ecosystem carbon sequestration existed in forest ecosystem sites.

Due to the large regional heterogeneity existing in terrestrial ecosystem of China, more EC flux sites involved in the future are essential to reveal the impacts of diffuse radiation on terrestrial ecosystem LUE in China. Furthermore, the cloudiness index could be incorporated as an environmental regulator into LUE models for regional GPP simulations in China.

Author Contributions

Conceived and designed the experiments: KH SW. Performed the experiments: KH SW LZ. Analyzed the data: KH SW LZ. Contributed reagents/materials/analysis tools: KH SW LZ HW JZ JY LZ YW PS. Wrote the paper: KH SW. Discussed the manuscript: KH SW LZ HW JZ JY LZ YW PS.

References

1. Nemani RR, Keeling CD, Hashimoto H, Jolly WM, Piper SC, et al. (2003) Climate-driven increases in global terrestrial net primary production from 1982 to 1999. Science 300: 1560–1563.

2. Monteith JL (1972) Solar Radiation and Productivity in Tropical Ecosystems. J Appl Ecol 9: 747–766.

3. Cai WW, Yuan WP, Liang SL, Zhang XT, Dong WJ, et al. (2014) Improved estimations of gross primary production using satellite-derived photosynthetically active radiation. J Geophys Res-Biogeo 119: 110–123.

4. Raupach MR, Canadell JG, Le Quere C (2008) Anthropogenic and biophysical contributions to increasing atmospheric CO_2 growth rate and airborne fraction. Biogeosciences 5: 1601–1613.

5. Wu CY, Munger JW, Niu Z, Kuang D (2010) Comparison of multiple models for estimating gross primary production using MODIS and eddy covariance data in Harvard Forest. Remote Sens Environ 114: 2925–2939.

6. Xiao JF, Zhuang QL, Baldocchi DD, Law BE, Richardson AD, et al. (2008) Estimation of net ecosystem carbon exchange for the conterminous United States by combining MODIS and AmeriFlux data. Agric For Meteorol 148: 1827–1847.

7. Canadell JG, Mooney HA, Baldocchi DD, Berry JA, Ehleringer JR, et al. (2000) Carbon metabolism of the terrestrial biosphere: A multitechnique approach for improved understanding. Ecosystems 3: 115–130.

8. Wang Y, Zhou GS (2012) Light Use Efficiency over Two Temperate Steppes in Inner Mongolia, China. Plos One 7.

9. Piao SL, Fang JY, Ciais P, Peylin P, Huang Y, et al. (2009) The carbon balance of terrestrial ecosystems in China. Nature 458: 1009–U1082.

10. He MZ, Ju WM, Zhou YL, Chen JM, He HL, et al. (2013) Development of a two-leaf light use efficiency model for improving the calculation of terrestrial gross primary productivity. Agric For Meteorol 173: 28–39.

11. Ogutu BO, Dash J (2013) Assessing the capacity of three production efficiency models in simulating gross carbon uptake across multiple biomes in conterminous USA. Agric For Meteorol 174: 158–169.

12. Monteith JL, Moss CJ (1977) Climate and the Efficiency of Crop Production in Britain [and Discussion]. Philosophical Transactions of the Royal Society of London B, Biological Sciences 281: 277–294.

13. Running SW, Nemani RR, Heinsch FA, Zhao MS, Reeves M, et al. (2004) A continuous satellite-derived measure of global terrestrial primary production. BioScience 54: 547–560.

14. Xiao XM, Hollinger D, Aber J, Goltz M, Davidson EA, et al. (2004) Satellite-based modeling of gross primary production in an evergreen needleleaf forest. Remote Sens Environ 89: 519–534.

15. Yuan WP, Liu S, Zhou GS, Zhou GY, Tieszen LL, et al. (2007) Deriving a light use efficiency model from eddy covariance flux data for predicting daily gross primary production across biomes. Agric For Meteorol 143: 189–207.

16. Wu CY, Niu Z, Gao SA (2010) Gross primary production estimation from MODIS data with vegetation index and photosynthetically active radiation in maize. J Geophys Res-Atmos 115.

17. Veroustraete F, Sabbe H, Eerens H (2002) Estimation of carbon mass fluxes over Europe using the C-Fix model and Euroflux data. Remote Sens Environ 83: 376–399.

18. Yang YT, Shang SH, Guan HD, Jiang L (2013) A novel algorithm to assess gross primary production for terrestrial ecosystems from MODIS imagery. J Geophys Res-Biogeo 118: 590–605.

19. Sims DA, Rahman AF, Cordova VD, El-Masri BZ, Baldocchi DD, et al. (2008) A new model of gross primary productivity for North American ecosystems based solely on the enhanced vegetation index and land surface temperature from MODIS. Remote Sens Environ 112: 1633–1646.

20. Garbulsky MF, Penuelas J, Papale D, Ardo J, Goulden ML, et al. (2010) Patterns and controls of the variability of radiation use efficiency and primary productivity across terrestrial ecosystems. Global Ecol Biogeogr 19: 253–267.

21. Gu LH, Baldocchi DD, Wofsy SC, Munger JW, Michalsky JJ, et al. (2003) Response of a deciduous forest to the Mount Pinatubo eruption: Enhanced photosynthesis. Science 299: 2035–2038.

22. Turner DP, Ritts WD, Styles JM, Yang Z, Cohen WB, et al. (2006) A diagnostic carbon flux model to monitor the effects of disturbance and interannual variation in climate on regional NEP. Tellus B 58: 476–490.

23. Mercado LM, Bellouin N, Sitch S, Boucher O, Huntingford C, et al. (2009) Impact of changes in diffuse radiation on the global land carbon sink. Nature 458: 1014–U1087.

24. Oliphant AJ, Dragoni D, Deng B, Grimmond CSB, Schmid HP, et al. (2011) The role of sky conditions on gross primary production in a mixed deciduous forest. Agric For Meteorol 151: 781–791.

25. Roderick ML, Farquhar GD, Berry SL, Noble IR (2001) On the direct effect of clouds and atmospheric particles on the productivity and structure of vegetation. Oecologia 129: 21–30.

26. Gu LH, Baldocchi D, Verma SB, Black TA, Vesala T, et al. (2002) Advantages of diffuse radiation for terrestrial ecosystem productivity. J Geophys Res-Atmos 107.

27. Misson L, Lunden M, McKay M, Goldstein AH (2005) Atmospheric aerosol light scattering and surface wetness influence the diurnal pattern of net ecosystem exchange in a semi-arid ponderosa pine plantation. Agric For Meteorol 129: 69–83.

28. Heald CL, Henze DK, Horowitz LW, Feddema J, Lamarque JF, et al. (2008) Predicted change in global secondary organic aerosol concentrations in response to future climate, emissions, and land use change. J Geophys Res-Atmos 113.

29. Kim SW, Jefferson A, Yoon SC, Dutton EG, Ogren JA, et al. (2005) Comparisons of aerosol optical depth and surface shortwave irradiance and their

effect on the aerosol surface radiative forcing estimation. J Geophys Res-Atmos 110.

30. Feddema JJ, Oleson KW, Bonan GB, Mearns LO, Buja LE, et al. (2005) The importance of land-cover change in simulating future climates. Science 310: 1674–1678.

31. Schiermeier Q (2006) Oceans cool off in hottest years. Nature 442: 854–855.

32. Ren XL, He HL, Zhang L, Zhou L, Yu GR, et al. (2013) Spatiotemporal variability analysis of diffuse radiation in China during 1981–2010. Ann Geophys-Germany 31: 277–289.

33. Baldocchi D, Falge E, Gu LH, Olson R, Hollinger D, et al. (2001) FLUXNET: A new tool to study the temporal and spatial variability of ecosystem-scale carbon dioxide, water vapor, and energy flux densities. B Am Meteorol Soc 82: 2415–2434.

34. Yu GR, Wen XF, Sun XM, Tanner BD, Lee XH, et al. (2006) Overview of ChinaFLUX and evaluation of its eddy covariance measurement. Agric For Meteorol 137: 125–137.

35. Yu GR, Zhu XJ, Fu YL, He HL, Wang QF, et al. (2013) Spatial patterns and climate drivers of carbon fluxes in terrestrial ecosystems of China. Glob Change Biol 19: 798–810.

36. Zhang M, Yu G-R, Zhuang J, Gentry R, Fu Y-L, et al. (2011) Effects of cloudiness change on net ecosystem exchange, light use efficiency, and water use efficiency in typical ecosystems of China. Agric For Meteorol 151: 803–816.

37. Guan DX, Wu JB, Zhao XS, Han SJ, Yu GR, et al. (2006) CO2 fluxes over an old, temperate mixed forest in northeastern China. Agric For Meteorol 137: 138–149.

38. Wen XF, Wang HM, Wang JL, Yu GR, Sun XM (2010) Ecosystem carbon exchanges of a subtropical evergreen coniferous plantation subjected to seasonal drought, 2003–2007. Biogeosciences 7: 357–369.

39. Yu GR, Zhang LM, Sun XM, Fu YL, Wen XF, et al. (2008) Environmental controls over carbon exchange of three forest ecosystems in eastern China. Glob Change Biol 14: 2555–2571.

40. Fu YL, Yu GR, Wang YF, Li ZQ, Hao YB (2006) Effect of water stress on ecosystem photosynthesis and respiration of a Leymus chinensis steppe in Inner Mongolia. Sci China Ser D 49: 196–206.

41. Fu YL, Yu GR, Sun XM, Li YN, Wen XF, et al. (2006) Depression of net ecosystem CO_2 exchange in semi-arid Leymus chinensis steppe and alpine shrub. Agric For Meteorol 137: 234–244.

42. Shi PL, Sun XM, Xu LL, Zhang XZ, He YT, et al. (2006) Net ecosystem CO2 exchange and controlling factors in a steppe - Kobresia meadow on the Tibetan Plateau. Sci China Ser D 49: 207–218.

43. Fu Y, Zheng Z, Yu G, Hu Z, Sun X, et al. (2009) Environmental influences on carbon dioxide fluxes over three grassland ecosystems in China. Biogeosciences 6: 2879–2893.

44. Hu ZM, Yu GR, Fu YL, Sun XM, Li YN, et al. (2008) Effects of vegetation control on ecosystem water use efficiency within and among four grassland ecosystems in China. Glob Change Biol 14: 1609–1619.

45. Webb EK, Pearman GI, Leuning R (1980) Correction of flux measurements for density effects due to heat and water vapour transfer. Quarterly Journal of the Royal Meteorological Society 106: 85–100.

46. Carrara A, Kowalski AS, Neirynck J, Janssens IA, Yuste JC, et al. (2003) Net ecosystem CO_2 exchange of mixed forest in Belgium over 5 years. Agric For Meteorol 119: 209–227.

47. Falge E, Baldocchi D, Olson R, Anthoni P, Aubinet M, et al. (2001) Gap filling strategies for defensible annual sums of net ecosystem exchange. Agric For Meteorol 107: 43–69.

48. Yu GR, Song X, Wang QF, Liu YF, Guan DX, et al. (2008) Water-use efficiency of forest ecosystems in eastern China and its relations to climatic variables. New Phytol 177: 927–937.

49. Turner DP, Urbanski S, Bremer D, Wofsy SC, Meyers T, et al. (2003) A cross-biome comparison of daily light use efficiency for gross primary production. Glob Change Biol 9: 383–395.

50. Fu P, Rich PM (1999) Design and implementation of the Solar Analyst: an ArcView extension for modeling solar radiation at landscape scales. Proceedings of the 19th annual ESRI user conference, San Diego, USA.

51. King DA, Turner DP, Ritts WD (2011) Parameterization of a diagnostic carbon cycle model for continental scale application. Remote Sens Environ 115: 1653–1664.

52. Farquhar GD, Roderick ML (2003) Atmospheric science: Pinatubo, diffuse light, and the carbon cycle. Science 299: 1997–1998.

53. Urban O, Janous D, Acosta M, Czerny R, Markova I, et al. (2007) Ecophysiological controls over the net ecosystem exchange of mountain spruce stand. Comparison of the response in direct vs. diffuse solar radiation. Glob Change Biol 13: 157–168.

54. Letts MG, Lafleur PM, Roulet NT (2005) On the relationship between cloudiness and net ecosystem carbon dioxide exchange in a peatland ecosystem. Ecoscience 12: 53–59.

55. Niyogi D, Chang H-I, Chen F, Gu L, Kumar A, et al. (2007) Potential impacts of aerosol–land–atmosphere interactions on the Indian monsoonal rainfall characteristics. Nat Hazards 42: 345–359.

16

Threshold Responses of Forest Birds to Landscape Changes around Exurban Development

Marcela Suarez-Rubio[1]*, **Scott Wilson**[2], **Peter Leimgruber**[3], **Todd Lookingbill**[4]

1 University of Maryland Center for Environmental Science, Appalachian Laboratory, Frostburg, Maryland, United States of America, 2 Canadian Wildlife Service, Environment Canada, Saskatoon, Canada, 3 Smithsonian Conservation Biology Institute, Front Royal, Virginia, United States of America, 4 University of Richmond, Department of Geography and the Environment, Richmond, Virginia, United States of America

Abstract

Low-density residential development (i.e., exurban development) is often embedded within a matrix of protected areas and natural amenities, raising concern about its ecological consequences. Forest-dependent species are particularly susceptible to human settlement even at low housing densities typical of exurban areas. However, few studies have examined the response of forest birds to this increasingly common form of land conversion. The aim of this study was to assess whether, how, and at what scale forest birds respond to changes in habitat due to exurban growth. We evaluated changes in habitat composition (amount) and configuration (arrangement) for forest and forest-edge species around North America Breeding Bird Survey (BBS) stops between 1986 and 2009. We used Threshold Indicator Taxa Analysis to detect change points in species occurrence at two spatial extents (400-m and 1-km radius buffer). Our results show that exurban development reduced forest cover and increased habitat fragmentation around BBS stops. Forest birds responded nonlinearly to most measures of habitat loss and fragmentation at both the local and landscape extents. However, the strength and even direction of the response changed with the extent for several of the metrics. The majority of forest birds' responses could be predicted by their habitat preferences indicating that management practices in exurban areas might target the maintenance of forested habitats, for example through easements or more focused management for birds within existing or new protected areas.

Editor: Justin G. Boyles, Southern Illinois University, United States of America

Funding: This work was supported by Appalachian Laboratory, University of Maryland Center for Environmental Science. The funders had no role in study design, data collection and analysis, decision to publish, or preparation of the manuscript.

Competing Interests: The authors have declared that no competing interests exist.

* E-mail: marcela.suarezrubio@boku.ac.at

¤ Current address: Institute of Zoology, Department of Integrative Biology and Biodiversity Research, University of Natural Resources and Life Sciences, Vienna, Austria

Introduction

The expansion of human settlement along the urban-rural fringe has received considerable global attention in recent decades [1–5]. In the United States, conversion of privately owned rural lands into low-density residential development (i.e., exurban development) has increased five- to sevenfold between 1950 and 2000 [6]. In the Mid-Atlantic region of the United States, the dispersed, isolated housing units typical of exurban areas are embedded within a forest matrix, often close to protected areas [7] and natural amenities [8,9]. Understanding the impacts of exurban development on wildlife and biodiversity is crucial to better understand long-term effects of exurban development and to develop successful land use and conservation planning [10,11].

Humans generally remove natural habitats by building settlements, which can serve to fragment the landscape [12–14]. Both habitat loss and fragmentation modify the spatial pattern of remnant habitats, creating smaller and isolated fragments, thus compromising habitat quality and quantity. Wildlife responds in a variety of ways depending on species traits and life histories [15,16]. Some species thrive in these environments whereas others, such as forest birds, decline rapidly (e.g., [17,18]). Possible reasons for long-term reductions of forest-bird species in these environ-

ments include predation [19], brood parasitism [20], and competition with human-adapted species [21]. Forest birds have been shown to be particularly susceptible to human settlement even at housing densities as low as 0.095 house/ha [22–27].

Understanding how exurban development alters forest birds' habitat over time is a conservation priority given the unprecedented rates of exurban development in eastern temperate forests of the Mid-Atlantic [6,28]. Forest bird abundance is generally positively related to proportion of forest cover (e.g., [29,30]), but the spatial distribution of suitable habitat also affects forest birds' occurrence and fecundity [31,32]. Declines of forest birds have been well documented in eastern North America, and these declines have been associated with habitat loss and fragmentation due to roads, power lines, and residential development [11,33,34]. However, few studies have examined the response of species through time as residential development progresses [18].

Species may respond nonlinearly to habitat loss and fragmentation (reviewed by [35]). Nonlinear responses of species to habitat loss and fragmentation may complicate our ability to determine the response of biodiversity to exurban development. Theoretical models predict the existence of a change point or threshold in which an abrupt reduction in occupancy occurs despite the presence of sufficient suitable habitat [36–39]. Some studies show empirical evidence for threshold existence in birds [40–43], although others have not found any evidence to support threshold responses [44]. It is uncertain whether threshold declines in forest birds apply to exurban development. If these relationships are appropriately characterized by threshold models, determining the range at which exurban development induces population crashes may provide guidance for landscape planning, management, and conservation.

The aim of this study was to assess whether and how forest birds respond to changes in habitat due to exurban growth. We evaluated habitat composition (amount) and configuration (arrangement) for selected bird species (i.e., forest and forest-edge species) around North America Breeding Bird Survey stops between 1986 and 2009. The approach accounted for year-to-year variability in species abundances and investigated species responses to both habitat loss and fragmentation as exurban development increased over time. In addition, we assessed whether selected bird species showed thresholds in both occurrence frequency and relative abundance. We used Threshold Indicator Taxa Analysis [45] to detect change points in species occurrence. We evaluated two spatial extents (400-m and 1-km radius buffer) to determine if species responded differently to changes at the local and landscape scales. We expected that forest species would exhibit a strong negative response to exurban development at both extents, whereas forest-edge species would respond positively to high levels of exurban land cover.

Methods

Study area

The study area encompassed nine counties in north-central Virginia (Clarke, Culpeper, Fauquier, Frederick, Madison, Page, Rappahannock, Shenandoah, and Warren) and two in western Maryland (Washington and most of Frederick; Figure 1). The region has experienced a remarkable population growth. For example, counties included in the study area had growth rates ranging from 4% (Page County) to 40% (Culpeper County) between 2000 and 2009 [46]. Concomitant with this population growth, the region has also experienced an increase in exurban area from 2.3% in 1986 to 7.3% in 2009 [28]. One reason for the increased exurban development is the easy access and well-maintained transportation infrastructure to the metropolitan Washington, DC area which provides employment opportunities [47].

Breeding bird survey

We used the North America Breeding Bird Survey (BBS; [48,49]) to gather relative abundance data. The BBS is a large-scale annual roadside survey to monitor the status and trend of breeding bird populations in the United States and southern Canada since 1966. The survey is performed along secondary roads by experienced volunteer observers in late May to early July, the peak of the breeding season. Routes are 39.4 km long and consist of 50 survey stops located at 0.8 km intervals. During the survey, observers record all birds heard or seen within 0.4 km in a 3-min period. We focused our analysis on survey stops instead of the entire route because our interest was on local characteristics of breeding habitats and routes might vary in local environmental conditions [50,51]. We chose all routes located in the study area and from them we uniformly selected at most 10 survey stops per route (every fifth stop along the route). We only considered survey stops that had detailed direction descriptions and fell within the study region (125 survey points in total; Figure 1). This information was important for geocoding and characterizing

site-specific features of selected survey stops. A maximum of 10 stops per route was chosen to reduce overlap between circular areas around survey stops and decrease the likelihood of spatial autocorrelation.

We focused on 11 forest-nesting passerine species whose habitat preferences included forest –Ovenbird (*Seiurus aurocapilla*), Red-eyed Vireo (*Vireo olivaceus*), American Redstart (*Setophaga ruticilla*), Wood Thrush (*Hylocichla mustelina*), Scarlet Tanager (*Piranga olivacea*), Eastern Wood-Pewee (*Contopus virens*), Eastern Phoebe (*Sayornis phoebe*); and forest-edge –Eastern Towhee (*Pipilo erythrophthalmus*), Gray Catbird (*Dumetella carolinensis*), Northern Cardinal (*Cardinalis cardinalis*), and Indigo Bunting (*Passerina cyanea*) [52]. We defined forest species as birds that utilized a wide variety of deciduous and mixed deciduous-coniferous forest types and that may favor interior forested habitats. Forest-edge species are those species that are strongly associated with forest edges and open habitats [53]. These species were selected to represent contrasting habitat preferences (forest vs. edge) and because they were detected on at least 5% of surveys during the 1986–2009 interval. In addition, many of these species are reported to have experienced population declines or reduced fecundity in their distribution range due to habitat loss or fragmentation [32,54–56]. Our study was designed to determine if the specific land conversion process of exurban development corresponded with abundance changes for these species.

Landscape structure around Breeding Bird Survey stops

We established circular areas of 400-m and 1-km radius around selected BBS stops. These areas were chosen to characterize both breeding bird territories [57,58], which were assumed to be in the immediate surroundings of survey stops, and areas feasibly visited during bird daily movements [59,60]. To quantify landscape structure around selected survey stops over time at these two extents, we used Landsat 5 TM imagery for 1986, 1993, 2000, and 2009. We performed standard pre-processing procedures (atmospheric and topographic correction) prior to image classification.

We used aerial photos to generate a training dataset to supervise a classification of areas of exurban development. Exurban development was defined as areas with housing densities between 1 unit per 0.4 ha and 1 unit per 16.3 ha (e.g., 6 - 250 houses per km^2) [6]. We used both spectral and spatial characteristics to define and identify exurban areas [28]. Spectral characteristics were derived from spectral mixture analysis [61] of corrected Landsat images to estimate the fractional cover of vegetation, substrate, non-photosynthetic vegetation, and shade within each image. We built decision trees based on spectral mixture analysis outputs to classify exurban development between 1986 and 2009. We used morphological spatial pattern analysis to further analyze terminal nodes from the decision trees that could not discriminate between exurban and urban areas based on spectral characteristics alone [62,63]. Scattered, isolated pixels were regarded as spatial characteristics typical of exurban development. This procedure allowed us to distinguish exurban areas from forest and urban areas and create a land-cover map that was used to characterize areas around survey stops.

We used FRAGSTATS 3.3 [64] and GUIDOS 1.3 [62,63] to estimate both landscape composition and configuration within the two circular areas around selected survey stops for 1986, 1993, 2000, and 2009. Landscape composition variables described the amount of habitat and included proportion of area occupied by forest and exurban development. Landscape configuration variables described the arrangement of forest habitat and included area-weighted average patch size, number of forest patches greater than 0.45 ha, and proximity index [65]. Proximity index is a

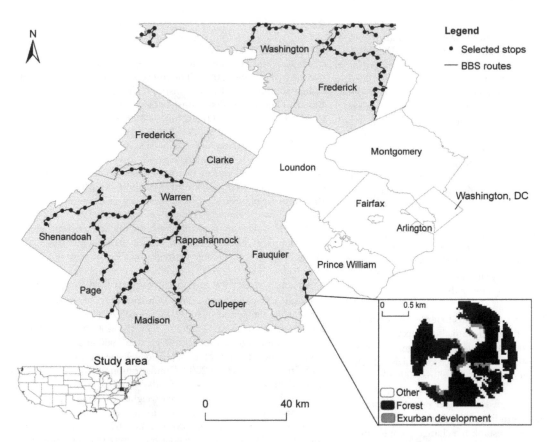

Figure 1. Study region (shaded area). It includes nine counties in north-central Virginia and two in western Maryland. Circles represent 125 North American Breeding Bird Survey (BBS) stops that were uniformly selected from routes. Zoom-in window shows example of a landscape within a 1-km radius of a selected survey stop.

measure of isolation that considers both patch size and proximity of a focal patch to all forest patches around. We only considered forest patches ≥ 100 ha within 2500 m of the focal patch. A 2500 m range was selected to reflect dispersal patterns of most songbirds (dispersal median distance range: 0.3 – 7.3 km; [66]). The proximity index increases as the neighborhood is increasingly occupied by forest patches and as those patches become closer and more contiguous or less isolated. GUIDOS was used because it identifies and graphically depicts the different types of landscape elements created by the fragmentation process [63]. The software package analyzes map geometry by applying mathematical morphological operators to allocate each pixel to one of a mutually exclusive set of classes. We quantified changes in the proportion of forest interior (core class), forest fragments (islet class), and forest edge (edge and perforation classes).

Although some of these variables are not independent, many have been shown to affect abundance of birds [32,67–70] and represent different aspects of potential habitat alteration.

Analysis

BBS data have unknown precision due to observer differences [71], first-year observers' skills [72,73], environmental conditions [74], and habitat features [50]. We used a hierarchical Bayesian model to adjust BBS counts and account for these limitations. We modeled count data as hierarchical over-dispersed Poisson variables and fit models using Markov Chain Monte Carlo (MCMC) methods in WinBUGS 1.4.3 [75]. Hierarchical Bayesian models are frequently applied to BBS data [76–78] and are better

able to account for variability in complex time series than other methods [79]. We specified C_{it} as the count for each species on stop i and time t where $i = 1,..., N$; $t = 1,..., T$; and N and T were the number of stops and the number of years species were observed, respectively. Conditioned on the model, counts (C_{it}) were independent across years and stops, and these conditional distributions for C_{it} were assumed to be Poisson with mean μ_{it}:

$$C_{it} \sim \text{Pois}(\mu_{it}) \qquad (1)$$

The full model was then:

$$\log(\mu_{it}) = \beta_{0stop} + \beta_{1stop} \times Year_t + \beta_2 \times FirstYear_{it} + Route_{it} + Observer_{it} + Noise_{it} \qquad (2)$$

where each stop was assumed to have a separate intercept (β_0) and time trend (β_1). The model also included several sources of variability including unknown route-level effects ($Route_{it}$), observer effects ($Observer_{it}$), and an additional noise component ($Noise_{it}$) to help account for over-dispersion in the data. BBS observers tend to over or under-record certain species in their first year relative to subsequent years [77,80] and to incorporate this effect we treated an individual's first year ($FirstYear_{it}$) as a binary indicator variable (β_2). The precision parameters (τ^2) for β_{0-2}, observer, route, and noise effects were assigned vague inverse gamma prior distributions [81] with parameters (0.001, 0.001).

We used two Markov chains for each model and examined model convergence and performance through Gelman-Rubin diagnostics and individual parameter histories [82,83]. Time to convergence varied among species depending on the amount of data for that species (30,000 – 200,000 iterations required). Once convergence was reached, we obtained derived estimates of the count at each stop and in each year, and these adjusted counts were then used for the threshold analysis. In addition, we estimated for each selected species the linear trend coefficient (i.e., the slope of abundance over time on a log scale) and percent annual change (the expected count in the last year divided by the expected count in the first year raised to 1/number of years). For trend coefficients (slope and percent annual change), we interpreted significance based on values with 95% credible intervals not overlapping zero.

We examined the relationship between landscape variables and selected species adjusted counts by fitting a non-parametric locally weighted polynomial regression (loess; [84]). When the loess regression highlighted nonlinearity in the relationship, then a change-point analysis to test for nonlinear threshold response was used.

We estimated potential species thresholds to landscape variables in space and time using Threshold Indicator Taxa ANalysis (TITAN; [45]). TITAN identifies abrupt changes in both occurrence frequency and relative abundance of individual taxa along an environmental gradient. It is able to distinguish responses of individual taxa with low occurrence frequencies or highly variable abundances and does not assume linear response along all or part of an environmental gradient. TITAN uses normalized indicator species taxa scores (z) to establish a change-point location that separates the data into two groups and maximizes association of each taxon with one side of the partition. Z scores measure the association of taxon abundance weighted by their occurrence and is normalized to facilitate cross-taxa comparison. Thus, TITAN distinguishes negative (z-) and positive (z+) indicator response taxa.

To measure quality of the indicator response and assess uncertainty around change-point locations, TITAN bootstraps the original dataset and recalculates change points with each simulation. Uncertainty is expressed as quantiles of the change-point distribution. Narrow intervals between upper and lower change-point quantiles (i.e., 5 and 95%) indicate nonlinear response in taxon abundance whereas broad quantile intervals are characteristic of taxa with linear or more gradual response. Diagnostic indices of the quality of the indicator response are purity and reliability. Purity is the proportion of bootstrap replicates that agree with the direction of the change-point for the observed response. Pure indicators (purity ≥ 0.95) are those that consistently assign the same response direction during the resampling procedure. Reliability is the proportion of change-point individual value scores (IndVal) among the bootstrap replicates that consistently have p-values below defined probability levels (0.05). Reliable indicators (reliability ≥ 0.95) are those with consistently large IndVal. Because purity and reliability indices did not differ for most metrics, we only reported the reliability index. We ran TITAN for the 11 selected bird species and each of the landscape variables in R 2.11.1 [85]. We used the minimum number of observations on each side of the threshold split that is required by TITAN (n = 5). Because our data set was very large, we specified 250 permutations to compute z scores and diagnostic indices as suggested by Baker and King [45].

Results

Breeding Bird Survey

There were 2481 detections on the 125 selected survey stops between 1986 and 2009. The most common species was the Indigo Bunting (1108 detections) and the least common was the Eastern Phoebe (190 detections; Table 1). Forest-edge species were the more abundant group (average of 1094 individuals per species) compared to the forest species (525 individual counts per species). Annual mean adjusted abundances (i.e., posterior means) showed population trends of selected species between 1986 and 2009 accounting for differences in route, observer, and detection year (Figure 2). The Gray Catbird, Northern Cardinal, American Redstart, Ovenbird, and Red-eyed Vireo showed significant increases in estimated abundance between 1986 and 2009 (Table 1). American Redstart had the highest percent change per year (3.1%). For the other six species, the estimated abundance did not significantly change through the 24-year period.

Landscape structure around Breeding Bird Survey stops

Landscape composition and configuration changed through time during the period of study, except for 20% of BBS that were inside protected areas (Table 2). For the 400-m radius buffer, amount of forest decreased from 49.2% in 1986 to 41.2% in 2009; whereas, the amount of exurban development increased from 1.7% in 1986 to 6.0% in 2009. Configuration of forest patches also differed among years. Although the number of forest patches remained nearly constant, area-weighted average patch size decreased by a mean of 2.1 ha in the last time period. This decrease in patch size was accompanied by a decrease in forest edge from 1986 to 2009, a decrease in forest interior, an increase in forest fragments, and a decrease in the proximity index. In general, all metrics changed much more in later time periods than early years reflecting the increasing rate of exurban development in the study region.

Similar patterns were observed for the 1-km radius buffer (Table 2). Those differences that did exist can largely be explained by the area effect of the larger buffer. More forest patches were found in the larger 1-km radius buffer, and these patches were generally larger (e.g., area-weighted average patch size in 2009 of 111.6 ha for the 1-km buffer vs. 18.5 ha for 400-m buffer). The larger buffer also contained fewer forest fragments (19.9 vs. 31.9% in 2009), but underwent a greater loss in forest interior from 1986 to 2009 (6.5% for 1-km buffer vs. 4.4% for 400-m buffer).

Threshold response of bird species to landscape structure

Scatterplots of adjusted counts fitted with a non-parametric locally weighted polynomial regression (loess) model indicated a nonlinear relationship between several of the landscape variables and selected bird species (see examples in Figure 3). In general, forest species exhibited threshold responses to both landscape composition and configuration (Figure 4). For the 400-m radius buffer, most of the forest species were positive indicator taxa for the amount of forest (mean change point: 24.3%), forest interior (15.4%), area-weighted average patch size (5.7 ha), and proximity index (9078). Most of the forest species were negative indicator taxa for the amount of exurban development (0.2%) and proportion of forest fragments (19.7%). American Redstart was the only forest species that responded negatively to forest edge (change point: 29.1%), whereas the rest of the forest species responded positively (mean change point: 16.6%). Eastern Phoebe was the only forest species that declined with amount of forest, proportion of forest interior, and area-weighted average patch size.

Table 1. Hierarchical-model estimates based on Breeding Bird Survey stops for forest and forest- edge species.

Species	Number of total detections (% of surveys)	Mean adjusted abundance	Trend coefficient	Percent change/year
Forest species				
American Redstart (AMRE)	225 (9.1)	0.132±0.015	**0.042**	**3.10**
Ovenbird (OVEN)	248 (10.0)	0.137±0.016	**0.029**	2.70
Red-eyed Vireo (REVI)	632 (25.5)	0.373±0.027	**0.024**	**2.70**
Eastern Phoebe (EAPH)	190 (7.7)	0.090±0.014	0.005	1.80
Wood Thrush (WOTH)	618 (24.9)	0.396±0.027	0.008	1.10
Scarlet Tanager (SCTA)	364 (14.7)	0.180±0.018	−0.004	0.30
Eastern Wood-Pewee (EAWP)	490 (19.8)	0.237±0.018	−0.001	−0.20
Forest-edge species				
Gray Catbird (GRCA)	509 (20.5)	0.401±0.048	**0.025**	**2.80**
Northern Cardinal (NOCA)	808 (32.6)	0.461±0.027	**0.022**	**1.50**
Eastern Towhee (EATO)	526 (21.2)	0.313±0.025	0.007	1.00
Indigo Bunting (INBU)	1108 (44.7)	0.657±0.031	−0.006	0.50

American Ornithologists Union alpha codes for English common names are in parenthesis. Trend coefficient represents the slope on a log scale of abundance over time. Values in bold indicate 95% credible intervals.

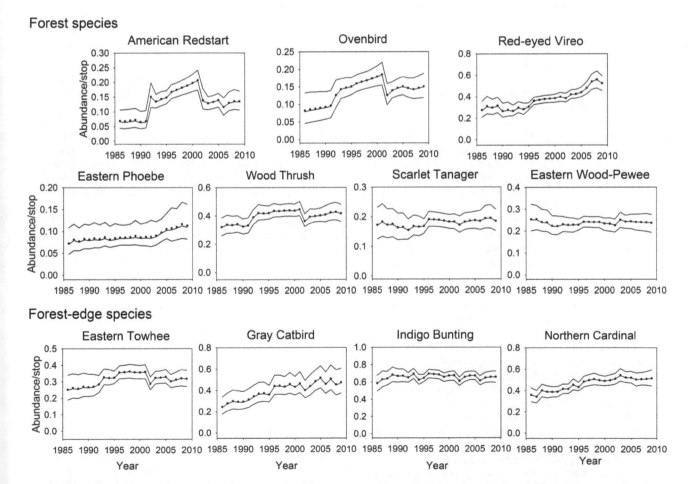

Figure 2. Time series of mean abundance adjusted for missing observations and observer differences. Lines indicate posterior median (line nearly coincident with the circles) with 95% confidence intervals.

Table 2. Landscape structure surrounding selected Breeding Bird Survey stops (n = 125) at 400-m and 1-km radius buffer (mean ± sd) for 1906, 1993, 2000, and 2009.

Variables	1986	1993	2000	2009
400-m radius buffer				
Forest (%)	49.2±39.3	48.3±39.3	46.2±39.4	41.2±39.2
Exurban development (%)	1.7±2.5	2.1±2.6	3.1±3.4	6.0±6.8
Forest interior (%)	39.8±32.2	38.1±31.9	35.8±31.8	29.3±32.4
Area- weighted average patch size (ha)	22.2±20.8	21.7±20.7	20.6±20.6	18.5±20.5
Forest fragments (%)	23.4±35.7	23.5±35.6	25.1±37.9	31.9±40.9
Number of forest patches (> 0.45 ha)	1.7±1.1	1.7±1.2	1.6±1.2	1.6±1.4
Forest edge (%)	24.1±14.7	24.3±14.8	24.5±16.4	20.7±16.2
Proximity index	25156.8±071.5	23165.1±749.6	14763.0±2712.3	9884.6±4949.1
1-km radius buffer				
Forest (%)	51.0±35.7	50.0±35.6	47.9±35.7	42.7±35.8
Exurban development (%)	1.8±1.6	2.2±1.9	3.2±2.6	6.2±5.6
Forest interior (%)	55.6±28.9	53.1±28.9	49.4±30.2	40.1±32.4
Area- weighted average patch size (ha)	134.4±123.5	131.8±123.1	123.2±121.7	111.6±121.3
Forest fragments (%)	10.2±17.8	11.2±19.6	14.4±24.5	19.9±28.8
Number of forest patches (> 0.45 ha)	5.0±4.2	5.0±4.2	5.3±4.3	5.4±4.4
Forest edge (%)	23.6±11.3	24.5±11.8	24.4±12.6	22.5±12.8
Proximity index	25957.0±205.7	23906.8±060.7	15272.4±1243.6	10533.3±4917.0

This species also responded positively to the proportion of forest fragments, though some relationships for this species were of lower reliability (Appendix S1).

Forest-edge species had strong threshold responses to landscape composition and most of the configuration metrics at both extents (Figure 4). For the 400-m radius buffer, for example, all forest-edge species responded positively to the number of forest patches (mean change point: 0.6 patches). Gray Catbird and Northern Cardinal increased sharply with amount of exurban development, proportion of forest fragments, and forest edge (although forest edge was not a reliable indicator for Gray Catbird). These two species responded negatively to the amount of forest, forest interior, area-weighted average patch size, and proximity index (Figure 4). However, Eastern Towhee and Indigo Bunting were positive indicator taxa for the amount of forest, forest interior, area-weighted average patch size, proximity index, and forest edge, and were negative indicator taxa for the proportion of forest fragments. Eastern Towhee was the only forest-edge species that responded negatively to the amount of exurban development and had similar change points to those exhibited by forest species.

Similar patterns in threshold response were observed for the two buffer widths (Figure 4 comparison of top and bottom panels) except for number of forest patches and proportion of forest edge. For these two variables, the direction of the response for roughly half of the species changed with buffer width. For most of the species, the direction of the response was positive for the 400-m radius buffer but negative for the 1-km radius buffer. However, the quality of indicators for the proportion of forest edge was less reliable for the 1-km radius buffer.

The quality of the indicator and confidence around change-point locations varied by extent and by landscape structure variable. For example, the forest species Red-eyed Vireo responded positively to the amount of exurban development. However, the indicator was only moderately reliable for the 400-m

radius buffer (reliability = 0.70; Appendix S1). Reliability also changed with extent of analysis for some species and indicators. For example, the reliability of the response of the forest species Eastern Phoebe to the proximity index was higher within the 400-m radius buffer (reliability = 0.74) than for the 1-km radius buffer (reliability = 0.38). Gray Catbird, an edge species, had a positive response to the number of forest patches within the 400-m radius buffer and a negative response within the 1-km radius buffer. However, the reliability for the 1-km radius buffer was poor (reliability = 0.32). In general, where there were differences in reliability at different extents, the 400-m relationships were more reliable.

Forest species had relatively narrow bootstrapped change-point distributions for most landscape structure characteristics indicating confidence about the existence of a threshold (Figure 4). However, for some landscape structure characteristics, forest species exhibited variable width in the bootstrapped change-point distributions. For example, some species (e.g., Eastern Wood-Pewee) had a sharp response to the amount of forest whereas others (e.g., Red-eyed Vireo) had a more gradual response. In general, forest-edge species (except for Eastern Towhee) had broad bootstrapped change-point distribution suggesting a more gradual response for most landscape structure characteristics.

Discussion

Our results support the existence of nonlinear responses to habitat loss and fragmentation [37,41,43] and variation in sensitivity to alteration of landscape structure due to exurban development depends on species habitat specificity (Figure 4; [41,86]). For example, species that positively responded to the amount of exurban development (e.g., Northern Cardinal) are often found throughout a range of habitats from shrubby sites in logged and second-growth forests to plantings around buildings

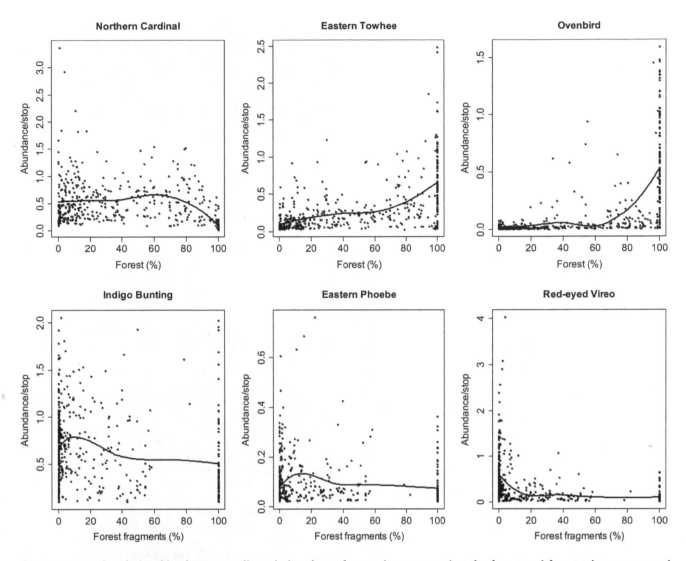

Figure 3. Example relationships between adjusted abundance for species representing the forest and forest-edge groups and selected landscape variables. Landscape composition (e.g., proportion of forest) and landscape configuration (e.g., proportion of forest fragments) in 400-m radius circular areas are depicted. The line represents non-parametric locally weighted polynomial regression curve (loess).

[87]. Sensitive species who responded negatively to amount of exurban development (e.g., Wood Thrush) are more frequently found in well-developed deciduous and mixed forests [88].

Despite loss of forest and increase of exurban development, bird sightings significantly increased during the 24-year period for five of the 11 species analyzed. The detection of two of the forest-edge species (Northern Cardinal and Gray Catbird) increased between 1986 and 2009. These species are found in forest edges and clearings, fencerows, abandoned farmland, or residential areas [87,89]. Thus, more sightings in exurban areas may indicate that these species have been taking advantage of the increased availability of suitable habitats and supplemental feeding provided by landowners [90]. The species also had broad change-point distributions indicating gradual responses to the land-cover change. Although we did not expect to find a threshold response, the direction of the response showed by these species corresponded with their habitat preferences. In other words, these species were indicators of habitat fragmentation due to exurban development (e.g., increased in abundance with increase in forest fragments and decrease in forest interior).

The other three species that experienced abundance increases were forest birds (American Redstart, Red-eyed Vireo, and Ovenbird). This was surprising given documented population declines in other studies for the Red-eyed Vireo and the Ovenbird due to habitat loss and fragmentation (e.g., [32,56]). American Redstart and Red-eyed Vireo are forest birds but seem to occur more frequently in early and mid-successional forest habitats and even start to decline as forests mature [91–93]. Thus, the type of forest disturbance associated with exurban development may benefit these species. The larger temporal and spatial scale regional regrowth of eastern forests due to farmland abandonment since the early twentieth century [94–96] also may explain the slight increase in abundance of these species. However, all three of the species showed a strong threshold response to amount of forest, suggesting that they are sensitive to reduced forest cover. It is important to note that the amount of forest of more than 45% of survey stops in 2009 were above the identified thresholds at both extents for American Redstart, Red-eyed Vireo, and Ovenbird. Thus, it seems that abundance increase is occurring disproportionately in relatively intact forests (e.g., protected areas)

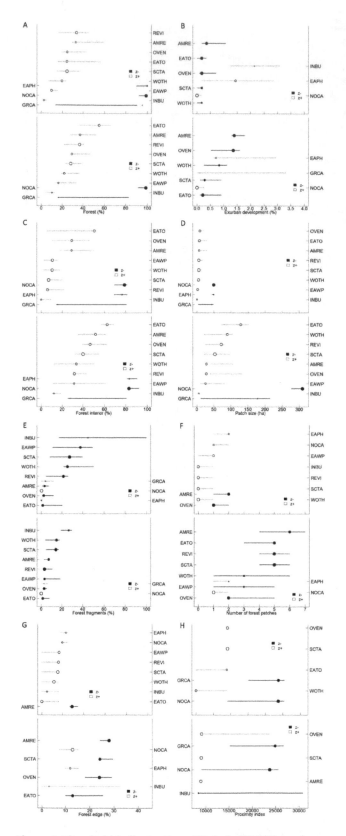

Solid circles correspond to negative (z-) indicator taxa (with corresponding species labels on the left axes) and open circles correspond to positive (z+) indicator taxa (with corresponding species labels on the right axes). Circles are sized in proportion to z scores. Lines overlapping each circle represent 5 and 95% percentiles among 250 bootstrap replicates. Landscape variables evaluated were (A) forest, (B) exurban development, (C) forest interior, (D) area-weighted averaged patch size, (E) forest fragments, (F) number of forest patches, (G) forest edge, and (H) proximity index. Taxa IDs correspond to the American Ornithologists Union alpha codes for English common names.

confounding any negative effects that forest decline [32] in exurban areas may have, though further assessment is required to confirm this assertion.

Although species showed similar response patterns at both extents, for two of the landscape configuration variables (number of patches and forest edge), the direction of the response changed with the extent. Similar results were found by Smith [97] who demonstrate that fragmentation effects depend on the landscape extent considered. Thus, the extent should be explicitly accounted for when evaluating the effects of these two metrics on forest birds. In general, the 400-m relationships were deemed more reliable by the TITAN threshold analysis indicating that more local change processes had a greater effect on species occurrence and relative abundances.

Although the majority of species responses were consistent with our classification regarding habitat preferences, there were two species (Eastern Phoebe and Eastern Towhee) whose response did not correspond to the assigned group. Eastern Phoebe is generally a woodland species [98] and was classified as a forest species. However, this species had threshold responses similar to those exhibited by forest-edge species for most of the landscape structure variables. This may be explained by nest placement preferences. Eastern Phoebe is mostly constrained by availability of suitable nest sites [98] and nests are often located on bridges, culverts, buildings, and rock outcrops in the vicinity of water [99]. Change in landscape structure due to exurban development may benefit this species, but further monitoring of its population is recommended. In contrast, Eastern Towhee exhibited a response similar to those showed by forest species. This species is thought of as an edge-associated generalist and places its nests on or above ground, usually at 1.5 m in shrubby areas [100]. However, these results suggest that Eastern Towhees may be more sensitive to habitat change due to exurban development than previously expected. Alternatively, Eastern Towhees might be more susceptible to increased predation pressure from free-ranging domestic cats common in exurban development [101,102].

The threshold responses that we detected for selected forest bird species indicate that species were affected in a nonlinear fashion by changes in landscape composition and configuration. However, the thresholds observed may not necessarily be similar for forest bird communities as a whole. In addition, threshold responses detected should not be used as a point below which a population will not persist [103] but rather as guidelines for management practices in areas prone to exurban development.

Given the wide range of threshold values observed in this study (e.g., threshold response to the amount of forest ranged between 9.6 and 33.9% for the 400-m radius buffer), it is problematic to suggest generic recommendations on how to best conserve forest birds in exurban areas. Exurban development is creating habitats that suit forest-edge species, and the main risk is at the other end of the spectrum for the forest species that require large amount of continuous forest cover. Incorporating threshold response in conservation planning might focus on maintaining forested

Figure 4. Threshold Indicator Taxa ANalysis (TITAN). Landscape variables were used as predictors of threshold changes in individual bird species in 400-m (top panel) and 1-km radius circular areas (bottom panel) between 1986 and 2009 in north-central Virginia and western Maryland. Only indicator taxa (purity ≥ 0.95 and reliability ≥ 0.95) are plotted in increasing order with respect to their observed change point.

habitats targeted towards the most sensitive species. For example, exurban areas can be managed to retain forest conditions close to the identified thresholds in species occurrence and relative abundance for the most sensitive of selected forest birds such as Red-Eyed Vireo, and in this way other forest birds would also be protected.

It is important to note that the BBS is poor for surveying sensitive forest species with large area requirements. As a result, this analysis considered species that are dependent on forests, but not some of those species that might have been especially sensitive to forest loss (e.g., Kentucky Warbler). Therefore, management efforts targeting the maintenance of larger forest patches as exurban development continues will also benefit some of these other sensitive forest-dependent species. This could be achieved through easements or more focused management for forest birds within existing or new protected areas. The value of high-quality potential source habitat is suggested by the unchanged or even increasing abundance of many of the forest species, although they exhibit negative threshold responses to many of the predictor variables when the entire spatial-temporal dataset is considered. Additional monitoring work, perhaps within the region's protected areas [104], could expand beyond the BBS roadside surveys to account for some of the limitations of its design.

Conclusion

Rural private lands are being converted to exurban development at high rates in the Mid-Atlantic region and around the world [28], and this trend is likely to continue into the future [5]. Our results show that exurban development is altering forest habitats. Forest birds exhibited a threshold response to landscape structure alteration at both local and landscape extents. The majority of forest birds' responses could be predicted by their habitat preferences indicating that management practices in exurban areas might target the maintenance of forested habitats (e.g., through easements or more focused management for birds within existing or new protected areas) lest risk broad-scale changes in bird community composition within these landscapes.

Acknowledgments

We would like to thank the thousands of volunteers who have collected Breeding Bird Survey Data and D. Ziolkowski and K. Pardieck (USFWS) for providing the bird data, the topographic maps, and the description of the BBS stops. We are grateful to J. B. Churchill, R. Hildebrand, R. Gardner, A. Elmore, L. Wainger, and P. Marra for technical assistance and fruitful discussions about the design of this study. We also thank the editor and reviewers for their valuable comments.

Author Contributions

Conceived and designed the experiments: MSR PL TL. Performed the experiments: MSR. Analyzed the data: MSR SW. Wrote the paper: MSR SW. Revised the paper critically for important intellectual content: PL TL.

References

1. Burnley IH, Murphy PA (1995) Exurban development in Australia and the United-States - through a glass darkly. J Plan Educ Res 14: 245–254.
2. Struyk RJ, Angelici K (1996) The Russian dacha phenomenon. Housing Stud 11: 233–261.
3. van den Berg L, Wintjes A (2000) New rural 'live style estates' in The Netherlands. Landsc Urban Plan 48: 169–176.
4. Liu J, Daily GC, Ehrlich PR, Luck GW (2003) Effects of household dynamics on resource consumption and biodiversity. Nature 421: 530–533.
5. Theobald DM (2005) Landscape patterns of exurban growth in the USA from 1980 to 2020. Ecol Soc10: 32–66.
6. Brown DG, Johnson KM, Oveland TR, Theobald DM (2005) Rural land-use trends in the conterminous United States, 1950–2000. Ecol Appl 15: 1851–1863.
7. Wade AA, Theobald DM (2010) Residential development encroachment on US protected areas. Conserv Biol 24: 151–161.
8. McGranahan DA (1999) Natural amenities drive population change. Washington, DC, USA: Food and Rural Economics Division, Economic Research Service, US Department of Agriculture.
9. Kwang-Koo K, Marcouiller DW, Deller SC (2005) Natural amenities and rural development: understanding spatial and distributional attributes. Growth Change 36: 273–297.
10. Miller JR, Hobbs RJ (2002) Conservation where people live and work. Conserv Biol 16: 330–337.
11. Hansen AJ, Knight RL, Marzluff JM, Powell S, Brown K, et al. (2005) Effects of exurban development on biodiversity: patterns, mechanisms, and research needs. Ecol Appl 15: 1893–1905.
12. Donnelly R, Marzluff JM (2006) Relative importance of habitat quantity, structure, and spatial pattern to birds in urbanizing environments. Urban Ecosystems 9: 99–117.
13. McKinney ML (2008) Effects of urbanization on species richness: a review of plants and animals. Urban Ecosystems 11: 161–176.
14. Evans KL, Newson SE, Gaston KJ (2009) Habitat influences on urban avian assemblages. Ibis 151: 19–39.
15. Marzluff JM (2001) Worldwide urbanization and its effects on birds. In: Marzluff JM, Bowman R, Donnelly R, editors. Avian ecology and conservation in an urbanizing world. Boston, USA: Kluwer Academic Publishers. pp. 19–47.
16. McDonnell M, Hahs A (2008) The use of gradient analysis studies in advancing our understanding of the ecology of urbanizing landscapes: current status and future directions. Landsc Ecol 23: 1143–1155.
17. Blair RB (2001) Birds and butterflies along urban gradients in two ecoregions of the US. In: Lockwood JL, McKinney ML, editors. Biotic Homogenization. Norwell, Massachusetts, USA: Kluwer. pp. 33–56.
18. Chace JF, Walsh JJ (2006) Urban effects on native avifauna: a review. Landsc Urban Plan 74: 46–69.
19. Newhouse MJ, Marra PP, Johnson LS (2008) Reproductive success of house wrens in suburban and rural landscapes. Wilson J Ornithol 120: 99–104.
20. Chace JFWJJ, Cruz A, Prather JW, Swanson HM (2003) Spatial and temporal activity patterns of the brood parasitic brown-headed cowbird at an urban/wildland interface. Landsc Urban Plan 64: 173–190.
21. Engels TM, Sexton CW (1994) Negative correlation of Blue jays and Golden-cheeked Warblers near an urbanizing area. Conserv Biol 8: 286–290.
22. Friesen LE, Eagles PFJ, Mackay RJ (1995) Effects of residential development on forest-dwelling Neotropical migrant songbirds. Conserv Biol 9: 1408–1414.
23. Engle DM, Criner TL, Boren JC, Masters RE, Gregory MS (1999) Response of breeding birds in the Great Plains to low density urban sprawl. Great Plains Res 9: 55–73.
24. Odell EA, Knight RL (2001) Songbird and medium-sized mammal communities associated with exurban development in Pitkin County, Colorado. Conserv Biol 15: 1143–1150.
25. Fraterrigo JM, Wiens JA (2005) Bird communities of the Colorado Rocky Mountains along a gradient of exurban development. Landsc Urban Plan 71: 263–275.
26. Merenlender AM, Reed SE, Heise KL (2009) Exurban development influences woodland bird composition. Landsc Urban Plan 92: 255–263.
27. Suarez-Rubio M, Renner SC, Leimgruber P (2011) Influence of exurban development on bird species richness and diversity. J Ornithol 152: 461–471.
28. Suarez-Rubio M, Lookingbill TR, Elmore AJ (2012) Exurban development from 1986 to 2009 surrounding the District of Columbia, USA. Remote Sens Environ 124: 360–370.
29. Pidgeon AM, Radeloff VC, Flather CH, Lepczyk CA, Clayton MK, et al. (2007) Associations of forest bird species richness with housing and landscape patterns across the USA. Ecol Appl 17: 1989–2010.
30. Valiela I, Martinetto P (2007) Changes in bird abundance in eastern North America: urban sprawl and global footprint? BioScience 57: 360–370.
31. Jones KB, Neale AC, Nash MS, Riitters KH, Wickham JD, et al. (2000) Landscape correlates of breeding bird richness across the United States Mid-Atlantic region. Environ Monit Assess 63: 159–174.
32. Donovan TM, Flather CH (2002) Relationships among North American songbird trends, habitat fragmentation, and landscape occupancy. Ecol Appl 12: 364–374.
33. Askins RA (1995) Hostile landscapes and the decline of migratory songbirds. Science 267: 1956–1957.
34. Mancke RG, Gavin TA (2000) Breeding bird density in woodlots: effects of depth and buildings at the edges. Ecol Appl 10: 598–611.

35. Swift TL, Hannon SJ (2010) Critical thresholds associated with habitat loss: a review of the concepts, evidence, and applications. Biol Rev 85: 35–53.

36. Gardner RH, Milne BT, Turner MG, O'Neill RV (1987) Neutral models for the analysis of broad-scale landscape pattern. Landsc Ecol 1: 19–28.

37. Andrén H (1994) Effects of habitat fragmentation on birds and mammals in landscapes with different proportions of suitable habitat: a review. Oikos 71: 355–366.

38. With KA, Crist TO (1995) Critical thresholds in species' responses to landscape structure. Ecology 76: 2446–2459.

39. Fahrig L (2001) How much habitat is enough? Biol Conserv 100: 65–74.

40. Radford JQ, Bennett AF, Cheers GJ (2005) Landscape-level thresholds of habitat cover for woodland-dependent birds. Biol Conserv 124: 317–337.

41. Betts MG, Forbes GJ, Diamond AW (2007) Thresholds in songbird occurrence in relation to landscape structure. Conserv Biol 21: 1046–1058.

42. Poulin J-F, Villard M-A, Edman M, Goulet PJ, Eriksson A-M (2008) Thresholds in nesting habitat requirements of an old forest specialist, the Brown Creeper (Certhia americana), as conservation targets. Biol Conserv 141: 1129–1137.

43. Zuckerberg B, Porter WF (2010) Thresholds in the long-term responses of breeding birds to forest cover and fragmentation. Biol Conserv 143: 952–962.

44. Lindenmayer DB, Fischer J, Cunningham RB (2005) Native vegetation cover thresholds associated with species responses. Biol Conserv 124: 311–316.

45. Baker ME, King RS (2010) A new method for detecting and interpreting biodiversity and ecological community thresholds. Methods Ecol Evol 1: 25–37.

46. U.S. Census Bureau (2010) State and County quick facts. Available: http://quickfacts.census.gov. Accessed 2010 June 1.

47. Weldon Cooper Center (2010) Final 2008 and provisional 2009 population estimates for Virginia Counties and cities. Available: http://www.coopercenter.org/demographics/data. Accessed 2010 June 1.

48. Peterjohn BG, Sauer JR (1994) Population trends of woodland birds from the North American Breeding Bird Survey. Wild Soc Bull 22: 155–164.

49. Sauer JR, Fallon JE, Johnson R (2003) Use of North American Breeding Bird Survey data to estimate population change for bird conservation regions. J Wildl Manage 67: 372–389.

50. Sauer JR, Pendleton GW, Orsillo S (1995) Mapping of bird distributions from point count surveys. USA: USDA Forest Service, Pacific Southwest Research Station. 151– 160 p.

51. Veech JA, Crist TO (2007) Habitat and climate heterogeneity maintain beta-diversity of birds among landscapes within ecoregions. Global Ecol Biogeogr 16: 650–656.

52. Poole AE (2005) The Birds of North America Online. Ithaca, NY, USA: Cornell Laboratory of Ornithology. Available: http://bna.birds.cornell.edu.bnaproxy.birds.cornell.edu/BNA. Accessed 2011 Nov 3.

53. Mikusiński G, Gromadzki M, Chylarecki Pa (2001) Woodpeckers as indicators of forest bird diversity. Conserv Biol 15: 208–217.

54. Hagan JM, III (1993) Decline of the Rufous-sided Towhee in the Eastern United States. Auk 110: 863–874.

55. Sherry TW, Holmes RT (1997) American Redstart (Setophaga ruticilla). The Birds of North America Online. Ithaca, NY, USA Cornell Lab of Ornithology Available: http://bna.birds.cornell.edu.bnaproxy.birds.cornell.edu/bna/species/277. Accessed 2011 Nov 3.

56. U.S. NABCI Committee (2009) The State of the Birds, United States of America, 2009.

57. Bowman J (2003) Is dispersal distance of birds proportional to territory size? Can J Zool 81: 195–202.

58. Mazerolle DF, Hobson KA (2004) Territory size and overlap in male Ovenbirds: contrasting a fragmented and contiguous boreal forest. Can J Zool 82: 1774–1781.

59. Krementz DG, Powell LA (2000) Breeding season demography and movements of Eastern Towhees at the Savanna River site, South Carolina. Wilson Bull 112: 243–248.

60. Lang JD, Powell LA, Krementz DG, Conroy MJ (2002) Wood Thrush movements and habitat use: effects of forest management for Red-cockaded Woodpeckers. Auk 119: 109–124.

61. Adams JB, Smith MO, Johnson PE (1986) Spectral mixture modeling: a new analysis of rock and soil types at the Viking Lander 1 site. J Geophys Res 91: 8098–8112.

62. Soille P (2003) Morphological image analysis: principles and applications. Berlin, Germany: Springer-Verlag.

63. Vogt P, Riitters K, Estreguil C, Kozak J, Wade T, et al. (2007) Mapping spatial patterns with morphological image processing. Landsc Ecol 22: 171–177.

64. McGarigal K, Cushman SA, Neel MC, Ene E (2002) FRAGSTATS: spatial pattern analysis program for categorical maps. Available: http://www.umass.edu/landeco/research/fragstats/fragstats.html. Accessed 2011 Nov 3.

65. Gustafson EJ, Parker GR (1992) Relationships between landcover proportion and indices of landscape spatial pattern. Landsc Ecol 7: 101–110.

66. Sutherland GD, Harestad AS, Price K, Lertzman KP (2000) Scaling of natal dispersal distances in terrestrial birds and mammals. Conserv Ecol 4: 16. Available: http://www.consecol.org/vol14/iss11/art16/. Accessed 2011 Dec 31.

67. Ambuel B, Temple SA (1983) Area-dependent changes in the bird communities and vegetation of southern Wisconsin forests. Ecology 64: 1057–1068.

68. Blake JG, Karr JR (1987) Breeding birds of isolated woodlots: area and habitat relationships. Ecology 68: 1724–1734.

69. van Dorp D, Opdam PFM (1987) Effects of patch size, isolation and regional abundance on forest bird communities. Landsc Ecol 1: 59–73.

70. Robinson SK, Thompson Iii FR, Donovan TM, Whitehead DR, Faaborg J (1995) Regional forest fragmentation and the nesting success of migratory birds. Science 267: 1987–1990.

71. Sauer JR, Peterjohn BG, and Link WA (1994) Observer differences in the North American Breeding Bird Survey. Auk 111: 50–62.

72. Erskine AJ (1978) The first ten years of the cooperative breeding bird survey in Canada. Can Wildl Serv Report Ser 42: 1–61.

73. Kendall WL, Peterjohn BG, Sauer JR (1996) First time observer effects in the North American Breeding Bird Survey. Auk 113: 823–829.

74. Robbins CS, Bystrak D, Geissler PH (1986) The Breeding Bird Survey: its first fifteen years, 1965–1979.

75. Lunn DJ, Thomas A, Best N, Spiegelhalter D (2000) WinBUGS a Bayesian modeling framework: concepts, structure and extensibility. Stat Comput 10: 325–337.

76. LaDeau SL, Kilpatrick AM, Marra PP (2007) West Nile virus emergence and large-scale declines of North American bird populations. Nature 447: 710–713.

77. Link WA, Sauer JR (2002) A hierarchical analysis of population change with application to Cerulean Warblers. Ecology 83: 2832–2840.

78. Sauer JR, Link WA (2011) Analysis of the North American breeding bird survey using hierarchical models. Auk 128: 87–98.

79. Clark JS (2005) Why environmental scientists are becoming Bayesians. Ecol Lett 8: 2–14.

80. Link WA, Sauer JR (2007) Seasonal components of avian population change: joint analysis of two large-scale monitoring programs. Ecology 88: 49–55.

81. Berger JO (1985) Statistical decision theory and Bayesian analysis. New York, New York, USA: Springer-Verlag.

82. Gelman A, Carlin JB, Stern HS, Rubin DB (2004) Bayesian data analysis. Boca Raton, Florida, USA: Chapman and Hall/CRC.

83. Link WA, Barker RJ (2010) Bayesian inference with ecological applications. London, UK: Academic Press.

84. Cleveland WS, Devlin SJ (1988) Locally-weighted regression: an approach to regression analysis by local fitting. J Am Stat Assoc 83: 596–610.

85. R Development Core Team (2011) R: a language and environment for statistical computing v 2.11.1 ed. R Foundation for Statistical Computing, Vienna, Austria.

86. Halkin SL, Linville SU (1999) Northern Cardinal (Cardinalis cardinalis). The Birds of North America Online. Ithaca, NY, USA: Cornell Lab of Ornithology. Available: http://bna.birds.cornell.edu.bnaproxy.birds.cornell.edu/bna/species/440. Accessed 2011 Nov 3.

87. Evans M, Gow E, Roth RR, Johnson MS, Underwood TJ (2011) Wood Thrush (Hylocichla mustelina). The Birds of North America Online. Ithaca, NY, USA: Cornell Lab of Ornithology. Available: (http://bna.birds.cornell.edu.bnaproxy.birds.cornell.edu/bna/species/246. Accessed 2011 Nov 3.

88. Andrén H, Delin D, Seiler A (1997) Population response to landscape changes depends on specialization to different landscape elements. Oikos 80: 193–196.

89. Smith RJ, Hatch MI, Cimprich DA, Moore FR (2011) Gray Catbird (Dumetella carolinensis). The Birds of North America Online. Ithaca, NY, USA: Cornell Lab of Ornithology. Available: http://bna.birds.cornell.edu/bna/species/167. Accessed 2011 Nov 3.

90. Lepczyk CA, Mertig AG, Liu J (2004) Assessing landowner activities related to birds across rural-to-urban landscapes. Environ Manage 33: 110–125.

91. Graber JW, Graber RR, Kirk EL (1985) Illinois birds: vireos. Biological Notes no 68 Illinois Natural History Survey.

92. Hunt PD (1998) Evidence from a landscape population model of the importance of early successional habitat to the American Redstart. Conserv Biol 12: 1377–1389.

93. Holmes RT, Sherry TW (2001) Thirty-year bird population trends in an unfragmented temperate deciduous forest: importance of habitat change. Auk 118: 589–609.

94. Matlack GR (1997) Four centuries of forest clearance and regeneration in the hinterland of a large city. J Biogeogr 24: 281–295.

95. Smith WB, Miles PD, Visage JS, Pugh SA (2004) Forest resources of the United States, 2002. St. Paul, Minneapolis, USA.: USDA Forest Service, North Central Forest Experiment Station. 137 p.

96. Bowen ME, McAlpine CA, House APN, Smith GC (2007) Regrowth forests on abandoned agricultural land: a review of their habitat values for recovering forest fauna. Biol Conserv 140: 273–296.

97. Smith AC, Fahrig L, Francis CM (2011) Landscape size affects the relative importance of habitat amount, habitat fragmentation, and matrix quality on forest birds. Ecography 34: 103–113.

98. Hill SR, Gates JE (1988) Nesting ecology and microhabitat of the Eastern Phoebe in the central Appalachians. Am Midland Nat J 120: 313–324.

99. Weeks HP, Jr (2011) Eastern Phoebe (Sayornis phoebe) The Birds of North America Online. Ithaca, NY, USA Cornell Lab of Ornithology. Available: http://bna.birds.cornell.edu.bnaproxy.birds.cornell.edu/bna/species/094. Accessed 2011 Nov 3.

100. Greenlaw JS (1996) Eastern Towhee (Pipilo erythrophthalmus). The Birds of North America Online. Ithaca, NY, USA: Cornell Lab of Ornithology. Available: http://bna.birds.cornell.edu.bnaproxy.birds.cornell.edu/bna/species/262. Accessed 2011 Nov 3.

101. Dauphiné N, Cooper RJ. Impacts of free-ranging domestic cats (*Felis catus*) on birds in the United States: a review of recent research with conservation and management recommendations In: Rich TD, Arizmendi C, Demarest DW, Thompson C, editors; 2009. pp. 205–219.

102. Cooper C, Loyd K, Murante T, Savoca M, Dickinson J (2012) Natural history traits associated with detecting mortality within residential bird communities: can citizen science provide insights? Environ Manage 50: 11–20.

103. Betts MG, Hagar JC, Rivers JW, Alexander JD, McGarigal K, et al. (2010) Thresholds in forest bird occurrence as a function of the amount of early-serial broadleaf forest at landscape scales. Ecol Appl 20: 2116–2130.

104. Wakamiya SW (2012) Breeding bird monitoring: Mid-Atlantic Network 2011 summary report. Natural Resource Data Series NPS/MIDN/NRDS—2012/331. National Park Service, Fort Collins, Colorado.

17

A Framework for Assessing Global Change Risks to Forest Carbon Stocks in the United States

Christopher W. Woodall[1]*, Grant M. Domke[1], Karin L. Riley[2], Christopher M. Oswalt[3], Susan J. Crocker[1], Gary W. Yohe[4]

1 US Department of Agriculture, Forest Service, Northern Research Station, St. Paul, Minnesota, United States of America, 2 College of Forestry and Conservation, University of Montana, Missoula, Montana, United States of America, 3 US Department of Agriculture, Forest Service, Southern Research Station, Knoxville, Tennessee, United States of America, 4 Wesleyan University, Middletown, Connecticut, United States of America

Abstract

Among terrestrial environments, forests are not only the largest long-term sink of atmospheric carbon (C), but are also susceptible to global change themselves, with potential consequences including alterations of C cycles and potential C emission. To inform global change risk assessment of forest C across large spatial/temporal scales, this study constructed and evaluated a basic risk framework which combined the magnitude of C stocks and their associated probability of stock change in the context of global change across the US. For the purposes of this analysis, forest C was divided into five pools, two live (aboveground and belowground biomass) and three dead (dead wood, soil organic matter, and forest floor) with a risk framework parameterized using the US's national greenhouse gas inventory and associated forest inventory data across current and projected future Köppen-Geiger climate zones (A1F1 scenario). Results suggest that an initial forest C risk matrix may be constructed to focus attention on short- and long-term risks to forest C stocks (as opposed to implementation in decision making) using inventory-based estimates of total stocks and associated estimates of variability (i.e., coefficient of variation) among climate zones. The empirical parameterization of such a risk matrix highlighted numerous knowledge gaps: 1) robust measures of the likelihood of forest C stock change under climate change scenarios, 2) projections of forest C stocks given unforeseen socioeconomic conditions (i.e., land-use change), and 3) appropriate social responses to global change events for which there is no contemporary climate/disturbance analog (e.g., severe droughts in the Lake States). Coupling these current technical/social limits of developing a risk matrix to the biological processes of forest ecosystems (i.e., disturbance events and interaction among diverse forest C pools, potential positive feedbacks, and forest resiliency/recovery) suggests an operational forest C risk matrix remains elusive.

Editor: Ben Bond-Lamberty, DOE Pacific Northwest National Laboratory, United States of America

Funding: The authors have no support or funding to report.

Competing Interests: The authors have declared that no competing interests exist.

* E-mail: cwoodall@fs.fed.us

Introduction

As the current carbon (C) stocks of forests in the United States (US) store an amount of C approximately equal to 25 years' worth of US fossil fuel CO_2 emissions at their current rate [1], the status and fate of these C stocks in the face of global change is an area of emerging concern [2,3,4,5]. The loss and subsequent partial recovery of forests in the US following the exploitive harvests and land-use conversion of the late 19th and early 20th century are a past event that can frame current discussions regarding managing/monitoring forest C stocks across large-scales [6]. The relatively rapid change in the status of forests in the US – from a steady state of minimal CO_2 emission/sequestration to major CO_2 emitter – during this time period (Fig 1) [7] may offer a cautionary tale of how quickly the source/sink status of large-scale forest C stocks can change. Despite this transcontinental degradation of forests in the US and concomitant emission of CO_2, global atmospheric CO_2 concentrations only slightly increased above pre-industrial levels during this period [8,9].

The advent of modern forestry in the US, conservation movements, and urbanization of the populace rapidly shifted the status of forests in the US so that they once again provided a net sequestration of C during the mid-20th century to the present day [7,10]. While it was a direct human disturbance (e.g., logging and land-use conversion) that precipitated the last US forest status change to a net emission source, could climate change (CC) and its potential to increase the probability of large-scale disturbance events and/or alterations to forest C cycles [11,12,13,14] result in a similar shift in sink/source status now that global CO_2 atmospheric concentrations approaching 400 ppm [9]? There is growing evidence that large-scale natural disturbances contribute to substantial C emissions from forests on an annual basis [3,15]. Given forests are the largest terrestrial C sink on earth [16] much of this C may be at risk. One of the most critical future consequences of CC on forest C stocks is the potential change in their source/sink status (i.e., net annual sequestration to net annual emission). Although forests currently sequester more C than they emit on an annual basis globally [16], the ability of forests to continue this trend in the future may be limited [7,17].

Developing a conceptual framework for assessing CC risks to forest ecosystem C stocks may enable efficient allocation of efforts to monitor and mitigate CC effects while informing future

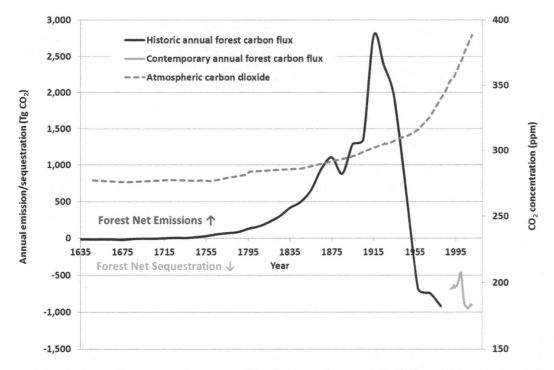

Figure 1. Historic annual rates of forest ecosystem and harvested wood product carbon dioxide net emissions/sequestration in US forests (black line: Birdsey et al. [7], green line: EPA [10]) and global atmospheric CO₂ concentration (Etheridge et al. [8], ESRL [9]), 1635 to 2010.

research into refined risk probabilities. In general, risk has been organized around two components, likelihood and consequence(s) (i.e., expectation of random variable) [18]. The exact metrics used to assess risk vary based on available data and information deemed important by stakeholders [19]. These metrics can be organized using a matrix where the magnitude of consequences is plotted against the relative likelihood of the event occurring [20]. Within the matrix, a series of grids can be developed to represent functional levels of risk which translate into monitoring strategies aimed at managing the risk [21,22]. This approach has been adopted by Iverson et al. [20] to evaluate the risk of CC on forested habitats. The creation of the risk matrices was not intended for use in decision making; rather the matrices were designed to focus the evaluation of risk in terms of likelihood and consequence [21,22]. Although Iverson et al.'s [20] focus was on risk of habitat change for tree species, the opportunity exists to extend the methodology to other forest ecosystem attributes and services (e.g., water or C). In the context of CC and forest C dynamics, risk may be conceptualized as the magnitude of C stock change due to CC and/or CC-induced natural disturbances multiplied by the associated probability that C stocks might change due to said events. Given the need to broadly assess global change risks to forest C stocks, the goal of this study was to develop and evaluate a general framework (i.e., risk matrix) for forest C stocks in the US using estimates from the National Greenhouse Gas Inventory (NGHGI) in the US and associated forest inventory data to parameterize the initial matrix across broad climate zones.

Methods

The basic premise of this study's proposed forest C risk matrix is that risk is a combination [20,22] of 1) the likelihood of a forest C pool's emission and 2) the consequence of such an emission. Within the risk framework, the consequence is equated with the

relative size (i.e., mass) of each forest C pool and plotted along the y-axis. The consequences of a forest C pool shifting from a sink to a source is postulated as being closely related to its population estimate over a large region of interest, in this case, the conterminous US. The likelihood of such a shift is equated with each forest C pool's variability across climate zones of the US – higher variability equates to a greater likelihood of change. In this analysis, total forest ecosystem C was apportioned into five pools (live above- and belowground biomass, dead wood, soil organic C, and forest floor) as broadly delineated by the United Nations Framework Convention on Climate Change [10]. The combination of each C pool's consequence and likelihood arrays itself within a matrix of societal responses [20] to CC including combinations of monitoring, mitigation, and adaptation. Given the complexities of human-forest-climate interactions, the size and arrangement of societal responses within the risk framework would vary with scale, geographic area, socioeconomic drivers and constraints, forest types, and existing monitoring infrastructure, among other factors.

The magnitude of C stocks (i.e., consequence) in each pool was estimated from the US Department of Agriculture's Forest Inventory and Analysis (FIA) program [23], which informs the NGHGI of the US [10]. For the purposes of FIA's inventory, forest is defined as at least 36.6 m wide and 0.4 ha in size with at least 10 percent cover by live trees. The FIA program employs a three phase inventory. Phase one is a variance reduction process where satellite imagery is used to assign individual field plots to strata (e.g., forest canopy cover classes). FIA's plot network contains over 125,000 plots which are systematically distributed approximately every 2,428 ha across the conterminous US. During the second phase of the inventory, if a sample point falls in a forested area then field crew visit the plot. Each forested plot is comprised of a series of smaller sub-plots where tree- and site-level attributes – such as diameter and tree height – are measured at regular temporal intervals [23]. During the third phase of the inventory, a subset of phase two plots are measured for additional variables related to forest health attributes (e.g., downed woody materials, understory vegetation, and soils). The FIA program does not directly measure forest C stocks. Instead, a combination of empirically derived C estimates (e.g., standing live and dead trees) [24] and models (e.g., forest floor C stocks related to stand age and forest type) are used to estimate forest C stocks [10]. Estimates of C stocks by pool were developed for 2010 using FIA's current inventory. For illustrative purposes, current C stocks were projected to the year 2100 using annual estimates of forest C stocks over the UNFCCC reporting period (1990-present) [10] and simple linear regression techniques by pool. Although it is not expected that forests will continue to sequester C at current rates to the year 2100 (especially for soil organic C), this parsimonious approach facilitates interim evaluation of the risk matrix.

To estimate the likelihood of forest C stock changes, current (1976–2000; referred to hereafter as 2010) and future projected Köppen-Geiger climate zones (2076–2100; referred to hereafter as 2100) from Rubel and Kotek [25] were used. The A1F1 emission scenario and associated projected climate zones (2100) were selected from Rubel and Kotek [25] as it demonstrated the largest future climate shift for illustration purposes in this study. The FIA program's plot network was overlaid with current (2010) and A1F1 Köppen-Geiger climate zones (2100) [25] for estimating the coefficient of variation of the median C stock densities among the various climates by pool for each year (2010 and 2100). In order to project coefficients of variation in the year 2100, a basic imputation approach was employed. If a FIA plot's current climate zone changed between 2010 and 2100 [25], then the plot was assigned the median C density of that projected future climate using that climate's current median C density by pool. Coefficients of variation were then re-calculated using combined data from current (i.e., no change in climate zone) and imputed (i.e., change in climate zone) values. For example, if half of FIA's forested plots experienced a shift in climate zones from 2010 to 2100, then half of the plot-level 2100 C densities would be based on 2010 C empirical estimates while half of the plots would be assigned the median C density associated with the new climate zone in which the plots are located. Hence, future forest C stock variability would be based on knowledge of the current distribution of C stock densities among climates assuming that current climates would still exist in 2100 but at different locations (e.g., arid zone with different spatial extent). For more information on: 1) population estimation procedures used by the FIA program please refer to Bechtold and Patterson [23], 2) forest C pool models and inventory data specific to 2010 please refer to the EPA [10], and 3) current and projected Köppen-Geiger climate classifications please see Rubel and Kottek [25].

Results

Current forest C stocks were estimated for the year 2010 by pool. Soil organic C was the largest stock at 17,572 Tg and dead wood the smallest at 2,627 Tg (Table 1). Associated univariate statistics suggest substantial variability in plot-level estimates of forest C density (Mg·ha^{-1}) by pool with the standard deviation exceeding the mean for live aboveground, live belowground, and dead wood in 2010.

As an initial appraisal of empirical variation in C stocks across the various climate zones of the US in 2010, the coefficients of variation (percent) of median C densities (based on individual plot-level estimates of C stocks) were calculated by pool and climate

Table 1. Estimates of total forest ecosystem C stocks (Tg), mean and associated standard deviation (SD) of carbon density, and associated univariate statistics (Q1: first quartile, median, Q3: third quartile; Mg·ha^{-1}), across the national forest inventory by carbon pool in the US, 2010.

Carbon pool	Total stocks	Mean	SD	Q1	Median	Q3
Live AG	14,541	38.73	40.57	10.47	27.35	54.76
Live BG	2,876	8.02	8.66	2.14	5.69	11.26
Dead Wood	2,627	8.43	10.70	3.71	6.17	9.61
Forest Floor	4,941	16.08	11.85	7.20	10.20	24.20
SOC	17,572	71.38	48.25	41.7	53.1	94.80

Note: estimates do not include Hawaii, Alaska, or trees on non-forest land (e.g., agricultural trees and urban parks).
*AG = aboveground, BG = belowground, dead = standing and downed dead wood, SOC = soil organic carbon.

zone across the US and are ordered as: soil organic C (70.9), dead wood (55.2), forest floor (53.4), live belowground (44.6), and live aboveground (43.1) (Table 2). The highest median C density (Mg·ha^{-1}) by pool across climates in 2010 was live aboveground (33.83), live belowground (7.95), dead wood (11.36), forest floor (33.11), and soil organic C (158.01) in the warm temperate (summer dry), warm temperate (summer dry), warm temperate (summer dry), snow (summer dry), and equatorial, respectively.

For illustrative purposes, C stocks and associated coefficients of variation among climate zones were projected to the year 2100 (Table 3). Projected C stocks increased in all forest C pools over the 90 year period (2010–2100), with the largest increases occurring in live aboveground biomass (49 percent; 7,116 Tg) and belowground biomass (49 percent; 1,406 Tg) followed by dead wood (25 percent; 671 Tg), soil organic C (8 percent; 1,331 Tg), and forest floor (7 percent; 334 Tg) (Table 3). The projected coefficients of variation of C pools among climate zones in 2100 increased for all pools. The largest increase (absolute change in coefficient percentage, percent) was for forest floor (4.2) followed sequentially by soil organic C (2.1), live belowground (0.5), dead wood (0.4), and live aboveground (0.2).

The estimates of current US forest C stocks by pool in conjunction with coefficients of variation among climate zones were used to array the forest C pools in a risk matrix (Fig 2). Pools were arrayed within the risk matrix as a combination of their associated stock size (i.e., magnitude of C sink/source) and C density variability among climates (i.e., likelihood of change under future CC). Because both the stocks and coefficients of variation were projected to the year 2100, the combination of population estimates (2010 and 2100) could be plotted within the risk matrix (i.e., general societal response key) which allowed for broad assessment of potential trends in forest C stocks and discussion of monitoring strategies. Soil organic C had the largest stock estimate in 2010 combined with a highest level of variation across climate zones and thus fell within the "initiate adaptation/mitigation" response category (Fig 2). As the remainder of pools had coefficients of variation within the same narrow range (43.1–55.2 percent), their assignment to response categories was largely dependent on their current stock size (i.e., mass). As the live aboveground pool had the second largest stock it was assigned to the "annually monitor/develop strategies" response category. The three pools of live belowground, forest floor, and dead wood had relatively small estimates of C stocks combined with moderate to

Table 2. Estimates of median forest carbon density (Mg·ha^{-1}) by carbon pool and Köppen-Geiger Climate Classifications [25] with coefficients of variation (CV) determined across climates for each carbon pool, 2010.

Pool	Climate Classification						CV (%)
	Equatorial	Arid	Warm Temperate, fully humid	Warm Temperate, summer dry	Snow, fully humid	Snow, summer dry	
	median carbon density (Mg·ha^{-1})						
Live AG	14.02	8.05	29.25	33.83	28.56	29.63	43.1
Live BG	2.85	1.68	6.00	7.35	5.94	6.60	44.6
Dead Wood	3.95	1.46	5.85	11.36	6.78	9.02	55.2
Forest Floor	7.26	21.09	7.84	28.14	20.28	33.11	53.4
SOC	158.01	24.12	45.95	49.80	94.76	44.12	70.9

*AG = aboveground, BG = belowground, dead = standing and downed dead wood, SOC = soil organic carbon.

low levels of variability among climates, thus their response category assignments were limited to periodic monitoring activities.

For the purpose of discussion, C stocks and associated variability among climates was projected to the year 2100 and arrayed in the risk matrix (Fig 2). All pools moved in varying degrees towards more proactive response categories (i.e., from periodically monitor to robust mitigation). For some pools, such as the forest floor and soil organic C, their movement towards a potential source was largely due to the increase in their coefficients of variability among climate zones in 2100. For other pools, such as live aboveground, there was minimal increase in risk of emission due to increases in their respective stock sizes.

Discussion

Empirically derived estimates (e.g., population totals and associated variability across climates) of forest C pools across the US may be used to develop an initial CC risk matrix. This risk matrix is not intended to be used to develop local strategies to mitigate CC and disturbances to forest C cycles and stocks; however, it may provide a common framework for discussing risks across large spatial/temporal scales. As the risk matrix attempts to mesh empirical estimates with societal response, a thorough

Table 3. Linear projections (year 2100) of total forest carbon stocks (Tg) based on contemporary baselines by carbon pool and projections of future coefficients of variation (CV) of carbon pools determined across A1F1 Köppen-Geiger Climate Classifications (years 2076–2100) [25].

Pool	Total Carbon Stock (Tg)	CV (%)
Live AG	21,657	43.3
Live BG	4,282	45.1
Dead Wood	3,298	55.6
Forest Floor	5,275	57.6
SOC	18,903	73.0

*AG = aboveground, BG = belowground, dead wood = standing and downed dead wood, SOC = soil organic carbon.

understanding of its limits, implications, and potential refinements is needed.

As the basic premise of the risk matrix is that the risk of C pool status change (i.e., C source or sink) relates to 1) the likelihood (i.e., coefficients of variation for each individual forest pool's median C density across climate zones) and 2) the consequence of such a shift (i.e., forest C pools' current stock). Coefficients of variation were used as an initial metric of change likelihood as it facilitated comparison across diverse forest C pools and has been used in risk assessments in the biological sciences [26] and finance [27]. If indeed CC occurred such that a forest experienced a climate shift from "temperate" to "equatorial," then the contemporary range in variation in C densities between those climates may indicate likelihood of C emission. Because societal responses can only be conjectured in this study, the proposed C risk framework suggests C source/sink status change and a tipping point where forest C emissions may exacerbate CC impacts and positive feedbacks (e.g., boreal forest heterotrophic respiration, [28]) as metrics critical to society. In practice, the size and arrangement of societal responses must align with socioeconomic drivers and constraints, the forest types and attributes that may be at risk, and existing monitoring infrastructure, among other factors.

In a manner similar to previous work [20], the risk matrix in this study is not intended to be used in decision making; rather it is designed to focus attention on the short- and long-term risks to forest C stocks. This is particularly true given the large amount of uncertainty associated with C stocks and stock changes associated with CC and intensified forest disturbances. Time is not explicitly part of the risk matrix to allow consideration of potentially long-term (100+ years), low-probability high-impact events (e.g., system collapse of boreal forests) that are often overlooked when considering risks [29] that may be beyond our contemporary frame of reference. There is growing evidence that CC-influenced natural disturbances are major drivers of C dynamics in forest ecosystems and contribute to substantial C emissions annually [3,15], while placing infrastructure and human dwellings at risk (a set of consequences not considered here). When the consequences (even absent economic losses) and likelihoods of forest C stocks shifting from sinks to sources are viewed together, a critical need for a cohesive approach to monitoring and managing risk emerges.

As much uncertainty is associated with the fate of forest ecosystem C following natural disturbance and the potential feedbacks between forests and climate [11,12,30] new questions emerge. What is the likelihood that US forests will once again

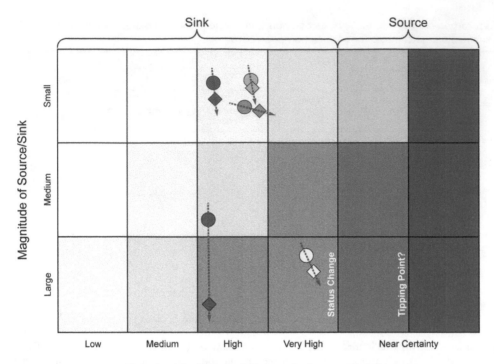

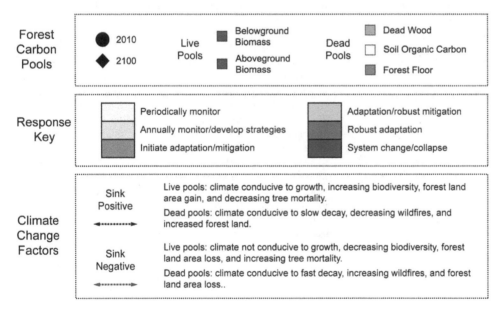

Figure 2. Climate change risk matrix for forest ecosystem carbon pools in the US. Likelihood of change in carbon stocks is based on the coefficient of variation of median forest carbon stock densities among Köppen-Geiger climate regions (i.e., x-axis) based on the national forest inventory plot network. Size of carbon stocks are based on the US National Greenhouse Gas Inventory (i.e., y-axis). Societal response (e.g., immediate adaptive response or periodic monitoring) to climate change events depends on the size and relative likelihood of change in stocks. Year 2100 projections are based on linear extrapolations of current carbon stocks and imputing current median carbon pool densities by climate region to projected future climate regions for calculation of coefficients of variation. The soil organic carbon pool exhibits the highest variability among climate regions and therefore may be most affected by climate change or climate change induced disturbance events. In contrast, the dead wood pool has a relatively small stock with low variability among climate regions. Explicit climate change effects are not incorporated into this matrix as they represent a number of complex feedbacks both between stocks (e.g., live aboveground biomass transitioning to the dead wood pool) and the atmosphere (e.g., forest floor decay).

become net emitters of CO_2? Is there a tipping point [31] at which CC and associated disturbances fundamentally alter forest ecosystem processes (e.g., regeneration and decay) such that the current system collapses with a concomitant large release of C?

This scenario can be viewed hypothetically wherein a forest ecosystem as it proceeds through time within a natural range of variability (i.e., disturbance and subsequent recovery) regarding its C source/sink status (Fig 3). As seen in the US, forest ecosystems and their status as a sink of C partially recovered despite a transcontinental forest disturbance (i.e., 1700–1910 widespread harvest and land-use conversion) that shifted the forest's source/sink status past what might be considered a natural range of variability [7] (Fig 1). The ability of forests to recover (i.e., sequester and accumulate forest C) following disturbance has been widely documented [32]. However, there may be a tipping point where specific disturbances and/or CC push a forest ecosystem beyond the point from which it can recover [31], potentially resulting in new systems (e.g., shrub/steppe). Conversely, disturbance regimes and changing climates that are conducive to forest ecosystems could result in the invasion of non-forest systems by forest communities (Fig 3). Beyond conjecture, framing these risks with empirical observations may refine mitigation/adaptation efforts while directing future technical refinements (i.e., calculation of risk).

The empirical parameterization of the risk matrix in this study highlights numerous knowledge gaps: 1) robust measures of forest C stock change likelihood under climate change scenarios, 2) projections of forest C stocks given unforeseen socioeconomic possibilities (i.e., land-use change), and 3) appropriate social responses to global change events for which there is no contemporary climate/disturbance analog.

First, the initial metric of C stock change likelihood in this study (coefficient of variation across climatic regions) was for illustrative purposes and falls short in various situations (e.g., coarse spatial resolution of future climate projections can inadvertently reduce coefficients of variation). There is the artifact of modeled versus empirical C estimates that affects the risk matrix. Stochastic disturbance events may not be accurately reflected in the coefficients of variation as some pools are modeled as a function of stand attributes (e.g., forest type) that may not be similarly impacted as the modeled pool. Woodall et al. [33] found that modeled standing dead tree C stocks across the US may not accurately reflect the empirical variation inherent with stochastic mortality events (e.g., pine beetle mortality). If forest C inventories

adopt more empirically-based assessments of C stocks (e.g., increased sample intensity or advanced remotely sensed imagery techniques) then perhaps the risk matrix would be more responsive to CC-induced disturbances. Although adequately gauging uncertainty (i.e., variation of forest C stocks) is an essential component of greenhouse gas inventories that facilitates societal policy discussions [34], there yet remains a robust method that could be employed in this study. If indeed measures of likelihood of forest C stock change are needed for society to appropriately value forest management actions (i.e., mitigation and adaptation activities) [35] then perhaps this remains the largest knowledge gap identified in this study.

Second, the size of future forest C stocks was linearly extrapolated to the year 2100 in this study which we acknowledge has a low probability of occurring [36,37]. Although the projected trends in forest ecosystem C stocks from 2010 to 2100 provide some insight into the direction and extent of potential change at a continental scale, it does not provide the level of detail to forecast, for example, the location and extent of fluctuations in the live biomass pool following a conversion of forest lands to grasslands due to changes in precipitation cycles, wildfires, and/or tree regeneration failure (i.e., system collapse). A hurdle in using a risk framework to guide social response is how far a given pool would move within the framework after a CC-induced event (i.e., how far and in what direction the sink negative/positive arrows extend). For example, recent evidence supports suggestions that CC will result in more frequent higher-intensity storms in the future [38]. While debate still surrounds the prediction of an increase in the number of storms, there appears to be considerable agreement that CC is likely to increase storm intensity and duration [11,13,14,39]. Beyond CC induced changes in forest C stock magnitude, there is perhaps the larger question regarding future socioeconomic conditions that often guide changes in forest C stocks. The economic recession of 2008 had a deleterious effect on forest and housing industries in the US [40], but assuming pre-2008 land-use trends to 2030 suggested reduction in the rate of US

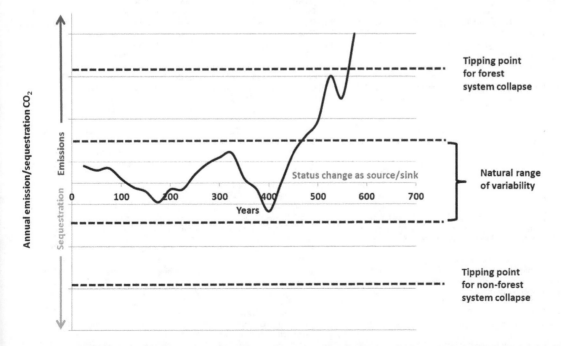

Figure 3. Hypothetical trend in forest ecosystem CO$_2$ emissions/sequestration over a multi-century time period in the context of a natural range of variability and potential tipping points between forest ecosystems and other systems (e.g., grasslands).

forest C sequestration [36]. Sharp changes in socioeconomic dynamics can have greater effect on forest C stocks than CC thus confounding technical advances to project forest ecosystem attributes under various climate scenarios.

Third, as this study only examined a limited set of societal responses to global change considered by previous studies [20,21], alternative response categories particular to forest C management should be considered. Despite the founding of the United Nation's Intergovernmental Panel on Climate Change nearly 25 years ago followed by a bevy of research and societal response (e.g., Reducing Emissions from Deforestation and Forest Degradation), the global CO_2 concentration continues to increase at an increasing rate [9]. Risk matrices, such as the one proposed in this study, may need to be constructed to accept evolving responses to CC such as emphasizing adaptation as opposed to mitigation. Furthermore, the cost metric of source/sink status of forest C stocks in this study may need to be changed to one of C baselines commonly used in C monitoring (e.g., 1990 baseline year) [10]. Perhaps likelihood could be defined as the likelihood of a certain percentage reduction in net C balance below a baseline projection or even historic benchmark. If adaptation is favored in lieu of mitigation activities then perhaps a cost-metric of gain/loss of forest land area or tree regeneration stocking could be used. If uncertainty of a forest C risk matrix (both future stock magnitude and likelihood of change) remains relatively high then it should follow that society's response and cost metrics should be flexible within a risk matrix.

Conclusions

Overall, the risk framework promulgated in this study not only offers an approach to identifying knowledge gaps associated with forest C dynamics in the context of global change, but also a generalized path to prioritizing research and monitoring, given risks associated with future CC. The risk of any particular forest C stock becoming a net atmospheric emission is related to its particular stock attributes (e.g., stock size and pool) and surrounding climate (e.g., equatorial versus temperate). These factors form a complex matrix that CC may inherently alter in unforeseen ways. The empirical parameterization of such a risk matrix highlights numerous major knowledge gaps: 1) robust measures of forest C stock change likelihood under climate change scenarios, 2) projections of forest C stocks given unforeseen socioeconomic possibilities (i.e., land-use change), and 3) appropriate social responses to global change events for which there is no climate/disturbance analog. Given the finite ability of our society to alter future climate trajectories and disturbances associated with global change, using a risk framework to address the greatest risks to forest C stocks may provide one path to future forest sustainability. Despite the qualitative nature and research gaps within the forest C stock risk framework, this approach provides a conceptual way forward for identifying priority research needs for mitigating or adapting to potential CC-induced events.

Author Contributions

Conceived and designed the experiments: CWW GMD GWY. Performed the experiments: CWW GMD. Analyzed the data: CWW GMD. Contributed reagents/materials/analysis tools: CWW GMD. Wrote the paper: CWW GMD KLR CMO SJC GWY.

References

1. Woodall CW, Skog K, Smith JE, Perry CH (2011) Criterion 5: Climate change and global carbon cycles. In National Report on Sustainable Forests – 2010. Editors: Robertson G, Gaulke P, McWilliams R, LaPlante S, Guldin R. FS-979. Washington DC: U.S. Department of Agriculture, Forest Service.
2. Davidson EA, Janssens IA (2006) Temperature sensitivity of soil carbon decomposition and feedbacks to climate change. Nature 440: 165–173.
3. Kurz WA, Stinson G, Rampley GJ, Dymond CC, Neilson ET (2008) Risk of natural disturbances makes future contribution of Canada's forests to the global carbon cycle highly uncertain. Proc Natl Acad Sci U S A 105: 1551–1555.
4. Melillo JM, Butler SM, Johnson JE, Mohan JE, Lux H, et al. (2011) Soil warming, carbon-nitrogen interactions and forest carbon budgets. Proc Natl Acad Sci USA 108: 9508–9512.
5. Anderegg WRL, Berry JA, Smith DD, Sperry JS, Anderegg LDL, et al. (2012) The roles of hydraulics and carbon stress in a widespread climate-induced forest die-off. Proc Natl Acad Sci USA 109: 233–237.
6. Houghton RA, Hackler JL, Lawrence KT (1999) The U.S. Carbon Budget: Contributions from Land-Use Change. Science 5427: 574–578.
7. Birdsey R, Pregitzer K, Lucier A (2006) Forest carbon management in the United States: 1600–2100. J Environ Qual 35: 1461–1469.
8. Etheridge DM, Steele LP, Langenfelds RL, Francey RJ, Barnola JM, et al. (1998) Historical CO_2 records from the Law Dome DE08, DE08-2, and DSS ice cores. 1998. In Trends, "A compendium of data on global change," Carbon Dioxide Information Analysis Center, Oak Ridge National Laboratory, Oak Ridge, TN. http://cdiac.esd.ornl.gov/trends/co2/lawdome.html.
9. ESRL (2012) Trends in atmospheric carbon dioxide. U.S. Department of Commerce, National Oceanic and Atmospheric Administration, Earth System Research Laboratory, Global Monitoring Division. http://www.esrl.noaa.gov/gmd/ccgg/trends/.
10. EPA (2011) Inventory of U.S. greenhouse gas emissions and sinks: 1990–2009. Chapter 7. Land use, land-use change, and forestry. Annex 3.12. Methodology for estimating net carbon stock changes in forest land remaining forest lands. Washington, DC: U.S. Environmental Protection Agency. #430-R-11-005.
11. Dale VH, Joyce LA, McNulty S, Neilson RP, Ayres MP, et al. (2001) Climate change and forest disturbance. BioScience 51: 723–734.
12. Running SW (2008) Ecosystem disturbance, carbon, and climate. Science 321: 652–653.
13. Turner MG (2010) Disturbance and landscape dynamics in a changing world. Ecology 91: 2833–2849.
14. Seidl R, Schelhaas MJ, Lexer MJ (2011) Unraveling the drivers of intensifying forest disturbance regimes in Europe. Glob Chang Biol 17: 2842–2852.
15. Zeng HC, Chambers JQ, Negron-Juarez RI, Hurtt GC, Baker DB, et al. (2009) Impacts of tropical cyclones on US forest tree mortality and carbon flux from 1851 to 2000. Proc Natl Acad Sci USA 106: 7888–7892.
16. Pan Y, Birdsey RA, Fang J, Houghton R, Kauppi PE, et al. (2011) A large and persistent carbon sink in the world's forests. Science 333: 988–993.
17. Reich PB (2011) Taking stock of forest carbon. Nat Clim Chang 1: 346–347.
18. Raiffa H, Schlaiffer R (2000) Applied statistical decision theory. Wiley Classics, New York.
19. Turner II BL, Kasperson RE, Matson PA, McCarthy JJ, Corell RW, et al. (2003) A framework for vulnerability analysis in sustainability science. Proc Natl Acad Sci USA 100: 8074–8079.
20. Iverson LR, Matthews SN, Prasad AM, Peters MP, Yohe G (2012) Development of risk matrices for evaluating climatic change responses of forested habitats. Clim Change 114: 231–243.
21. Yohe G (2010) Risk assessment and risk management for infrastructure planning and investment. Bridge 40: 14–21.
22. Yohe G, Leichenko R (2010) Adopting a risk-based approach. Ann N Y Acad Sci 1196: 29–40.
23. Bechtold WA, Patterson PJ (2005) The enhanced Forest Inventory and Analysis program—national sampling design and estimation procedures. USDA For. Serv. Gen. Tech. Rep. SRS-80. 85 p.
24. Woodall CW, Heath LS, Domke GM, Nichols M, Oswalt CM (2011) Methods and models for estimating volume, biomass, and C for forest trees in the U.S's national inventory, 2010. USDA For. Serv. Gen Tech. Rep. NRS-88. 30 p.
25. Rubel F, Kottek M (2010) Observed and projected climate shifts 1901–2100 depicted by world maps of the Köppen-Geiger climate classification. Meteorol. Z. 19: 135–141.
26. Pimm SL, Jones HL, Diamond J (1988) On the risk of extinction. Am Nat. 132: 757–785.
27. Brief RP, Owen J (1969) A note on earnings risk and the coefficient of variation. Finance 24: 901–904.
28. Grosse G, Harden J, Turetsky M, McGuire AD, Camill P, et al. (2011) Vulnerability of high-latitude soil organic carbon in North America to disturbance. J Geophys Res 116: G00K06.
29. Weitzman ML (2011) Fat-tailed uncertainty in the economics of catastrophic climate change. Rev Environ Econ Policy 5: 275–292.
30. O'Halloran TL, Law BE, Goulden ML, Wang Z, Barr JG, et al. (2012) Radiative forcing of natural forest disturbances. Glob Chang Biol 18: 555–565.
31. Walker G (2006) The tipping point of the iceberg: Could climate change run away with itself? Nature 441: 802–805.

32. Amiro BD, Barr AG, Barr JG, Black TA, Bracho R, et al. (2010) Ecosystem carbon dioxide fluxes after disturbance in forests of North America. J Geophys Res 115: G00K02.

33. Woodall CW, Domke GM, MacFarlane DW, Oswalt CM (2012) Comparing field- and model-based standing dead tree carbon stock estimates across forests of the United States. Forestry 85: 125–133.

34. Jonas M, Marland G, Winiwarter W, White T, Naborski Z, et al. (2010) Benefits of dealing with uncertainty in greenhouse gas inventories: introduction. Clim Change 103: 3–18.

35. Hurteau MD, Hungate BA, Koch GW (2009) Accounting for risk in valuing forest carbon offsets. Carbon Balance Manag 4: 1.

36. U.S. Department of Agriculture, Forest Service (2012) Future of America's forest and rangelands: Forest Service 2010 Resources Planning Act Assessment. Gen. Tech. Rep. WO-87. Washington, DC. 198 p.

37. Zhao M, Running SW (2010) Drought-induced reduction in global terrestrial net primary production from 2000 through 2009. Science 329: 940–943.

38. Elsner JB (2006) Evidence in support of the climate change–Atlantic hurricane hypothesis. J Geophys Res 33: L16705.

39. Knutson TR, McBride JL, Chan J, Emanuel K, Holland G, et al. (2010) Tropical cyclones and climate change. Nat Geosci 3: 147–163.

40. Woodall CW, Ince PJ, Skog KE, Aguilar FX, Keegan CE, et al. (2012) An overview of the forest products sector downturn in the US. For Prod J 61: 595–603.

18

Using LiDAR Data to Measure the 3D Green Biomass of Beijing Urban Forest in China

Cheng He[1], Matteo Convertino[2,3], Zhongke Feng[4], Siyu Zhang[1]*

1 Nanjing Forest Police College, Nanjing, China, **2** School of Public Health-Division of Environmental Health Sciences, University of Minnesota Twin-Cities, Minnesota, United States of America, **3** Institute on the Environment, University of Minnesota Twin-Cities, Minnesota, United States of America, **4** Institute of GIS, RS&GPS, College of Forestry, Beijing Forestry University, Beijing, China

Abstract

The purpose of the paper is to find a new approach to measure 3D green biomass of urban forest and to testify its precision. In this study, the 3D green biomass could be acquired on basis of a remote sensing inversion model in which each standing wood was first scanned by Terrestrial Laser Scanner to catch its point cloud data, then the point cloud picture was opened in a digital mapping data acquisition system to get the elevation in an independent coordinate, and at last the individual volume captured was associated with the remote sensing image in SPOT5(System Probatoired'Observation dela Tarre)by means of such tools as SPSS (Statistical Product and Service Solutions), GIS (Geographic Information System), RS (Remote Sensing) and spatial analysis software (FARO SCENE and Geomagic studio11). The results showed that the 3D green biomass of Beijing urban forest was 399.1295 million m^3, of which coniferous was 28.7871 million m^3 and broad-leaf was 370.3424 million m^3. The accuracy of 3D green biomass was over 85%, comparison with the values from 235 field sample data in a typical sampling way. This suggested that the precision done by the 3D forest green biomass based on the image in SPOT5 could meet requirements. This represents an improvement over the conventional method because it not only provides a basis to evalue indices of Beijing urban greenings, but also introduces a new technique to assess 3D green biomass in other cities.

Editor: Matteo Convertino, University of Florida, United States of America

Funding: This work was supported by the state forestry administration project "948" (2013-4-65); National Project "863" (2009AA12Z327, 2008AA121305-4); Beijing Natural Science Foundation (09D0297); National Natural Science Foundations of China (30872038, 30671696). The funders had no role in study design, data collection and analysis, decision to publish, or preparation of the manuscript.

Competing Interests: The authors have declared that no competing interests exist.

* E-mail: siyu85878817@163.com

Introduction

It is hard to estimate the amount of urban green space due to its characteristics of diverse structure and scattered distribution [1,2]. Therefore, 3D green biomass could be vividly defined as a 3D volume of the stems and leaves of all plants growing in the region [3], which can not only more accurately reflect the proportion of all vegetations in the region than such traditional 2D indicators as forest area and coverage, but also provide some ecological efficiency and green indexes suitable for the ecological assessment of the urban landscape, while playing an important role in planning the city and building the forestry discipline [4,5] .

In general, two methods can be used for the 3D Green Biomass estimation: the ground survey and the estimating with the remote sensing technology [6–9]. Actually, the ground survey is difficult to be done on a large scale even if the value can get a high accuracy because the green biomass can be acquired by the 3D volume measured by each tree's crown width and diameter at the breast height so that the systems need continuous field tests to be improved [10–21]. The remote sensing technology has been widely used in vegetation classification, forest fire monitoring and 3D green biomass measuring. Lv et al. [22] calculated the 3D volume by the crown height and width, how the crown width and height on the aerial photo was first measured. Cheng et al. [24] acquired the 3D volume by using the screen tracking vectorization

by means of GIS, who first made some field investigations to get the data of leaf area and vegetation coverage, and then combined the data with some high resolution images (IKONOS). Zhou et al. [25] succeeded in making an estimation of the 3D green biomass of Shanghai urban city forest by classifying the species on the aerial photos with high resolution, then simulating the stereo quantity by the plane quantity on the computer. Compared with the traditional ground work, the remote sensing techniques mentioned above have made greater improvements with such less cost as manpower, material and time, leading a fast calculation of 3D green biomass on a large scale. However, although the approaches can mitigate some problems, their precisions are not guaranteed because they are computered by the crown volume based on an appropriate formula suitable for the crown shape. Additionally, there are so timely and limited calculations that it is difficult to be widely promoted. Therefore, it is essential to find a more precise and generalized approach capable of achieving the 3D green biomass by means of remote sensing retrieval method today when the ecological environment is more and more important.

The spatial distribution of leaf area determines resource capture and canopy exchanges with the atmosphere. It is generally tedious and time-consuming to measure the spatial distribution of leaf area, even when 3D digital techniques are employed [21–26]. Many tree models, like light models, therefore choose individual canopies as a volume filled with leaf area. Simple shapes like ellipsoids or frustums have been extensively used to model tree shape [26–31]. More sophisticated parametric envelopes have been proposed by Cescatti (1997) to extend the range of modeled canopy shapes, and non-parametric envelopes like polygonal envelopes are expected to fit any tree shape [32]. However, all the envelopes showed that different shape models for the same tree may lead to large differences in crown volume [33–36]. None of these methods for tree crown volume estimation has been evaluated by comparison with direct measurements. Moreover, neither method accounts for the fractal nature of plants, because only one value of crown volume is computed (i.e., at the observation scale) and changes in crown volume with measurement scale are ignored.

On the other hand, airborne laser scanners can be used to acquire vertical and horizontal forest structure in detail as scanning targets with laser pulses. In particular, such vertical measurements enable the prediction of forest biomass and carbon storage. Furthermore, laser sensors can be used to accurately measure topographical information, the physical properties of a forest and other information. Therefore, ALS(Auto Scanner Laser System)has been recognized as a more efficient and precise instrument than field surveys and optical remote sensing techniques [37–43]. Since the early to mid 1980s, several studies using full waveform sensors have been performed for forest inventory, merchantable timber volume estimation [44], and forest canopy characterization [45–47]. Recently, several researchers have applied discretely emitted laser pulses for the individual- and stand-level tree height estimation [47–53]and height-based timber volume estimates [44–46].However, there is currently no effective approach in the methods mentioned above to resolve the problems on how to calculate the canopy volume accurately and quickly, especially for the volume on a large scale.

In this paper, the 3D green biomass of Beijing urban forest was calculated and analyzed based on the remote sensing retrieval model. This approach, in which to obtain the point cloud data of the crown, 30 different trees in size of each species from over 30 common tree species in Beijing urban area(like *arborvitae, cedar, pine, cypress, ginkgo, poplar, sycamore, willow tree, Sophora japonica, Ailanthus altissima, Koelreuteria, ash, maple, cork oak* and other about ten common tree species)were chosen for scanning with laser scanner FARO(FARO develops and markets portable CMMs (coordinate measuring machines) and 3D imaging devices to solve dimensional metrology problems.), was designed to calculate the 3D green biomass of single wood by CASS (CAD AID SURVEY SYSTEM) software(CASS mapping equipment in the South in AutoCAD 2004 to develop a new generation of digital terrain cadastral mapping software)which has been patented in China in 2011 [24–28], associated with the remote sensing image in SPOT5, and by means of SPSS, GIS, RS and other spatial analysis tools [46–49].

Materials and Methods

Ethics Statement

No specific permits were required for the described field studies, since the trees chosen in the study are owned and managed by the state including the sites for our sampling are not privately-owned or protected in any way and specific permission for non-profit research, therefore, is not required. The field studies were not involved in endangered or protected species in this area.

Data Acquisition

The study site was located in Beijing (39°26′40″N to 41°03′05″N, 115°25′45″E to 117°30′20″E), The SPOT5 remote

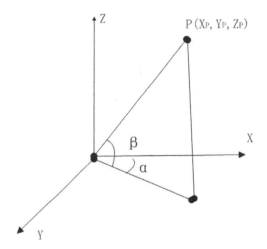

Figure 1. Schematic diagram of scanning point coordinate calculation. X_P——P abscissa values; Y_P——P ordinate values; Z_P——P Elevation Value; α——included angle of P was perpendicular to the YZ plane with X axis; β——included angle of P with XY plane.

sensing image data from four views in summer of 2009 in Beijing were selected in this paper, including resolution of 10 m multispectral band and 2.5 m panchromatic band. Besides, there is much supporting information, such as Beijing 1:250,000 administrative map, traffic road map, water maps, maps of forest resources, Worldview remote sensing images of 2008 in Beijing City, the latest Goolge Earth data and so on. The total area of Beijing is 16,800 km², of which mountainous areas occupy about 62% and plains take up the rest. Forestry areas is 104,609,637 m, including 65,891,408 m forestation-suitable, 557,631 m open forest, 30,580,843 m shrub, 2,110,388 m young forest and 5,469,367 m other forest. Geographically, Beijing is a transitional zone for southern and northern plants of China. Influenced by warm-temperate continental monsoon climate, its sub-natural flora generally belongs to warm temperate zone deciduous broad-leaved forest and coniferous forest, but due to serious destruction in early years, currently there are only small area forests with sporadically scattered trees. In some higher mountainous planted forest, the Larix principis-rupprechtii forests were originally

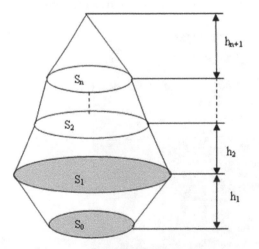

Figure 2. Mimic diagram of three-dimensional laser scanning method for measuring the crown volume.

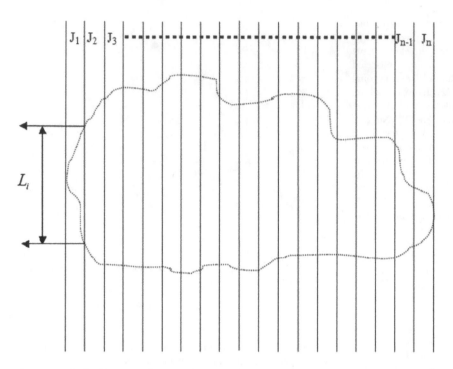

Figure 3. Mimic diagram of measuring crown cross-sectional area.

distributed as the sub-natural compositions, however, shrub and secondary forests are the most widely distributed zonal vegetations of Beijing, such as betula, populus, quercus.

In this study, at first over 1000 trees (30 species and over 30 trees in each species), like pinus bungeana, cedar and pinus tabulaeformis, planting in campuses, parks, roadsides, housing estates and mountain forests in Beijing city, were sampled and scanned systematically and representatively with terrestrial 3D laser scanner FARO LS880, and then their 3D coordinates were measured by means of some instruments like Trimble GPS(Global Positioning System) and Topcon total station. To acquire the point cloud data, we put up a platform, where some parameters of the scanner should be first set. The horizontal direction was 360°, vertical direction was 155° (from −90 to 65) and a resolution of 2 mm in 10 m. Next, three stations should be set up according to

its growth and relative terrain of the tree to be measured, because they can generally constitute an equilateral triangle in theory whose included angle is 120°. Additionally, at least three public spheres should be placed on a non-straight line between two stations and all could be scanned by the above 3 stations without any shelter. The target paper, which only acted as a reference for scanning, were finally posted on the trunk with a height of about 1.3 m above ground northwards (Figure 1). Only when all trees were scanned, could the target ball be removed, otherwise, they had to be re-scanned. To get the volume of the sampling forest, each single tree had to be scanned in about 10 minutes. For field application, scanning for single tree should be done in open forest because too many close planting trees with branches and leaves will lead to serious shadows. Meanwhile, some pictures of the trees

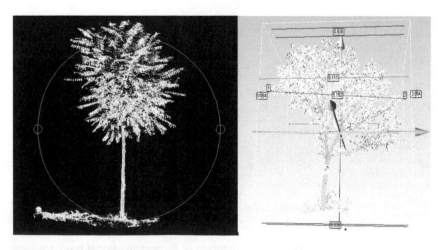

Figure 4. 3D point cloud data.

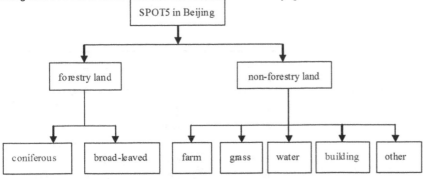

Figure 5. Classification flow diagram of the image information.

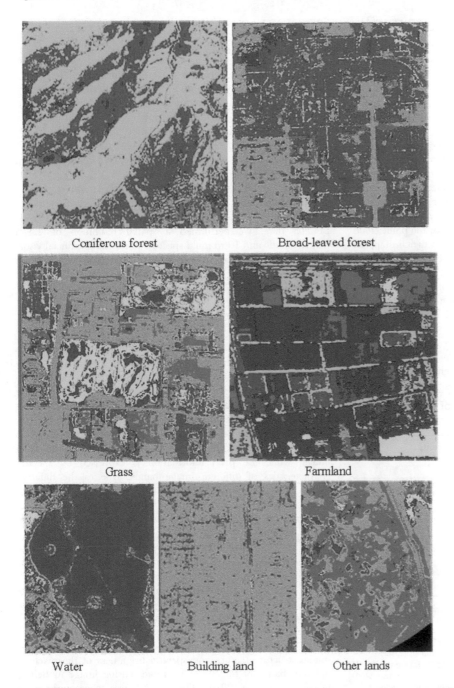

Figure 6. Interpretation signs. Coniferous forest, Broad-leaved forest, Grass, Farmland, Water, Building land and other lands are signed in the figure.

Table 1. Table of accuracy totals of classification.

	Reference Totals	Classified Totals	Number Correct	Producer's Accuracy/%	User's Accuracy/%	Kappa
Coniferous forest	14	7	6	42.86	85.71	0.834
Broadleaf forest	15	15	11	73.33	73.33	0.686
Grass	5	5	4	80.00	80.00	0.790
Farmland	11	15	10	90.91	66.67	0.626
Waters	2	2	2	100	100	1
Construction lands	39	39	37	94.87	94.87	0.916
Other lands	14	17	13	92.86	76.47	0.726
Σ	100	100	83			

to be measured as reference for post-processing point cloud trees should be taken, where the scanner was placed.

To achieve the crown 3D green biomass of single tree, we made a calculation on point cloud data of the crown using a CASS software which could computer a volume with digital elevation after a secondary development of CAD (Computer Aided Design). In processing, some pictures with different point clouds of the standing woods were first pieced together at a coordinate system by coordinate match by means of the 3D scanner's software, and then a 3D model of standing wood was developed and saved as .dxf which could be discerned by the digital mapping system after the data were pre-processed and extracted. Next, the point cloud picture captured by the scanner was opened in a digital mapping data acquisition system to set up an independent coordinate system and the elevation could be extracted (Figure 2).The area whose volume should be calculated was outlined with a closed compound line instead of fitting curve in a mesh with 3D triangles, because a fitted curve would be replaced by the broken line so that the precision of the results could be affected. Finally, the point cloud volume was calculated with the help of DTM (Digital Terrain Model) method of the system shown as follow: the points on the crown surface collected at the same height were linked with the smooth curve to form some contour lines which were then separated into some grids with a regular 2 cm cell size(Certainly, the length can be divided into any other size, but 2 cm here was just for convenience), so that the topmost elevation in each grid could be estimated by linear interpolation and marked at its top right, where the designed elevation was set 0; next, we calculated the volumes with some formulas shown as follows:(1) $V_{\text{cornerpoint}} = h*1/4\,S_{\text{grid}}$; (2) $V_{\text{edgepoint}} = h*2/4S_{\text{grid}}$;(3) $V_{\text{turningpoint}} = h*3/4\,S_{\text{grid}}$;(4) $V_{\text{midpoint}} = h*S_{\text{grid}}$. (5)$V_{\text{crown}} = nV_{\text{grid}}$, where h is a canopy height and n is the number of all grids (Figure 2 and 3).

The point cloud data are shown in Figure 4. In remote sensing SPOT5 data, we performed a spatial resolution of 2.5 m panchromatic and 10 m multi-spectral bands using artificial visual interpretation method to extract the vegetation classification information. And we added SPOT5 image gray value, its remote sensing factors and GIS factors into the independent values.

Remote sensing image classification and information extraction

The green biomass retrieval model can be developed by means of various information, methods, monitoring and manual interpretation. In supervised method, at first, some known characteristic parameters are extracted from the spectral features based on samples from training areas, and then other unspecified parameters which can be extracted and classified in sorts from images are analyzed according to prior probability of different kinds of objects. The ground information is demonstrated by pixels in image, while the pixel information is expressed by spectral characteristics of different image bands. Due to the different spatial resolutions and complex grounds, some mixed pixels can appear in images which will result in different objects with the same spectra characteristics or same spectrum with different objects. Therefore, the accuracy will be confined if the urban vegetation is classified only by supervised classification because misclassification and loss classification can arise in some specific classification and extraction. The present vegetation is classified by means of visual interpretation or computer-aided, however, it is of low automation, long hours and low efficiency or likely to be worse because of the unskilled labors or less educated operators (Figure 5 and Figure 6).

In this study, the first information was captured based on the spatial structure feature and spectral brightness of the pixels in different bands of the SPOT5 remote sensing image in Beijing urban regions in 2010. The second was characteristics of landscape, landform and forest resource distribution. The third

Table 2. Factors loading rotation matrix of varimax.

Variables	Remote sensing	principal components	
		1	2
1	B1	−0.716	0.696
2	B2	−0.751	0.658
3	B3	−0.741	0.670
4	SWIR	−0.726	0.686
5	NDVI	0.745	0.667
6	SAVI	0.744	0.668
7	MSAVI	0.742	0.670

Note: B1, B2 are visible bands, where B1 can detect absorption and reflectance of plant green hormone, and B2 belongs to the red light zone capable of distinguishing the color of different types of vegetation from the color difference; Where B3 is near-infrared bands, which can reflect the sensitivity of plants to chlorophyll by the correlation between some acquired strong information and factors like leaf area index and biomass; SWIR (short-wave (length) infrared (band)); NDVI (Normalized Difference Vegetation Index); SAVI (Soil-Adjusted Vegetation Index); MSAVI (Modified Soil-Adjusted Vegetation Index).

Table 3. Correlation coefficient of conifer volume and RS factors.

Variables	Band and combinations	Correlation coefficient with 3D green biomass
1	B1	0.92917
2	B2	0.93602
3	B3	0.93711
4	SWIR	0.93102

was about such maps as current vector, forest distribution, greening investigation data, contour and traffic each year.

Beijing is so large that a lot of random points must appear in the accuracy assessment of classification, thereby they should be chosen at absolute random instead of any mandatory rule in practice and the regional classification maps were assessed under the Accuracy Assessment module of ERDAS IMAGINE software(Table 1).

For classification of remote sensing images in some large areas, its accuracy has been able to meet the needs of the latter analysis and assessment.

Modeling

The Principal Component Analysis(PCA) with varimax rotation was used for factor analysis in this study. As two PCs can be shown in Table 2, the loading values of each independent variable in either the first PC or the second were little changed, indicating that there was only a small amount of correlation between grouping variables. If the eight variables were forcibly added into the model, it could not be guaranteed to high accuracy that some uncorrelated variables to 3D green biomass could be directly regressed by the model. So the factor involved into modeling could be acquired by automatic filtration with the help of stepwise regression modeling.

Based on the results interpreted by the remote sensing images and the green biomass data measured in the plots, the relative remote sensing factors and GIS factors were chosen as the independent values of a model by means of the spatial relationship of RS image rectified by GPS to ground samples, where the remote sensing factors were related to the image gray value in SPOT5 and its linear and nonlinear combination etc, and GIS factors included slope, elevation, aspect etc.(Some factors were not involved in modeling as independent variables like slope and aspect since the subjects in the study were mainly located in Beijing urban areas and topographic relief was not much changed.) The 3D green biomass was repeatedly regressed to model the conifer and broadleaf tree respectively on basis of the correlation between the factors and the green biomass values observed in the plots.

The 3D green biomass model of conifer is shown as follow:

$$V = 1.149 - 0.096\,B1 - 0.1\,B2 + 0.199\,SWIR \qquad (1)$$

The 3D green biomass model of broadleaf tree is shown as follow:

$$V = -40.290 - 0.236\,B1 - 0.188\,B2 + 0.487\,SWIR \qquad (2)$$

Where B1, B2 are visible bands, where B1 can detect absorption and reflectance of plant green hormone, and B2 belongs to the red light zone capable of distinguishing the color of different types of vegetation from the color difference; SWIR band which can well reflect the water feature in the plant leaves is a shortwave infrared zone, by which it is easy to make the identification and classification of vegetation, soil, and water.

Based on validation and accuracy assessment on the model and comparison with the actual measured data in the ground, the data for modeling should be systemized, and then 150 samples were added into modeling and 100 checking samples were chosen to test their accuracy.

It is possible to develop a reliable, scientific and operable model. Analysis on the correlation coefficient between the gray values of four bands in SPOT5 and the 3D green biomass of conifer and broadleaf tree by using EXCEL software, the results were shown in Table 3 and 4.

All tests were performed using version 18.0 of the SPSS software (SPSS Inc., Chicago, IL).

Results and Analysis

Classification results and analysis

All the 3D green biomass of 1015 trees scanned by 3D laser were added in the calibrated remote sensing image so that we could get the gray value at each sampling point, and then by means of remote sensing image Worldview 1 and Google carth with the resolution of 0.5 m, each sampling tree was soon located.

Table 4. Correlation coefficient of broadleaf tree volume and RS factors.

Variables	Band and combinations	Correlation coefficient with 3D green biomass
1	B1	0.86879
2	B2	0.80498
3	B3	0.805022
4	SWIR	0.87112

Table 5. Checking table of model precision.

Species	Correlation Coefficient	Square of Correlation Coefficient	Revised Square	Estimation Error	Statistical Analysis			
					F value	Degree of Freedom1	Degree of Freedom2	Significant Change
Conifer tree	0.951	0.904	0.902	2.21	411.651	3	147	0.0000
Broadleaf tree	0.894	0.799	0.792	7.67	622.611	3	478	0.0000

Based on supervised classification and visual interpretation, the total Beijing urban forest area of conifer and broadleaf tree was 275.08 km^2, of which the area of conifer tree was 54.47 km^2, while that of broadleaf tree was 220.61 km^2. The correlation coefficient between the 3D biomass of conifer tree and RS factors was over 0.9 and that of broadleaf tree was over 0.8, thus revealing that the linear correlation was very close [44–47], thereby a model of 3D green biomass could be developed with the direct remote sensing data by the multiple linear regression.

Model checking

After sampling and drying every organ of the tree, we converted them and got the biomass. Meanwhile, the stem and volume were accurately measured by means of sectional measurement (Table 1), As can be seen from Table 1, the overall classification accuracy reaches more than 80%, and Kappa factor of 0.780, exceeding the requirements of 0.7. After inducing and analyzing the eight variable factors of B1, B2, B3, SWIR(short-wave (length) infrared (band)), NDVI (Normalized Difference Vegetation Index), SAVI (Soil-Adjusted Vegetation Index) and RVI (Ratio Vegetation Index), we got the regression equations of conifer and broadleaf tree shown as Eq.1 and Eq.2. The accuracy of the model was tested with the correlation coefficient and F test to evaluate (Table 5).

It is known that the correlation coefficient of the multiple regression model of Beijing urban 3D green biomass was high(over 0.89) which correlated well with the 3D green biomass, remote sensing factors and GIS factors, and F values also showed that the significant differences existed in the model. The histogram of regression standardized residual in Figure 7 illustrated that it was an ideal bell-shaped normal distribution, and a better diagonal distribution was illustrated in the cumulative probability distribution (Figure 7 and Figure 8).

Accuracy analysis

We performed 241 samples to identify whether the actual measured values and estimated values significant differences existed among the treatments, and when we did, we used the (3D Green Biomass)TGB test to determine which specific combinations of values differed significantly (Table 6)

(1) The residual standard deviation is acquired by formula $S = \sqrt{\dfrac{\sum (TGB_s - TGB_g)^2}{n-2}}$, where S is residual standard deviation, TGB_s is the actual measured value of 3D green biomass, TGB_g is the estimated value of 3D green biomass and n is the number of sampling plots for accuracy test.

(2) The Standard error is acquired by formula $\delta_x = \dfrac{S}{\sqrt{n}}$, where δ_x is Standard error.

(3) The absolute error limit of 95% and 99% is calculated by formula $\Delta = \delta_x t_{n-2}^{\alpha}$, where Δ is the absolute error limit which acquired by t value distribution table difference.

(4) The relative error limit of 95% and 99% is calculated by formula $E = \dfrac{\Delta}{\bar{X}}$, where E is the relative error limit, and $\bar{X} = \dfrac{\sum TGB_g}{n}$ is the mean value of TGB.

(5) Precision C is calculated by formula $C = 100\% - E$.

The monitoring data in Table 6 demonstrated that the precision of the 3D green biomass of the sample based on the SPOT5 was over 85%, indicating that it could fully meet the requirements.

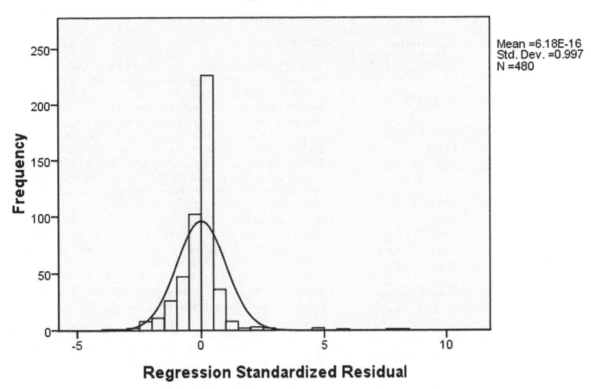

Figure 7. Histogram of regression standardized residual.

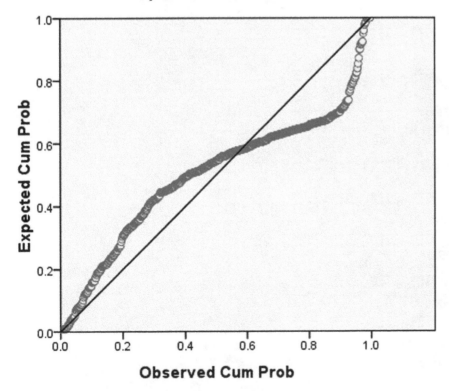

Figure 8. Cumulative probability distribution map.

Table 6. Testing Results of Precision Ratio of TGB Based-on SPOT5 Image.

	Confidence Level	
	a = 0.05	a = 0.01
Sample Number(n)	241	
Total of Actual Measured Values $\sum$	33687.621	
TGB_s Average $\bar{X}$	140.127	
Total of Estimated Values $\sum$	34788.741	
TGB_g Average $\bar{X}$	148.013	
$\sum(TGB_s - TGB_g)$	−1874.453	
$\sum(TGB_s - TGB_g)^2$	3513600.322	
Standard Deviation	122.804	
Standard Error δ_x	8.126	
t_{n-2}^a	1.989	2.620
Absolute Error Δ	15.873	20.958
Relative Error E/%	10.695	14.204
Precision C/%	89.302	85.796

In all, based on the remote sensing image gray values extracted from the model by means of ArcGIS 9.3 and all statistical data calculated on the remote sensing retrieval model of the 3D green biomass, the green biomass in each region of Beijing could be determined. The results shows that the 3D green biomass of Beijing urban forest was 399.1295 million m³, of which coniferous was 28.7871 million m³ and broad-leaf was 370.3424 million m³.

Discussion

As the above statistical data demonstrates, the case study described in this paper confirms that this is possible. Compared with the traditional 2D green indices in forest area, 3D green biomass represents an improvement over the conventional method because 3D index demonstrates that it can both accurately reflect the volume of the vegetation in the region, and scientifically assess the ecological environment of the city, while providing an important basis for urban planning and forest sciences development. The new approach of 3D green biomass illustrates that it is performed more accurately, efficiently, easily, and rapidly than the conventional, because the 3D green biomass not only involves the processing of remote sensing image including identification and classification, but also includes the investigation of forest vegetation on the ground, especially where the same species make great differences in different climatic zones.

The total 3D green biomass of Beijing urban areas can be acquired by the model and its grade distribution of biomass per unit can be also computerized in ArcGIS 9.3 shown in figure 9 [30,31].

Beijing is an important capital urban district of China. In order to fully implement the strategy of "Humanistic Beijing, Scientific Beijing, Green Beijing" and promote the development of urban eco-environment, the study can provide some materials for references in urban green lands.

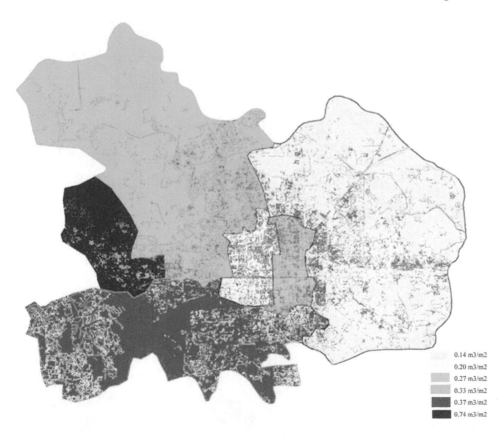

0.14 m3/m2
0.20 m3/m2
0.27 m3/m2
0.33 m3/m2
0.37 m3/m2
0.74 m3/m2

Figure 9. Grade distribution of 3D green biomass per unit area in Beijing.

The crown form is one of non-negligible factors in the calculation of biomass with the help of 3D laser scanner, for example, the shape of the crown will shake in the wind when the scanned point cloud may not fully reflect the true state of the involved crown volume. Therefore, the wood should be scanned at rest to ensure the accuracy of the volume. Meanwhile, only the biomass from the upper half of crowns were involved while the under part, shrubs and herbs were not to be considered. In the future we will focus on the relative research in the field.

The green biomass of 3D in Beijing was estimated by the interpretation and classification of remote sensing data and modeling. In this paper, the urban vegetation was extracted by artificial visual interpretation and computer-aided, or strictly speaking it was still semi-automated and time and labor consuming. The extraction of vegetation is still the hot spot researchers interested in. The resolution in SPOT5 remote sensing image data was not high so that the crown width was greater than 2.5 m like a pixel. However, the gray on the corresponding band can produce the deviation if it is less than a pixel. In all, the biomass acquired in the study work as the exploratory research and reference and more remote sensing images with higher resolution will be used later to study.

Author Contributions

Conceived and designed the experiments: CH SYZ ZKF. Performed the experiments: CH. Analyzed the data: CH ZKF. Contributed reagents/materials/analysis tools: CH SYZ ZKF. Wrote the paper: CH MC.

References

1. Pregitzer KS, Euskirchen ES (2004) Carbon cycling and storage in world forests: biomass patterns related to forest age. Global Change Biology 10: 2052–2077.
2. Running SW (2008) Climate change - Ecosystem disturbance, carbon, and climate. Science 321: 652–653.
3. Yamada T, Zuidema PA, Itoh A, Yamakura T, Ohkubo T, et al. (2007) Strong habitat preference of a tropical rain forest tree does not imply large differences in population dynamics across habitats. Journal of Ecology 95: 332–342.
4. Condit R, Ashton PS, Baker P, Bunyavejchewin S, Gunatilleke S, et al. (2000) Spatial patterns in the distribution of tropical tree species. Science 288: 1414–1418.
5. Pan Y, Birdsey RA, Fang J, Houghton R, Kauppi PE, et al. (2011) A large and persistent carbon sink in the world's forests. Science 333: 998–993.
6. Fang J, Wang G, Liu G, Xu S (1998) Forest biomass of China: An estimate based on the biomass-volume relationship. Ecological Applications 8: 1084–1091.
7. Shi L, Zhao S, Tang Z, Fang J (2011) The changes in China's forests: An analysis using the forest identity. PLoS ONE 6: e20778.
8. Zhang J, Ge Y, Chang J, Jiang B, Jiang H, et al. (2007) Carbon storage by ecological service forests in Zhejiang Province, subtropical China. Forest Ecology and Management 245: 64–75.
9. Keith H, Mackey B, Lindenmayer D (2009) Re-evaluation of forest biomass carbon stocks and lessons from the world's most carbon-dense forests. Proceedings of the National Academy of Sciences of the United States of America 106: 11635–11640.
10. Yang T, Song K, Da L, Li X, Wu J (2010) The biomass and aboveground net primary productivity of Schima superba-Castanopsis carlesii forests in east China. Science in China (Series C: Life Sciences) 53: 811–821.
11. Post WM, Kwon KC (2000) Soil carbon sequestration and land–use change: Processes and potential. Global Change Biology 6: 317–328.
12. Ruiz-Jaen MC, Potvin C (2011) Can we predict carbon stocks in tropical ecosystems from tree diversity? Comparing species and functional diversity in a plantation and a natural forest. New Phytologist 189: 978–987.
13. Kira T (1991) Forest ecosystems of east and southeast-Asia in a global perspective. Ecological Research 6: 185–200.
14. Du YJ, Mi XC, Liu XJ, Ma KP (2012) The effects of ice storm on seed rain and seed limitation in an evergreen broad-leaved forest in east China. Acta Oecologica 39: 87–93.
15. Zhong Z (1987) The typical subtropical evergreen broadleaved forest of China. Journal of Southwest China Normal University 3: 109–121.
16. Wang X, Kent M, Fang X (2007) Evergreen broad-leaved forest in Eastern China: Its ecology and conservation and the importance of resprouting in forest restoration. Forest Ecology and Management 245: 76–87.
17. Piao S, Fang J, Ciais P, Peylin P, Huang Y, et al. (2009) The carbon balance of terrestrial ecosystems in China. Nature 458: 1009–1013.
18. Malhi Y, Wood D, Baker TR, Wright J, Phillips OL, et al. (2006) The regional variation of aboveground live biomass in old-growth Amazonian forests. Global Change Biology 12: 1107–1138.
19. Stegen JC, Swenson NG, Enquist BJ, White EP, Phillips OL, et al. (2011) Variation in aboveground forest biomass across broad climatic gradients. Global Ecology and Biogeography 20: 744–754.
20. Baraloto C, Rabaud S, Molto Q, Blanc L, Fortunel C, et al. (2011) Disentangling stand and environmental correlates of aboveground biomass in Amazonian forests. Global Change Biology 17: 2677–2688.
21. Baker TR, Phillips OL, Malhi Y, Almeida S, Arroyo L, et al. (2004) Variation in wood density determines spatial patterns in Amazonian forest biomass. Global Change Biology 10: 545–562.
22. Lv ME, Pu YX, Huang XY (2000) Application of Remote Sensing on Urban Green, Chinese Landscape Architecture 16(5): 41–43.
23. Zhang LP, Zheng LF, Tong QX (1997) The Estimation of Vegetation Variables Based on High Resolution Spectra, Journal of Remote Sensing 1(2): 111–114.
24. Chen F, Zhou ZX, Wang PC, Li HF, Zhong YF (2006) Green space vegetation quantity in workshop area of Wuhan Iron and Steel Company, Chinese Journal of Applied Ecology 17(4):592–596.
25. Zhou JH, Sun TZ (1995) Study on Remote Sensing Model of Three-Dimensional Green Biomass and the Estimation of Environmental Benefits of Greenery, Remote Sensing of Environment 10(3):162–174.
26. Kra'l K, Jani'k D, Vrska T, Adam D, Hort L, et al. (2010) Local variability of stand structural features in beech dominated natural forests of Central Europe: Implications for sampling. Forest Ecology and Management 260: 2196–2203.
27. Brown S (2002) Measuring carbon in forests: current status and future challenges. Environmental Pollution 116: 363–372.
28. Chen B, Mi X, Fang T, Chen L, Ren H, et al. (2009) Gutianshan forest dynamic plot: Tree species and their distribution patterns. Beijing: China Forestry Publishing House.
29. Legendre P, Mi X, Ren H, Ma K, Yu M, et al. (2009) Partitioning beta diversity in a subtropical broad-leaved forest of China. Ecology 90: 663–674.
30. Yu M, Hu Z, Yu J, Ding B, Fang T (2001) Forest vegetation types in Gutianshan National Natural Reserve in Zhejiang. Journal of Zhejiang University (Agriculture and Life Science) 27: 375–380. (In Chinese).
31. Hu Z, Yu M, Suo F, Wu F, Liu Q (2008) Species diversity characteristics of coniferous broad-leaved forest in Gutian Moutain National Nature Reserve, Zhejiang province. Ecology and Environment 17: 1961–1964. (In Chinese).
32. Du G, Hong L, Yao G (1987) Estimate and analysis the aboveground biomass of a secondary evergreen broad-leaved forest in Northwest of Zhejiang. Journal of Zhejiang Forestry Science and Technology 7: 5–12. (In Chinese).
33. Chen W (2000) Study on the net productivity dynamic changes of the aboveground portion of Alniphyllum fortunei plantation. Journal of Fujian Forestry and Technology 27: 31–34. (In Chinese).
34. Chen Q, Shen Q (1993) Studies on the biomass models of the tree stratum of secondary Cyclobalanopsis glauca forest in Zhejiang. Acta Phytoecologica Sinica 17: 38–47. (In Chinese).
35. Caspersen JP, Pacala SW (2000) Successional diversity and forest ecosystem function. Ecological Research 16: 895–903.
36. Cardinale BJ (2011) Biodiversity improves water quality through niche partitioning. Nature 472: 86–89.
37. Loreau ML, Mouquet N, Gonzalez A (2003) Biodiversity as spatial insurance in hetegrogenous landscapes. Proceedings of the National Academy of Sciences of the United States of America 100: 12765–12770.
38. Canadell JG, Raupach MR (2008) Managing forests for climate change mitigation. Science 320: 1456–1457.
39. Imai N, Samejima H, Langner A, Ong RC, Kita S, et al. (2009) Co-Benefits of sustainable forest management in biodiversity conservation and carbon sequestration. PLoS ONE 4: e8267.
40. Mascaro J, Asner GP, Muller-Landau HC, van Breugel M, Hall J, et al. (2011) Controls over aboveground forest carbon density on Barro Colorado Island, Panama. Biogeosciences 8: 1615–1629.
41. Laurance WF, Fearnside PM, Laurance SG, Delamonica P, Lovejoy TE, et al. (1999) Relationship between soils and Amazon forest biomass: A landscape-scale study. Forest Ecology and Management 118: 127–138.
42. Tateno R, Takeda H (2003) Forest structure and tree species distribution in relation to topography-mediated heterogeneity of soil nitrogen and light at the forest floor. Ecological Research 18: 559–571.
43. Paoli GD, Curran LM, Slik JWF (2008) Soil nutrients affect spatial patterns of aboveground biomass and emergent tree density in southwestern Borneo. Oecologia 155: 287–299.
44. Ferry B, Morneau F, Bontemps JD, Blanc L, Freycon V (2010) Higher treefall rates on slopes and waterlogged soils result in lower stand biomass and productivity in a tropical rain forest. Journal of Ecology 98: 106–116.
45. McEwan RW, Lin Y, Sun I, Hsieh C, Su S, et al. (2011) Topographic and biotic regulation of aboveground carbon storage in subtropical broad-leaved forests of Taiwan. Forest Ecology and Management 262: 1817–1825.

46. Man X, Mi X, Ma K (2011) Effects of an ice strom on community structure of an evergreen broad-leaved forest in Gutianshan National Natural Reserve, Zhejiang Province. Biodiversity Science 19: 197–205. (In Chinese).

47. Malhi Y, Baker TR, Phillips OL, Almeida S, Alvarez E, et al. (2004) The aboveground coarse wood productivity of 104 Neotropical forest plots. Global Change Biology 10: 563–591.

48. Elser JJ, Bracken MES, Cleland EE, Gruner DS, Harpole WS, et al. (2007) Global analysis of nitrogen and phosphorus limitation of primary producers in freshwater, marine and terrestrial ecosystems. Ecology Letters 10: 1135–1142.

49. Paoli G, Curran L (2007) Soil nutrients limit fine litter production and tree growth in mature lowland forest of southwestern Borneo. Ecosystems 10: 503–518.

50. Schaik CPV (1985) Spatial variation in the structure and litterfall of a Sumatran rain forest. Biotropica 17: 196–205.

51. Zhang L (2010) The effect of spatial heterogeneity of environmental factors on species distribution and community structure. Beijing, China: PhD thesis. Institute of Botany, Chinese Academy of Sciences. (In Chinese).

52. Russo SE, Davies SJ, King DA, Tan S (2005) Soil-related performance variation and distributions of tree species in a Bornean rain forest. Journal of Ecology 93:879–889.

53. Vittoz P, Engler R (2007) Seed dispersal distances: a typology based on dispersal modes and plant traits. Botanica Helvetica 117: 109–124.

19

Social Insects Dominate Eastern US Temperate Hardwood Forest Macroinvertebrate Communities in Warmer Regions

Joshua R. King[1]*, Robert J. Warren[2], Mark A. Bradford[3]

1 Biology Department, University of Central Florida, Orlando, Florida, United States of America, 2 Biology Department, SUNY Buffalo State, Buffalo, New York, United States of America, 3 Yale School of Forestry and Environmental Studies, Yale University, New Haven, Connecticut, United States of America

Abstract

Earthworms, termites, and ants are common macroinvertebrates in terrestrial environments, although for most ecosystems data on their abundance and biomass is sparse. Quantifying their areal abundance is a critical first step in understanding their functional importance. We intensively sampled dead wood, litter, and soil in eastern US temperate hardwood forests at four sites, which span much of the latitudinal range of this ecosystem, to estimate the abundance and biomass m^{-2} of individuals in macroinvertebrate communities. Macroinvertebrates, other than ants and termites, differed only slightly among sites in total abundance and biomass and they were similar in ordinal composition. Termites and ants were the most abundant macroinvertebrates in dead wood, and ants were the most abundant in litter and soil. Ant abundance and biomass m^{-2} in the southernmost site (Florida) were among the highest values recorded for ants in any ecosystem. Ant and termite biomass and abundance varied greatly across the range, from <1% of the total macroinvertebrate abundance (in the northern sites) to >95% in the southern sites. Our data reveal a pronounced shift to eusocial insect dominance with decreasing latitude in a temperate ecosystem. The extraordinarily high social insect relative abundance outside of the tropics lends support to existing data suggesting that ants, along with termites, are globally the most abundant soil macroinvertebrates, and surpass the majority of other terrestrial animal (vertebrate and invertebrate) groups in biomass m^{-2}. Our results provide a foundation for improving our understanding of the functional role of social insects in regulating ecosystem processes in temperate forest.

Editor: Judith Korb, University of Freiburg, Germany

Funding: JRK was supported by National Science Foundation (NSF) grant DEB 1020415 and an In-House grant from the University of Central Florida. MAB and RJW were supported by NSF grants DEB-0823293 and DEB-0218001 to the Coweeta LTER Program (http://www.nsf.gov). The funders had no role in study design, data collection and analysis, decision to publish, or preparation of the manuscript.

Competing Interests: The authors have declared that no competing interests exist.

* E-mail: joshua.king@ucf.edu

Introduction

To a degree seldom grasped even by entomologists, the modern insect fauna has become predominantly social. – Bert Hölldobler and Edward O. Wilson, *The Ants*.

The conspicuous presence of social insects in almost all terrestrial ecosystems has captivated the imaginations of biologists, motivating more than a century's worth of ecological study of ants and termites [1], [2], [3]. Social insects appear most abundant – and most diverse – in the tropics, subtropics, and warm temperate latitudes [1], [2], [4]. Their ecological importance, however, is defined by their influence on nutrient cycling, decomposition, soil engineering, predation upon arthropods, and plant community turnover [5], [6], [7], [8], [9], [10]. Ants and termites are thus described as ecosystem engineers because, along with earthworms, they are typically the only physically large members of the soil invertebrate fauna that are presumed to have sufficient abundance and biomass to influence the formation and maintenance of soil structure and to regulate biological processes across landscapes [6], [7], [9], [11]. Understanding the magnitude of their influence on these ecosystem processes, however, is limited by a lack of data

regarding areal abundance (individuals m^{-2}) and biomass (grams dry mass m^{-2}) estimates [9], [12].

Whereas the engineering effects of earthworms are studied across many systems because their areal abundance is quantified, work on termites as engineers has focused mainly on a few sites in the humid tropics and some African savannas with estimates of areal abundance [2], [9], [13], [14]. Ant engineering effects are, in comparison, little studied as almost nothing is known about their areal abundance [1], [3], [5], [8], [9], [10], [13], [15], [16]. This paucity of data is cited as the reason for omitting ants and termites from syntheses of biogeographical patterns in belowground communities [17], but see also [18]. The shortage of social insect observations probably occurs because most soil fauna studies do not estimate social insect biomass: this requires searching for and collecting whole colonies [19], [20], [21].

The primary focus of our study is to provide areal abundance and biomass estimates of soil macroinvertebrates. That is, invertebrates >2 mm in body width and all ants and termites. This primarily excludes mites and Collembola. We emphasize improving estimates of social insect abundance and biomass across a latitudinal gradient in eastern US temperate hardwood forest [22] to stimulate further investigations into their role in ecosystem processes, paralleling work for salamanders in temperate forest [23]. These forests have a broad geographical range and are important to humans for recreation, carbon storage, timber production and wildlife conservation [24].

Ants and termites are generally warm-loving and numerically dominant in the tropics (and thus described as thermophilic; [1], [9], [13]). So we hypothesized that the range of temperate forests that our sampling covered would show that social insects would increase in areal abundance from northern to southern latitudes, but we expected them to be subdominant to other taxa in these communities, and especially so in northern, cooler regions [2], [25], [26], [27].

Materials and Methods

Study Sites

The study was conducted in mid-late August (Connecticut, North Carolina and Georgia sites) and early September (Florida site) of 2011 in four locations spanning ~12° latitude along the eastern US in second-growth hardwood forests. In Connecticut, the northernmost site, sampling was conducted at Yale Myers Forest in Windham and Tolland Counties (41°57′N 72°07′W). In North Carolina, sampling was conducted at Coweeta Hydrologic Laboratory in the Nantahala Mountain Range in western North Carolina (35°03′N 83°25′W). In Georgia, sampling was conducted at Whitehall Forest, located in the piedmont region of Clarke and Oconee Counties, Georgia (33°53′N 83°21′W). In north Florida, the southernmost site, sampling was conducted at San Felasco Hammock which is in San Felasco State Park, Alachua County, Florida (29°43′N 82°26W). Permits and approval for the work was obtained from the Florida Department of Environmental Protection for permission to work on protected public land at San Felasco Hammock State Park, the US Forest Service and Coweeta LTER for permission to work on the protected public land at the Coweeta Hydrologic Laboratory, the Warnell School of Forestry for permission to work on private land at Whitehall Forest, and the Yale School of Forests for permission to work on private land at Yale Myers Forest.

Yale Myers Forest is managed for timber and is comprised primarily of even-aged northern hardwood species with understory dominated by mountain laurel (*Kalmia latifolia*), gently rolling topography with slopes rarely exceeding 40%, elevation at or below 300 m above sea-level, and temperate climate (mean summer 20°C, winter −4°C, 110 cm annual rainfall [28]). Coweeta Hydrologic Laboratory is a long term site of the USDA Forest Service, Southern Research Station. Elevations range from 675–1592 m and annual climate is temperate (mean summer 21.6°C, winter 1.7°C, 180 cm annual rainfall). Slopes are steep, ranging from 30–100%. Timber was harvested until the 1920 s and currently the forest is largely comprised of even-aged mixed southern hardwood species with a frequently dense understory cover of *Rhododendron* and *Kalmia* species [29]. At Whitehall Forest, elevation ranges from 150–240 m above sea level. The forest is evenly aged (60–70 year old) southern hardwood. Climate is temperate (mean summer 25.6°C, winter 6.7°C, 125 cm annual rainfall). San Felasco Hammock has a topography that is slightly rolling and elevation ranges from approximately 43–52 m above sea level. Climate is southern temperate (mean summer 26.9°C, winter 12.9°C, 132 cm annual rainfall). The forest is secondary growth (selectively logged prior to 1937), even-aged southern hardwood forest with a high diversity of tree and understory species and a well-developed litter layer [30].

Two of the study sites were geographically close and near the center of the latitudinal range (Coweeta and Whitehall) and we recorded the highest average soil temperatures at Coweeta during sampling periods, in spite of its elevation (see Results, below). Thus, to verify that the study sites were representative of a temperate latitudinal gradient we determined the estimated annual temperatures of each site and the above ground productivity of the forest ecosystems present at each site. We used general elevation and latitude lapse rates to generate a temperature index for each site, following Warren and references therein [31]. The relative productivity (annual above ground productivity) of each site was estimated from published values in the literature [32], [33], [34], [35], [36].

Sampling

At each study site, two 10×10 m plots were established on two north and two south facing slopes (except at YMF, where slopes face East-West), for a total of 8 plots per study site and 32 100 m² plots across all locations. Along each slope, a transect line was randomly placed and one plot was established upslope and another was established downslope, with ~60 m separating them. Transects were separated by 100 m or more. Macroinvertebrates were sampled in the leaf litter, soil, and coarse woody material (CWM = all dead wood >10 cm dia.). The sampling approach was modified to combine traditional macroinvertebrate sampling (quadrat-based rapid collection of surface material to avoid escapes) while simultaneously sampling whole colonies (ants) or feeding groups (subterranean termites) of social insects in litter, CWM, and soil. Earthworms were captured, however, this study does not properly sample earthworm abundance (c.f. [37]) so earthworm data were not analyzed separately.

To sample litter (including fine woody material, FWM = all dead wood <10 cm dia.) and the top ~5 cm of soil, within each 100 m² plot, ten 25×25 cm quadrats were established in a regularly spaced pattern (Fig. 1). In each quadrat soil temperature was measured at 5 cm depth and volumetric soil moisture (Campbell Hydrosense™) to 12 cm depth. A machete was used to quickly cut (at quadrat edge) any FWM and the surface of the soil within the quadrat. The entirety of the litter sample in the quadrat down to the soil surface was then immediately collected and bagged. The remaining material from the quadrat and the top ~5–10 cm of soil were then immediately collected using a cordless vacuum (Dewalt™) and bagged. If ants or termites were

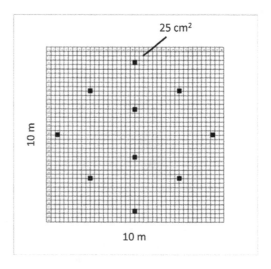

Figure 1. Arrangement of 25 cm² quadrat samples within 100 m² plots.

discovered after vacuuming, all soil to 30 cm depth was excavated to collect the colony. If necessary (i.e. the whole colony was not obviously collected), excavation followed the colony to the depth necessary to get the entire colony (typically no greater than ~1 m). Within quadrats every effort was made to collect whole colonies. The majority of ant species collected were monodomous (single nest, Table 1). Satellite nests of polydomous species (multiple nest sites per colony) outside quadrats were not collected. Any species observed were noted as collection proceeded if more than one species was present to assure collection of separate, whole colonies.

After collecting litter, within each 100 m^2 plot, all CWM was measured along the center axis for length, and at either end for diameter, to estimate the volume (cm^3). Any CWM falling on the edge of plots was either cut or measured approximately (for pieces too large to cut) and sampled so as to include only CWM inside plots. For CWM small enough to bag, pieces were returned to the lab for sorting. For larger pieces, over a tarp to prevent escapes, every 50 cm of material was inspected and macroinvertebrates were collected.

When ant colonies or termite infestation were encountered in large pieces that could not be returned to the lab (e.g. stumps, large trees), a 15 cm wide piece of wood was collected, at every 50 cm inspection point, for sorting in the lab up to 5 encounters. Using the cordless vacuum, all of the material surrounding the colony was suctioned to assure collection of colonies. Any encounters after the first five were scored as presence and colony size was estimated visually. There were less than ten visual estimates for the entire study. Additionally, five 15 cm wide slices of very large CWM (without visible ants or termites) were taken and the surrounding area vacuumed and sorted in the lab.

All field material was returned to the lab and frozen on the day of collection. All material was later hand sorted after thawing. All woody material was broken apart and all litter material was carefully sorted and inspected. All macroinvertebrates were sorted to Class or Order and termites and ants were sorted to species and counted. All specimens were dried at 65°C prior to weighing. Voucher specimens were taken and currently reside in the University of Central Florida's insect collection.

Statistical Analysis

The primary data consisted of the abundance and dry biomass of all macroinvertebrates collected in 100 m^2 plots and the number of colonies of social insects. Data were converted to m^{-2} and m^{-3} values, which represents an extrapolation of the smaller area sampled. For CWM, total abundance, number of colonies, and biomass could either be reported per unit volume (m^{-3}) by dividing by total volume of CWM in 100 m^2 plots or per unit area (m^{-2}) by dividing totals per 100 m^2 plot by 100. For litter samples,

Table 1. Species of ants and termites captured at the four study sites.

Site	Ant species	Termite species
Yale Myers Forest	Aphaenogaster picea (Wheeler)	
Coweeta Forest	Aphaenogaster fulva Roger	Reticulitermes flavipes(Kollar)
	Aphaenogaster picea	
	Camponotus chromaiodes Bolton	
	Camponotus pennsylvanicus (De Geer)*	
	Lasius alienus (Foerster)*†	
	Myrmecina americana Emery	
	Nylanderia concinna Trager *†	
	Nylanderia faisonensis (Forel)*†	
	Ponera pennsylvanica Buckley	
	Prenolepis imparis (Say)	
Whitehall Forest	Amblyopone pallipes (Haldeman)	Reticulitermes flavipes
	Aphaenogaster picea	
	Aphaenogaster rudis Enzmann	
	Camponotus castaneus (Latreille)	
	Nylanderia concinna*†	
	Nylanderia faisonensis*†	
	Pheidole dentata MR Smith	
	Ponera pennsylvanica	
	Prenolepis imparis	
	Temnothorax curvispinosus (Mayr)*†	
San Felasco Forest	Camponotus floridanus (Buckley)*	Reticulitermes flavipes
	Formica pallidefulva Latrielle	Reticulitermes hageni Banks
	Nylanderia faisonensis*†	
	Odontomachus brunneus (Patton)	
	Pheidole dentata	
	Pheidole dentigula MR Smith	

Ant species noted with an asterisk (*) are polydomous (multiple nests per colony). Ant species noted with a † are polygyne (multiple queens per colony).

Table 2. The total abundance and dry mass of macroinvertebrates, from all plots, listed alphabetically by Class or Order.

	Total number	Total dry mass (g)
Invertebrates		
Megadrilacea (Earthworms)	7	0.2941
Stylommatophora (Terrestrial Snails)	539	5.3695
Isopoda (Isopods)	33	0.0910
Chilopoda (Centipedes)	81	0.5766
Diplopoda (Millipedes)	216	7.5466
Araneae (Spiders)	96	0.7355
Opiliones (Harvestmen)	9	0.2778
Insects		
Blattaria (Roaches)	44	2.0600
Coleoptera (Beetles)	324	3.0643
Diptera (Flies)	5	0.0205
Formicidae (Ants)	28351	35.477
Hemiptera (Bugs)	18	0.2213
Hymenoptera (Sawflies, Wasps, Bees)	8	0.0542
Isoptera (Termites)	24605	14.464
Lepidoptera (Moths and Butterflies)	74	1.6179
Orthoptera (Crickets, Katydids, Grasshoppers)	6	0.3627
Zygentoma (Silverfish)	1	0.0023

Ants are listed separately, as a Family, from other Hymenoptera.

total abundance, number of colonies, and biomass were converted to per unit area (m^{-2}) estimates by multiplying totals from all quadrats collected within 100 m^2 plots by 1.6 [i.e. $10 \times (0.25 \text{ m}^2 \times 0.25 \text{ m}^2) \times 1.6 = 1 \text{ m}^2$].

Data were analyzed in SAS version 9 using a mixed-model ANOVA design with number and dry mass of invertebrates as dependent variables and sites (Yale Myers in Connecticut, Coweeta in North Carolina, Whitehall in Georgia, and San Felasco in Florida), sociality (ants and termites versus all other invertebrates), and habitat (litter versus CWM), as classification variables and transect assigned as a random variable. Count data were $\log_{10}+1$ transformed and biomass data were $\log_{10}+0.0001$ transformed to satisfy normality and homoscedasticity assumptions. As PROC MIXED uses restricted maximum likelihood to estimate unknown covariance parameters, it was necessary to select the best-fitting covariance structure model for the data [38]. The data in all cases were best fit by the most general form possible, an unstructured covariance matrix structure, which was then used to construct the tests for fixed effects.

Approximate Type III F-statistics for fixed effects were calculated in PROC MIXED using a general Wald-type quadratic form [38] which we report here as F-statistics and associated P-values for localities, social and non-social invertebrates, litter and CWM and all two-way interactions. Three-way interactions were not fit because the degrees of freedom were too limited once we accounted for the spatial design of the study. We used an alpha of 0.05 to indicate significance. Inferences for fixed effects in PROC MIXED allows for comparisons across dependent variables while simultaneously accounting for the underlying covariance structure using differences of least squares means. We examined the result of multiple comparisons here as P-values after Tukey-Kramer adjustment.

Results

In the $3{,}200 \text{ m}^2$ of forest floor that we surveyed, we found a wide diversity (16 higher taxa) and high abundance (54,417 individuals) of macroinvertebrates (Table 2). Ants (52% of all macroinvertebrates sampled) and termites (45% of all macroinvertebrates) were by far the most abundant organisms overall, becoming increasingly abundant in the southern localities [Georgia (Whitehall) and Florida (San Felasco); Table 2, Fig. 2)]. There were significant differences in the abundance of all invertebrates among sites (ANOVA, $F_{3,64} = 21.1$, $P<0.0001$), between social insects and other invertebrates (ANOVA, $F_{1,64} = 6.44$, $P = 0.01$), and between microhabitats (CWM vs. litter; ANOVA, $F_{1,64} = 108$, $P<0.0001$). There were also significant differences in the biomass of all invertebrates among sites (ANOVA, $F_{3,64} = 12.6$, $P<0.0001$), between social insects and other invertebrates (ANOVA, $F_{1,64} = 31.0$, $P<0.0001$), and between microhabitats (ANOVA, $F_{1,64} = 137$, $P<0.0001$).

There were significant two-way interactions between the main effects on abundance (ANOVA, site×habitat $F_{3,64} = 14.7$, $P<0.0001$; site×sociality $F_{3,64} = 10.5$, $P<0.0001$; habitat×sociality, $F_{1,64} = 46.7$, $P<0.0001$) and dry mass (ANOVA, site×habitat $F_{3,64} = 11.9$, $P<0.0001$; site×sociality $F_{3,64} = 9.43$, $P<0.0001$; habitat×sociality, $F_{1,64} = 68.3$, $P<0.0001$) of invertebrates (Fig. 2). These interactions demonstrate the interdependence of latitude, habitat (CWM or litter) and invertebrate type (social insects or other) on the abundance of these organisms. The social×habitat interaction was likely driven by the fact that litter samples produced a much greater abundance of macroinvertebrates than CWM samples, and that this was especially pronounced for non-social taxa (Fig. 2A–D). The site×habitat, and site×social, interactions presumably arose because from north to south there were large differences in macroinvertebrate abundance in litter (Fig. 2A, C) but not CWM (Fig. 2B, D). Social insects were most

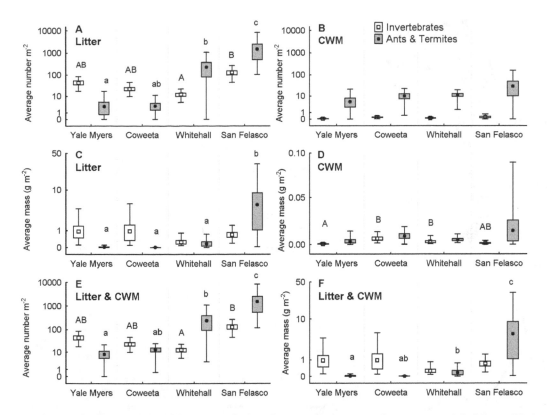

Figure 2. Macroinvertebrate biomass and abundance varied across sites, taxa, and habitats. (A) Average number of non-social invertebrates (not including ants and termites) and social insects (ants and termites) m^{-2} in litter samples and (B) in coarse woody material (CWM) samples. Average ants and termites m^{-2} differed among some sites in litter samples (A) but not in CWM (B). (C) Average dry mass of non-social invertebrates and social insects m^{-2} in all litter samples and (D) in CWM samples. Social insects were more abundant in San Felasco in litter samples (C) while non-social invertebrates only varied among some sites in CWM samples (D). In combined litter and CWM samples, the abundance (E) of both groups varied among sites, while only social insects varied in biomass (F). In both cases, the southern sites had higher numbers and masses of social insects (E and F). Points = mean, bars = +/− SE, and whiskers = range. The Y-axis is log$_{10}$ scaled. Letters above whiskers represent differences revealed through multiple comparisons. Shared letters of the same case (upper vs. lower) indicate no significant differences. Box plots without letters had no significant pairwise difference (Tukey-Kramer adjustment, $P>0.05$).

abundant at the two southern sites [Florida (San Felasco) and Georgia (Whitehall)] irrespective of habitat, but social insects were less abundant in the litter habitat at the two northern sites [North Carolina (Coweeta) and Connecticut (Yale Myers); Fig. 2A].

When litter and CWM samples were pooled, it was apparent that non-social invertebrate biomass did not differ markedly across locations, and its abundance was greatest in Florida (San Felasco) and least in Georgia (Whitehall) (Fig. 2E, F). The average dry mass of non-social invertebrates m^{-2} was only slightly less than 1 g in the two northernmost sites [0.94 g in Connecticut (Yale Myers), 0.98 g in North Carolina (Coweeta)], whereas it was 0.25 g and 0.69 g in Georgia and Florida, respectively (Fig 2F). In contrast, there were pronounced effects of site on social insect abundance and biomass (Fig. 2E, F), which increased as latitude decreased, resulting in the highest values for Florida, especially for ants (Fig. 3A, B). Termite abundance also increased with decreasing latitude but to less of an extent, and biomass varied only slightly across locations (Fig. 3A, B).

In Florida (San Felasco), average abundance of ants (Fig. 3A) was nearly ten times greater than that of other invertebrates (Fig. 2E) and average biomass more than ten times greater (Figs. 3B and 2F). Ant mass and abundance were, on average, a minimum of five times greater than for termites in all sites and peaked at nearly 500 times (dry mass) and 55 times (abundance) greater in Florida (Fig. 3B). In all sites except Connecticut (Yale Myers),

where termites were not found (Fig. 3), there were more ant than termite species (Table 1).

Though we did not observe termites in Connecticut, *Reticulitermes flavipes* has long been known to be present in the state, but uncommon [39], [40], and has been observed at the Yale Myers site (MAB, pers. obs.). In litter across the other sites, the relative abundance and biomass of termites m^{-2} was very low (average of <1%, Fig. 3A, B) whereas ants accounted for the vast majority of social insect biomass (Fig. 3B). In contrast, except for Connecticut, termites were a major component of the abundance and biomass of macroinvertebrates in CWM (Fig. 3C, D). Thus, measuring termites m^{-3} CWM likely provides the most accurate estimate of standing termite biomass in eastern US temperate forests. Ant colonies also were abundant and, combined with termites, made up a large majority of the abundance of invertebrates in CWM (Yale Myers: average ant abundance = 63%, average termite abundance = 0%; Coweeta: ants = 71%, termites = 21%, Whitehall: ants = 63%, termites = 36%; San Felasco: ants = 30%, termites = 56%; Table 3).

The number of social insect colonies m^{-2} collected from litter and CWM were higher in southern sites, especially Florida (extrapolations for Yale Myers and Coweeta: ~2 colonies m^{-2}, Whitehall: ~5 colonies m^{-2}, San Felasco: ~13 colonies m^{-2}). A majority (65%) of ant colonies collected included queens (were queenright) and colony size numbers (i.e. for non-queenright

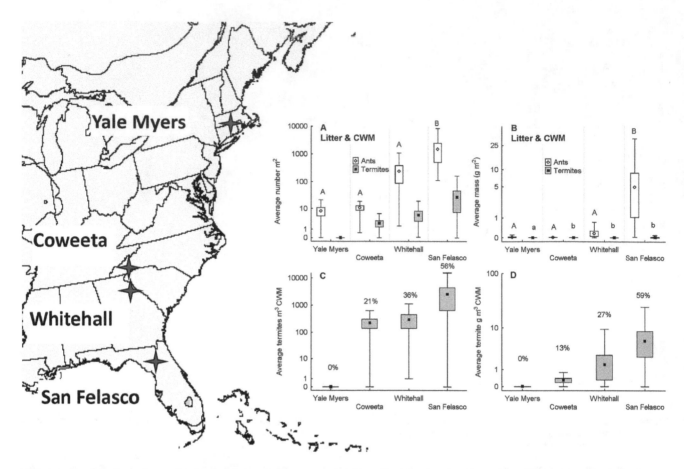

Figure 3. The location of sample sites and the average abundance m^{-2} (A) and g dry biomass m^{-2} (B) of ants and termites in combined litter and CWM samples. San Felasco (Florida) had a much greater abundance and biomass of ants than other sites, while termites did not vary in abundance. (C) The average g dry mass of termites m^{-3} and (D) the average number of termites m^{-3} in coarse woody material (CWM) in plots. Termite dry mass and numbers were zero at Yale Myers and did not differ significantly among the other sites. Points = mean, bars = +/− SE, and whiskers = range. The Y-axis is log$_{10}$ scaled. Percentages above whiskers in (C and D) represent the mean proportion of invertebrate numbers and biomass in CWM that termites comprised. Map image derived from http://upload.wikimedia.org/wikipedia/commons/d/de/Eastern_US_range_map_blank.png, created by Alan Rockefeller.

colony fragments) produced from this sampling method generally match other published estimates for these species [41], [42]. A few species were, by far, the most abundant both in terms of total abundance (numbers of individuals) and colony abundance, comprising greater than 50% of ant or termite workers in all sites (Table 4). Termites were almost entirely *R. flavipes*, although one colony of *R. hageni* was collected in Florida (Table 4 and Table 1).

A broad-scale pattern emerged among the most common species whereby *Aphaenogaster picea* was, by far, the most dominant ant in the two northernmost sites (Table 4). In Georgia (Whitehall), however, *Pheidole dentata* was the most common ant species (although a closely related *A. picea* congener, *A. rudis*, still was present in considerable numbers). *Pheidole dentata* had the most colonies in Florida, too (although the carpenter ant *Camponotus floridanus* had a higher number of workers and biomass in Florida due to sampling two large colonies – Table 4). The genus *Aphaenogaster* was not found in the Florida site although it is present in San Felasco State Park [43], whereas the species *A. rudis* and *A. picea* are not present in Florida [44]. The abundance of ants in CWM results from the predominance of *Aphaenogaster* and *P. dentata* colonies (Fig. 3, Table 4). These species opportunistically nest in rotting wood and will typically have portions of the nest extending into the soil (JRK, RJW, MAB pers. obs.). The highest diversity of

ant species occurred in North Carolina (Coweeta) and Georgia (Whitehall), each with ten species (Table 1). These sites also had five species in common (Table 1).

Average soil temperature at the time of sampling differed among sites (*ANOVA, P* < 0.0001), as did soil moisture (*ANOVA, P* < 0.0001). Connecticut (Yale Myers) was, on average, the coolest and wettest site (18.4°C, 13.2% soil moisture) and North Carolina (Coweeta) was the warmest and driest (24.6°C, 3.2% soil moisture). Georgia (Whitehall) and Florida (San Felasco) were similar (22.8°C, 9.6% soil moisture; 23.1°C, 5.6% soil moisture, respectively). Only termite abundance had a linear relationship with soil temperature across plots (ant abundance *P* = 0.25, R^2 = 0.04; termite abundance *P* = 0.02, R^2 = 0.17; invertebrate abundance *P* = 0.55, R^2 = 0.01) although it was weakly correlated. There was no relationship between soil moisture and abundance for any group (ant abundance *P* = 0.38, R^2 = 0.03; termite abundance *P* = 0.21, R^2 = 0.05; invertebrate abundance *P* = 0.36, R^2 = 0.03).

Both annual temperature and productivity show that our sites are representative of a latitudinal gradient where temperature and productivity increased with decreasing latitude. Using San Felasco Hammock (the southernmost and annually warmest site) as a baseline "0," assignment of general elevational and latitudinal

Table 3. The global reported ranges of numbers of individuals m^{-2} and biomass m^{-2} for ecosystem engineers and macroinvertebrates.

Source	Ants m^{-2}/g m^{-2}	Termites m^{-2}/g m^{-2}***	Earthworms m^{-2}/g m^{-2}***	Other macroinvertebrates m^{-2}/g m^{-2}***	Ants%/Termites % (maximum)**
This study					
Yale Myers (41° N)	0–22/0–0.102	0/0	0–3/0–0.300	18–83/0.108–4.003	2.5%/0%
Coweeta (35° N)	1–19/0.001–0.018	0–6/0–0.005	0/0	10–47/0.098–5.186	0.3%/0.09%
Whitehall (33° N)	2–1084/0.003–0.739	1–19/0–0.013	0/0	5–23/0.079–0.823	47%/0.8%
San Felasco (29° N)	111–8310/0.027–31.578	0–163/0–0.091	0/0	45–268/0.185–1.506	95%/0.3%
[68] Bignell & Eggleton					
Tropical forests (Africa, Asia, Neotropics)	NA	38–6957/0–33.264	NA	NA	NA
Tropical savannas (Africa)	NA	49–4402/0.216–2.990	NA	NA	NA
Temperate forests (Australia)	NA	NA/0.810–1.350	NA	NA	NA
Temperate scrub and grasslands (Australia, USA)	NA	NA/0.262–1.350	NA	NA	NA
[12] Wood & Sands					
Temperate forest (Australia)	NA	600/0.810	NA	NA	NA
Semi arid savanna and grasslands (North America, Africa)	NA	0–9127/0–5.997	NA	NA	NA
Tropical savannas (Africa, Australia)	NA	70–4402/0.459–2.997	NA	NA	NA
Tropical Forests (Africa, Southeast Asia, Neotropics)	NA	87–4450/0.027–2.970	NA	NA	NA
[69] Baroni-Urbani & Pisarski					
Various (mostly temperate Europe and USA)	0–115,825/NA	NA	NA	NA	NA
[21] Kaspari & Weiser					
Various (New World temperate to tropics)	NA/<0.010–<1.000	NA	NA	NA	NA
[9] Lavelle & Spain					
Various (worldwide "cold," temperate, and tropical)	NA	NA	~20–120/~0.6–~ 24.3	NA	NA
[14] Lavelle					
Tropical grasslands (Ivory Coast, Mexico)	500–1400/0.273–0.525	2–1200/<0.100–0.756	230–700/3.345–7.350	147–558/0.240–14.370	0.9%/1.4%
[37] Callaham & Hendrix					
Appalachian Piedmont (33° N, USA)	NA	NA	0–120/0–~ 8.250 *	NA	NA
[70] Shakir & Dindal					
Various temperate forests (43°N, USA)	NA	NA	37–200/0.375–4.785*	NA	NA
[71] Suarez et al.					
Temperate hardwood forest (42°N, USA)	NA	NA	22–99/0.9660–8.085 *	NA	NA
[72] Hendrix et al.					
Southeastern pine forest (30°N, USA)	NA	NA	2/0.900	NA	NA
[15] Petersen & Luxton					
Tundra	0/0	0/0	NA/0.330	NA/0.550	0%/0%
Temperate grasslands	NA/0.1	0/0	NA/3.100	NA/1.410	2%/0%
Tropical grasslands	NA/0.3	NA/1.000	NA/0.170	NA/0.075	19%/64%
Temperate coniferous forests	NA/0.01	0/0	NA/0.450	NA/0.570	1%/0%

Table 3. Cont.

Source	Ants m^{-2}/g m^{-2}	Termites m^{-2}/g m^{-2}***	Earthworms m^{-2}/g m^{-2}***	Other macroinvertebrates m^{-2}/g m^{-2}***	Ants%/Termites % (maximum)**
Temperate deciduous forests	NA/0.01	0/0	NA/0.200–5.300	NA/1.280	0.2–0.6%/0%
Tropical forests	NA/0.03	NA/1.000	NA/0.340	NA/0.060	2%/70%

*Majority exotic species.
**Percent of maximum biomass (all macroinvertebrates) reported.
***Conversion of fresh weights to dry weights (g) are estimates and followed that of [15]: termite fresh weight×0.27 = dry mass, earthworm fresh mass×0.15, ant fresh mass×0.23, and other macroinvertebrates fresh mass×0.30. These conversions do not apply to the invertebrates sampled in this study as those were dried and weighed.

lapse rates ($-6.5°$C per 1000 m increase in elevation and $-1°$C per 145 km north increase in latitude), showed that Whitehall Forest was annually 3.75°C cooler, Coweeta was 9.21°C cooler, and Yale Myers was 10.86°C cooler than the Florida site. Thus our sites comprise a climate gradient where temperature declined with latitude and Coweeta and Yale Myers sites (North Carolina and Connecticut) group as northern sites that were closer in annual temperature than the southern sites at Whitehall and San Felasco (Georgia and Florida). Above ground annual production by plants in each site was estimated as 1500 g m^{-2} yr^{-1} at San Felasco Hammock, 1050 g m^{-2} yr^{-1} at Whitehall Forest, 920 g m^{-2} yr^{-1} at Coweeta, and 840 g m^{-2} yr^{-1} at Yale Myers Forest.

Discussion

The relative abundance of ants in the southern temperate sites was unexpected and is comparable to the impressively high abundance and biomass of ants in tropical arboreal ecosystems (up to 70% of all arboreal arthropods, up to 50% of arboreal arthropod biomass; [45], [46], [47]) and higher than estimates for tropical ground-dwelling ants (e.g. [21], [48]). Furthermore, our average areal biomass estimate (4.87 g dry mass m^{-2})in Florida (San Felasco) equals or surpasses most other commonly abundant, terrestrial vertebrate, animal groups such as salamanders in northeastern temperate forest in the US (~0.05 g dry mass m^{-2}), British Virgin Island reptile communities (~0.0001 g dry mass m^{-2}), all large (greater than 500 g body weight, including elephants) mammals in equatorial rainforest in Gabon (~0.32 g dry mass m^{-2}), mammals in dry tropical forest in Thailand (0.73 g dry biomass m^{-2}), mammals in a variety of neotropical forest and

grassland (0.11–0.33 g dry biomass m^{-2}) and even pastures stocked with cattle (up to 2.28 g dry biomass m^{-2}) in Brazil (approximation of dry mass values = 0.3×fresh mass values reported in [23], [49], [50], [51], [52]).

In contrast to the patterns observed in the southern sites, termites and ants were subordinate within the macroinvertebrate communities in the northern part of the range of eastern temperate forests (Fig. 2E, F). This result supports the long-standing consensus that ants are thermophilic (warm-loving) and their abundance is greater in ecosystems with higher primary productivity, especially where temperatures are higher [1], [21], [26], [27]. Termites are also a thermophilic taxon and their abundance is greater in warmer regions [2]. Our data (Figs. 3C and D) and those of Vargo et al. [53] show a large increase in termite abundance in warmer regions within temperate zones. Temperate termite abundance may also be affected by the availability of standing and downed dead woody material (Fig. 3C, D), however, the more northern part of the temperate zone has higher CWM stocks than in the south [54], again suggesting that cooler climate is an important limit on their abundance (Fig. 3) [53], [55], [56].

An interesting example of biogeographic turnover occurred in the most abundant ant species in CWM from northern (*A. picea*) to southern (*P. dentata*) sites (Table 4). These two species are ecologically similar despite being in different genera. Both species opportunistically nest in decaying wood and soil, have very similar diets (both prey upon termites and both will take eliasome bearing seeds), appear to be weakly territorial or not territorial at all, and have colonies that are typically below 1,000 workers in size (JRK,

Table 4. The abundance of the most common species of ants and termites collected in each site.

Social insect	Site	Species	Number of colonies or occurrences	Average worker number	% of total abundance
Ants	Yale Myers	*Aphaenogaster picea*	10	414	100
Termites	Yale Myers	NA	0	0	0
Ants	Coweeta	*A. picea*	20	189	64
Termites	Coweeta	*Reticulitermes flavipes*	11	161	100
Ants	Whitehall	*P. dentata*	7	692	68
Termites	Whitehall	*R. flavipes*	13	181	100
Ants	San Felasco	*Camponotus floridanus*	2	2727	55
Termites	San Felasco	*R. flavipes*	12	1684	98

Colony numbers (ants) and occurrences in CWM (termites) as well as average number of workers are shown. Percent of total abundance was determined for ants and termites separately as a percentage of the total number of workers captured.

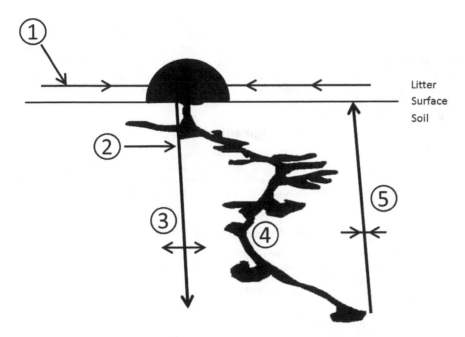

Figure 4. The myrmecosphere is centered upon ant nests constructed at the soil surface and below ground. (1) Prey and carrion, plant material, plant and insect exudates are brought into the colony. (2) Below-ground prey and carrion, plant material, plant and animal exudates are brought into the colony. (3) Materials brought into the colony are assimilated into the soil over time. (4) Feces, saliva, and other excretions are produced within the colony. (5) Soil, corpses, and midden material are returned to the soil surface.

RJW, MAB pers. obs.). The divergence in the ant communities that occurs somewhere between North Carolina and Georgia is almost certainly under climatic influence, with the cooler temperate species (*A. picea*), giving way to the southeastern coastal plain species (*P. dentata*). Termites showed no such pattern, with *R. flavipes* remaining the dominant species throughout the entire range of the study (Table 4), but termite biomass in CWM does increase markedly between North Carolina and Georgia.

Variation in sampling protocols for soil fauna complicates the comparison of our results with other studies (i.e. Table 3). We under-sampled earthworms in our study, and possibly other groups (e.g. fast-moving large spiders), although the remaining macroinvertebrate fauna are well-represented [15]. Our sampling approach thus appears useful for estimating areal abundance of social insects and most co-occurring macroinvertebrates. Notably though, our sampling design was likely effective at estimating areal biomass but not effective at capturing whole colonies of termites. This is because termites in the genus *Reticulitermes* are dead wood-feeding species of the "multiple-piece nesting" functional group, which means that they feed upon decaying wood away from the primary nest where the reproductive members of the colony reside [57], [58]. Nests are cryptic and the majority of above and belowground termite abundance represents feeding rather than nesting activity (nests include sexuals and some workers), though nests sometimes are located above ground [59]. Colonies tend to be simple family groups, comprising a single reproductive pair (a queen and king) and their offspring, and maintain foraging areas typically ≤100 m² in size in the southeastern US [53]. Inbreeding, larger territories, and extended family colonies become more common in the northern part of the range where abundance and colony density is also lower (Figs. 3C and D, [53]).

In contrast to termites, and due to the fact that the majority of the ant colonies collected were queenright (65%), monogyne (single queen), monodomous (single nest) species (Table 1), our sampling protocol appears to be effective at estimating both areal

biomass and colony abundance of ground-dwelling ants. Despite the high numbers and biomass we report, it is important to note that our areal abundances are still underestimates. At Yale Myers, Coweeta, and Whitehall we also searched for ants under rocks and at Yale Myers in fine woody material (FWM, <10 cm dia.). Using Yale Myers as an example, in addition to the 10 *A. picea* colonies in CWM we used for this analysis, we did not include seven colonies found under rocks and fourteen colonies found in FWM, which contained the only colonies with >1,000 workers. If this under-sampling holds across all our sites, then the true biomass and numbers of ants may be ~3-times larger than the already high values we report, with the obvious caveat that these are extrapolations. We did under-sample ant species diversity because of the relatively small number of samples and use of one, rather than multiple, sampling techniques [60], [61].

Maintaining ecosystem services is a critically important, central component of global biodiversity conservation strategies [62]. Lavelle [5], [7], [9], [63] and others [2], [5], [8], [10], [16] have called attention to the central importance of social insects, along with earthworms, in maintaining soil ecosystem function and all of the associated ecosystem services. If the importance of social insects for soil processes is at least partially dependent upon biomass [12], then the data we present here suggest ants and termites are among the most important macroinvertebrates in eastern US temperate forests, at least in the southern parts of the range and likely in other temperate systems (Table 3).

Lavelle et al. [64] identify four principal systems of biological regulation of decomposition and soil structure: the litter-superficial root system, the rhizosphere, the drilosphere, and the termitosphere. The drilosphere and the termitosphere are the processes under the influence of earthworm populations and termite populations, respectively, through their activities in the soil environment. These include intestinal contents, castings, and galleries for earthworms and mounds, galleries, woody material, and gut symbionts for termites. No such system of biological regulation has been identified for ants: a myrmecosphere, Given the huge abundance and biomass of ants we observed in the southern part of our system, the myrmecosphere might be considered a fifth system of biological regulation in soils. It's contribution to the soil ecosystem, particularly the physical structure and chemical make-up of the soil environment [5], [10], [65], would be an emergent property of the social organization of colonies and the nests they construct and maintain [66]. The nest is the organizational centerpiece of colonial living for ants, shaping the spatial arrangement of individuals and division of labor [67] as well as the movement of materials into and out of the colony (Fig. 4). The nest is thus the "building block" of the myrmecosphere (Fig. 4). More data on areal abundance (Table 3) and the belowground activities of ants [66], [67] are necessary to better quantify the functional role of a myrmecosphere in ecosystems.

The paucity of data on ant and termite abundances is cited as the reason for omitting them from syntheses of biogeographical patterns in belowground communities [17]. Our observations begin to redress this shortcoming and reveal a pronounced shift from social insect subordinance to dominance across decreasing latitude in a major, temperate forest ecosystem. Termites were the most abundant macroinvertebrates in dead wood and ants were the most abundant in litter and soil. Ant abundance and biomass m^{-2} in the southernmost site were among the highest values recorded for ants in any ecosystem, highlighting the potential importance of these faunal groups to the belowground functioning of temperate systems.

Acknowledgments

We thank Ella Bradford, Ben Gochnour, Lindsay Gustafson, Sarah Huber, Mary Schultz and Anna Wade for field and lab assistance. We thank Edward Vargo and one anonymous reviewer for comments that greatly improved an earlier version of the manuscript. This is the Termite Ecology And Myrmecology (TEAM) working group's publication number 1.

Author Contributions

Conceived and designed the experiments: JRK RJW MAB. Performed the experiments: JRK RJW MAB. Analyzed the data: JRK RJW MAB. Contributed reagents/materials/analysis tools: JRK RJW MAB. Wrote the paper: JRK RJW MAB.

References

1. Hölldobler B, Wilson EO (1990) The Ants. Harvard University Press, Cambridge, MA, USA.
2. Abe T, Bignell DE, Higashi M, eds. (2000) Termites: Evolution, Sociality, Symbioses, Ecology. Kluwer, Dordrecht, Netherlands.
3. Lach L, Parr C, Abbott K, eds. (2010) Ant Ecology. Oxford University Press, Oxford, UK.
4. Bolton B, Alpert G, Ward PS, Naskrecki P (2006) Bolton's Catalogue of the Ants of the World: 1758–2005. Harvard University Press, Cambridge, MA, USA.
5. Lobry de Bruyn LA, Conacher AJ (1990) The role of termites and ants in soil modification: a review. Australian Journal of Soil Research 28: 55–93.
6. Lavelle P, Bignell D, Lepage M, Wolters V, Roger P, et al. (1997) Soil function in a changing world: the role of invertebrate ecosystem engineers. European Journal of Soil Biology 33: 159–193.
7. Lavelle P, Decaëns T, Aubert M, Barot S, Blouin M, et al. (2006) Soil invertebrates and ecosystem services. European Journal of Soil Biology 42: S3–A15.
8. Folgarait PJ (1998) Ant biodiversity and its relationship to ecosystem functioning: a review. Biodiversity and Conservation 7: 1221–1244.
9. Lavelle P, Spain AV (2001) Soil Ecology. Kluwer, Dordrecht, Netherlands.
10. Cammeraat ELH, Risch AC (2008) The impact of ants on mineral soil properties and processes at different spatial scales. Journal of Applied Entomology 132: 285–294.
11. Edwards CA, ed. (1998) Earthworm Ecology. CRC Press, Boca Raton, FL, USA.
12. Wood TG, Sands WA (1978) The role of termites in ecosystems. In: Brian, M.V. (ed.), Production Ecology of Ants and Termites. Cambridge University Press, Cambridge, UK, 245–292.
13. Brian MV, ed. (1978) Production Ecology of Ants and Termites. Cambridge University Press, Cambridge, UK.
14. Lavelle P (1984) The soil system in the humid tropics. Biology International 9: 2–15.
15. Petersen H, Luxton M (1982) A comparative analysis of soil fauna populations and their role in decomposition processes. Oikos 39: 287–388.
16. Lobry de Bruyn LA (1999) Ants as bioindicators of soil function in rural environments. Agriculture, Ecosystems and Environment 74: 425–441.
17. Fierer N, Strickland MS, Liptzin DL, Bradford MA, Cleveland CC (2009) Global patterns in belowground communities. Ecology Letters 12: 1238–1249.
18. Brussaard L, Aanen Dk, Briones MJI, Decaëns T, De Deyn GB, et al. (2012) Biogeography and phylogenetic community structure of soil invertebrate ecosystem engineers: global to local patterns, implications for ecosystem functioning and services, and global environmental change impacts. In: Wall, D.H. et al. (eds.), Soil Ecology and Ecosystem Services. Oxford University Press, UK, 201–232.
19. Wheeler WM (1911) The ant colony as an organism. Journal of Morphology 22: 307–325.
20. Tschinkel WR (1991) Insect sociometry: a field in search of data. Insectes Sociaux 38: 77–82.
21. Kaspari M, Weiser MD (2012) Energy, taxonomic aggregation, and the geography of ant abundance. Ecography 35: 65–72.
22. Lugo AE, Brown SL, Dodson R, Smith TS, Shugart HH (1999) The Holdridge life zones of the conterminous United States in relation to ecosystem mapping. Journal of Biogeography 26: 1025–1038.
23. Burton TM, Likens GE (1975) Salamander populations and biomass in the Hubbard Brook Experimental Forest, New Hampshire. Copeia 1975: 541–546.
24. Irland LC (1999) The Northeast's Changing Forest. Harvard University Press, Petersham, MA, USA.
25. Brown WL (1973) A comparison of the Hylean and Congo-West African rain forest ant faunas. In: Meggers, B.J. et al. (eds.), Tropical Forest Ecosystems in Africa and South America: A Comparative Review. Smithsonian Institution Press, Washington, DC, USA, 161–185.
26. Andersen AN (1997) Functional groups and patterns of organization in North American ant communities: a comparison with Australia. Journal of Biogeography 24: 433–460.
27. Kaspari M, Alonso L, O'Donnell S (2000) Three energy variables predict ant abundance at a geographical scale. Proceedings of the Royal Society, B 267: 485–489.
28. Goodale E, Lalbhae P, Goodale UM, Ashton PMS (2009) The relationship between shelterwood cuts and crown thinnings and the abundance and distribution of birds in a southern New England forest. Forest Ecology and Management 258: 314–322.
29. Elliott KJ, Vose JM (2011) The contribution of the Coweeta Hydrologic Laboratory to developing an understanding of long-term (1934–2008) changes in managed and unmanaged forests. Forest Ecology and Management 261: 900–910.
30. Platt WJ, Schwartz MW (1990) Temperate hardwood forests. In: Myers RL, Ewel JJ (eds.), Ecosystems of Florida. University of Central Florida Press, Orlando, FL, USA, 194–229.
31. Warren RJ (2010) A test of temperature estimation from solar irradiation and a simple statistical method to integrate elevation into prediction models. Castanea 75: 67–77.
32. Bolstad PV, Vose JM, McNulty SG (2001) Forest productivity, leaf area and terrain in southern Appalachian deciduous forests.
33. Lugo AE, Gamble JF, Ewel KC (1978) Organic matter flows in a mixed-hardwood forest in north central Florida. In: Adriano DC, Brisbin, IL (eds.) Environmental Chemistry and Cycling Processes. DOE Symposium Series, CONF-760429, 790–800.
34. Mickler RA, Earnhardt TS, Moore JA (2002) Modeling and spatially distributing forest net primary production at the regional scale. Journal of the Air & Waste Management Association 52: 407–415.
35. Ollinger SV, Aber JD, Federer CA (1998) Estimating regional forest productivity and water yield using an ecosystem model linked to a GIS. Landscape Ecology 13: 323–334.
36. Turner MG (1987) Land use changes and net primary production in the Georgia, USA, landscape: 1935–1982. Environmental Management 11: 237–247.
37. Callaham MA, Hendrix PJF (1997) Relative abundance and seasonal activity of earthworms (Lumbricidae and Megascolecidae) as determined by hand-sorting and formalin extraction in forest soils on the southern Appalachian piedmont. Soil Biology and Biochemistry 29: 317–321.
38. Khattree R, Naik DN (1999) Applied Multivariate Statistics with SAS Software, 2nd edition. SAS Institute, Cary, NC, USA.
39. Emerson AE (1936) Termite distribution in the United States. Science 83: 410–411.
40. Packard CE (1936) Termite distribution in the United States. Science 83: 575.

41. Clark R, King JR (2012) The ant, *Aphaenogaster picea*, benefits from plant elaisomes when insect prey is scarce. Environmental Entomology 41: 1405–1408.

42. King JR (2010) Size-abundance relationships in Florida ant communities reveal how social insects break the energetic equivalence rule. Ecological Entomology 35: 287–298.

43. King JR, Porter SD (2007) Body size, colony size, abundance, and ecological impact of exotic ants in Florida's upland ecosystems. Evolutionary Ecology Research 9: 757–774.

44. Deyrup M (2003) An updated list of Florida ants (Hymenoptera: Formicidae). Florida Entomologist 86: 43–48.

45. Stork NE (1988) Insect diversity: facts, fiction, and speculation. Biological Journal of the Linnean Society 35: 321–337.

46. Tobin JE (1995) Ecology and diversity of tropical forest canopy ants. In: Lowman MD, Nadkarni NM (eds.), Forest Canopies. Academic press, NY, NY, USA, 129–147.

47. Davidson DW (1997) The role of resource imbalances in the evolutionary ecology of tropical arboreal ants. Biological Journal of the Linnean Society 61: 153–181.

48. Watt AD, Stork NE, Bolton B (2002) The diversity and abundance of ants in relation to forest disturbance and plantation establishment in southern Cameroon. Journal of Applied Ecology 39: 18–30.

49. Schaller GB (1983) Mammals and their biomass on a Brazilian ranch. Arquivos de Zoologia 31: 1–36.

50. Prins HHT, Reitsma JM (1989) Mammalian biomass in an African equatorial rain forest. Journal of Animal Ecology 58: 851–861.

51. Srikosamatara S (1993) Density and biomass of large herbivores and other mammals in a dry tropical forest, western Thailand. Journal of Tropical Ecology 9: 33–43.

52. Rodda DH, Perry G, Rondeau RJ, Lazell J (2001) The densest terrestrial vertebrate. Journal of Tropical Ecology 17: 331–338.

53. Vargo EL, Leniaud L, Swoboda LE, Diamond SE, Weiser MD, et al. (2013) Clinal variation in colony breeding structure and level of inbreeding in the subterranean termites *Reticulitermes flavipes* and *R. grassei*. Molecular Ecology DOI: 10.1111/mec.12166.

54. Chojnacky DC, Mickler RA, Heath LS, Woodall CW (2004) Estimates of down woody materials in eastern US forests. Environmental Management 33: S44–S55.

55. Cornwell WK, Cornelissen JHC, Allison SD, Bauhus J, Eggleton P, et al. (2009) Plant traits and wood fates across the globe: rotted, burned, or consumed? Global Change Biology 15: 2431–2449.

56. Vargo EL, Husseneder C (2009) Biology of subterranean termites: insights from molecular studies of *Reticulitermes* and *Coptotermes*. Annual Review of Entomology 54: 379–403.

57. Abe T (1990) Evolution of worker caste in termites. In: Veeresh, G.K. et al. (eds.), Social Insects and the Environment. Oxford and IBH, New Delhi, India, 29–30.

58. Korb J (2007) Termites. Current Biology 17: 995–999.

59. Thorne BL, Traniello JFA, Adams ES Bulmer MS (1999) Reproductive dynamics and colony structure of subterranean termites of the genus *Reticulitermes* (Isoptera: Rhinotermitidae): a review of the evidence from behavioral, ecological, and genetic studies. Ethology, Ecology and Evolution 11: 149–169.

60. King JR, Porter SD (2005) Evaluation of sampling methods and species richness estimators for ants in upland ecosystems in Florida. Environmental Entomology 34: 1566–1578.

61. Ellison AM, Record S, Arguello A, Gotelli NJ (2007) Rapid inventory of the ant assemblage in a temperate hardwood forest: species composition and assessment of sampling methods. Environmental Entomology 36: 766–775.

62. Millennium Ecosystem Assessment (2005) Ecosystems and Human Well-Being: Biodiversity Synthesis. World Resources Institute, Washington, DC, USA.

63. Lavelle P (1997) Faunal activities and soil processes: adaptive strategies that determine ecosystem function. Advances in Ecological Research 27: 93–132.

64. Lavelle P, Blanchart E, Martin A, Spain AV, Martin S (1992) The impact of soil fauna on the properties of soils in the humid tropics. In: Lal, R. and Sanchez, P.A. (eds.), Myths and Science of Soils of the Tropics, Soil Science Society of America, Madison, WI, USA, 157–185.

65. Jiménez JJ, Decaëns T, Lavelle P (2008) C and N concentrations in biogenic structures of a soil-feeding termite and a fungus-growing ant from the Colombian savannas - clues for modeling the impact of soil ecosystem engineers. Applied Soil Ecology 40: 120–128.

66. Tschinkel WR (2004) The nest architecture of the Florida harvester ant, *Pogonomyrmex badius*. Journal of Insect Science 4: 21.

67. Tschinkel WR (2005) The nest architecture of the ant, *Camponotus socius*. Journal of Insect Science 5: 9.

68. Bignell DE, Eggleton P (2000) Termites in ecosystems. In: Abe, T. et al. (eds.), Termites: Evolution, Sociality, Symbioses, Ecology. Kluwer, Dordrecht, Netherlands, 363–387.

69. Baroni-Urbani C, Pisarski B (1978) Appendix 1. In: Brian, M.V. (ed.), Production Ecology of Ants and Termites. Cambridge University Press, Cambridge, UK, 336–339.

70. Shakir SH, Dindal DL (1997) Density and biomass of earthworms in forest and herbaceous microecosystems in central New York, North America. Soil Biology and Biochemistry 29: 275–285.

71. Suarez ER, Fahey TJ, Yavitt JB, Groffman PM, Bohlen PJ (2006) Patterns of litter disappearance in a northern hardwood forest invaded by exotic earthworms. Ecological Applications 16: 154–165.

72. Hendrix PF, Callaham MA, Kirn L (1994) Ecology of Nearctic earthworms in the southern USA. II. Effects of bait harvesting on *Diplocardia* (Oligochaeta, Megascolecidae) populations in Apalachicola National Forest, north Florida. Megadrilogica 5: 73–76.

Permissions

The contributors of this book come from diverse backgrounds, making this book a truly international effort. This book will bring forth new frontiers with its revolutionizing research information and detailed analysis of the nascent developments around the world.

We would like to thank all the contributing authors for lending their expertise to make the book truly unique. They have played a crucial role in the development of this book. Without their invaluable contributions this book wouldn't have been possible. They have made vital efforts to compile up to date information on the varied aspects of this subject to make this book a valuable addition to the collection of many professionals and students.

This book was conceptualized with the vision of imparting up-to-date information and advanced data in this field. To ensure the same, a matchless editorial board was set up. Every individual on the board went through rigorous rounds of assessment to prove their worth. After which they invested a large part of their time researching and compiling the most relevant data for our readers.

The editorial board has been involved in producing this book since its inception. They have spent rigorous hours researching and exploring the diverse topics which have resulted in the successful publishing of this book. They have passed on their knowledge of decades through this book. To expedite this challenging task, the publisher supported the team at every step. A small team of assistant editors was also appointed to further simplify the editing procedure and attain best results for the readers.

Apart from the editorial board, the designing team has also invested a significant amount of their time in understanding the subject and creating the most relevant covers. They scrutinized every image to scout for the most suitable representation of the subject and create an appropriate cover for the book.

The publishing team has been an ardent support to the editorial, designing and production team. Their endless efforts to recruit the best for this project, has resulted in the accomplishment of this book. They are a veteran in the field of academics and their pool of knowledge is as vast as their experience in printing. Their expertise and guidance has proved useful at every step. Their uncompromising quality standards have made this book an exceptional effort. Their encouragement from time to time has been an inspiration for everyone.

The publisher and the editorial board hope that this book will prove to be a valuable piece of knowledge for researchers, students, practitioners and scholars across the globe.

List of Contributors

Wei Wei, Xiao Guang Qi, SongTao Guo and Pei Zhang
Key Laboratory of Resource Biology and Biotechnology in Western China, College of Life Sciences, Northwest University, Xi'an, China

Paul A. Garber
Anthropology Department, University of Illinois, Urbana, Illinois, United States of America

BaoGuo Li
Key Laboratory of Resource Biology and Biotechnology in Western China, College of Life Sciences, Northwest University, Xi'an, China
Institute of Zoology, Shaanxi Academy of Sciences, Xi'an, China

Yawei Wei, Xiangmin Fang and Wei Zhao
State Key Laboratory of Forest and Soil Ecology, Institute of Applied Ecology, Chinese Academy of Sciences, Shenyang, China
University of Chinese Academy of Sciences, Beijing, China

Maihe Li
State Key Laboratory of Forest and Soil Ecology, Institute of Applied Ecology, Chinese Academy of Sciences, Shenyang, China
Tree Physiology Group, Swiss Federal Research Institute WSL, Birmensdorf, Switzerland

Hua Chen
State Key Laboratory of Forest and Soil Ecology, Institute of Applied Ecology, Chinese Academy of Sciences, Shenyang, China
University of Illinois at Springfield, Springfield, Illinois, United States of America

Bernard J. Lewis, Dapao Yu, Li Zhou, Wangming Zhou and Limin Dai
State Key Laboratory of Forest and Soil Ecology, Institute of Applied Ecology, Chinese Academy of Sciences, Shenyang, China

Krista L. McGuire
Department of Biology, Barnard College, Columbia University, New York, New York, United States of America

Kathleen K. Treseder
Department of Ecology & Evolutionary Biology, University of California Irvine, Irvine, California, United States of America

Steven D. Allison
Department of Ecology & Evolutionary Biology, University of California Irvine, Irvine, California, United States of America
Department of Earth System Science, University of California Irvine, Irvine, California, United States of America

Noah Fierer
Department of Ecology & Evolutionary Biology, University of Colorado, Boulder, Colorado, United States of America
Cooperative Institute for Research in Environmental Sciences, University of Colorado, Boulder, Colorado, United States of America

Nerea Abrego
Department of Biological and Environmental Science, University of Jyväskylä, Jyväskylä, Finland
Department of Plant Biology and Ecology, University of the Basque Country (UPV/EHU), Bilbao, Spain

Gonzalo García-Baquero and Isabel Salcedo
Department of Plant Biology and Ecology, University of the Basque Country (UPV/EHU), Bilbao, Spain

Panu Halme
Department of Biological and Environmental Science, University of Jyväskylä, Jyväskylä, Finland
Natural History Museum, University of Jyväskylä, Jyväskylä, Finland

Otso Ovaskainen
Department of Biosciences, University of Helsinki, Helsinki, Finland

Jun Ma and Qin Qin
State Key Laboratory of Forest and Soil Ecology, Institute of Applied Ecology, Chinese Academy of Sciences, Shenyang, People's Republic of China
University of Chinese Academy of Sciences, Beijing, People's Republic of China

Yuanman Hu, Rencang Bu and Yu Chang
State Key Laboratory of Forest and Soil Ecology, Institute of Applied Ecology, Chinese Academy of Sciences, Shenyang, People's Republic of China

Huawei Deng
Shengli Oilfield's Shengli Engineering co., LTD, Dongying, People's Republic of China

Wenquan Zhu
State Key Laboratory of Earth Surface Processes and Resource Ecology, Beijing Normal University, Beijing, China
College of Resources Science and Technology, Beijing Normal University, Beijing, China

Guangsheng Chen
Environmental Sciences Division, Oak Ridge National Laboratory, Oak Ridge, Tennessee, United States of America

Nan Jiang, Jianhong Liu and Minjie Mou
College of Resources Science and Technology, Beijing Normal University, Beijing, China

Zhiyong Zhou and Chao Guo
Ministry of Education Key Laboratory for Silviculture and Conservation, Beijing Forestry University, Beijing, China
The Institute of Forestry and Climate Change Research, Beijing Forestry University, Beijing, China

He Meng
College of Forestry, Inner Mongolia Agriculture University, Hohhot, China

Leandro Da Silva Duarte, Rodrigo Scarton Bergamin and Guilherme Dubal Dos Santos Seger
Departamento de Ecologia, Universidade Federal do Rio Grande do Sul, Porto Alegre, Brazil

Vinícius Marcilio-Silva and Márcia Cristina Mendes Marques
Departamento de Botânica, Universidade Federal do Paraná, Curitiba, Brazil

Alvaro G. Gutiérrez
Department of Ecological Modeling, Helmholtz Centre for Environmental Research (UFZ), Leipzig, Germany
Forest Ecology Group, Institute of Terrestrial Ecosystems, Department of Environmental Sciences, Swiss Federal Institute of Technology (ETH Zürich), Zürich, Switzerland

Juan J. Armesto
Instituto de Ecología y Biodiversidad (IEB), Santiago, Chile
Departamento de Ecología, Facultad de Ciencias Biológicas, Universidad Catolica de Chile, Santiago, Chile

M. Francisca Díaz
Departamento de Ciencias Biológicas, Facultad de Ciencias Biológicas, Universidad Andrés Bello, Santiago, Chile

Andreas Huth
Department of Ecological Modeling, Helmholtz Centre for Environmental Research (UFZ), Leipzig, Germany

Konrad Martin
University of Hohenheim, Agroecology in the Tropics and Subtropics (380b), Stuttgart, Germany

Ling-Zeng Meng
Key Laboratory of Tropical Forest Ecology, Xishuangbanna Tropical Botanical Garden, Chinese Academy of Sciences, Yunnan, China
University of Hohenheim, Agroecology in the Tropics and Subtropics (380b), Stuttgart, Germany

Andreas Weigel
Rosalia Umweltmanagement, Am Schloßgarten 6, Wernburg, Germany

Xiao-Dong Yang
Key Laboratory of Tropical Forest Ecology, Xishuangbanna Tropical Botanical Garden, Chinese Academy of Sciences, Yunnan, China

Shufen Pan
International Center for Climate and Global Change Research, School of Forestry and Wildlife Sciences, Auburn University, Auburn, Alabama, United States of America
State Key Laboratory of Urban and Regional Ecology, Research Center for Eco-Environmental Sciences, Chinese Academy of Sciences, Beijing, China

Hanqin Tian, Shree R. S. Dangal, Jia Yang, Bo Tao, Chaoqun Lu, Wei Ren, Kamaljit Banger, Qichun Yang, Bowen Zhang and Xia Li
International Center for Climate and Global Change Research, School of Forestry and Wildlife Sciences, Auburn University, Auburn, Alabama, United States of America

Chi Zhang
State Key Laboratory of Desert and Oasis Ecology, Xinjian Institute of Ecology and Geography, Chinese Academy of Sciences, Urumqi, China

Zhiyun Ouyang and Xiaoke Wang
State Key Laboratory of Urban and Regional Ecology, Research Center for Eco-Environmental Sciences, Chinese Academy of Sciences, Beijing, China

Mioko Ataka and Makoto Tani
Laboratory of Forest Hydrology, Division of Environmental Science and Technology, Graduate School of Agriculture, Kyoto University, Kyoto, Japan

Yuji Kominami, Kenichi Yoshimura andTakafumi Miyama
Kansai Research Center, Forestry and Forest Products Research Institute (FFPRI), Kyoto, Japan

Mayuko Jomura
College of Bioresource Sciences, Nihon University, Fujisawa, Kanagawa, Japan

Heather Keith, David B. Lindenmayer, David Blair, Lauren Carter, Lachlan McBurney and Sachiko Okada
The Fenner School of Environment and Society, Australian National University, Building 48, Canberra, ACT, Australia

Brendan G. Mackey
Griffith Climate Change Response Program, Griffith University, Queensland, Australia

Tomoko Konishi-Nagano
Fujitsu Laboratories Ltd., Kawasaki, Japan

Kun Huang
Key Laboratory of Ecosystem Network Observation and Modeling, Institute of Geographic Sciences and Natural Resources Research, Chinese Academy of Sciences, Beijing 100101, China
University of Chinese Academy of Sciences, Beijing 100049, China

Shaoqiang Wang, Lei Zhou, Huimin Wang and Peili Shi
Key Laboratory of Ecosystem Network Observation and Modeling, Institute of Geographic Sciences and Natural Resources Research, Chinese Academy of Sciences, Beijing 100101, China

Junhui Zhang
Institute of Applied Ecology, Chinese Academy of Sciences, Shenyang 110016, China

Junhua Yan
South China Botanical Garden, Chinese Academy of Sciences, Guangzhou 510650, China

Liang Zhao
Northwest Plateau Institute of Biology, Chinese Academy of Sciences, Xining 810001, China

Yanfen Wang
University of Chinese Academy of Sciences, Beijing 100049, China

Marcela Suarez-Rubio
University of Maryland Center for Environmental Science, Appalachian Laboratory, Frostburg, Maryland, United States of America

Scott Wilson
Canadian Wildlife Service, Environment Canada, Saskatoon, Canada

Peter Leimgruber
Smithsonian Conservation Biology Institute, Front Royal, Virginia, United States of America

Todd Lookingbill
University of Richmond, Department of Geography and the Environment, Richmond, Virginia, United States of America

Christopher W. Woodall, Grant M. Domke and Susan J. Crocker
US Department of Agriculture, Forest Service, Northern Research Station, St. Paul, Minnesota, United States of America

Karin L. Riley
College of Forestry and Conservation, University of Montana, Missoula, Montana, United States of America

Christopher M. Oswalt
US Department of Agriculture, Forest Service, Southern Research Station, Knoxville, Tennessee, United States of America

Gary W. Yohe
Wesleyan University, Middletown, Connecticut, United States of America

Cheng He and Siyu Zhang
Nanjing Forest Police College, Nanjing, China

Matteo Convertino
School of Public Health-Division of Environmental
Health Sciences, University of Minnesota Twin-
Cities, Minnesota, United States of America
Institute on the Environment, University of
Minnesota Twin-Cities, Minnesota, United States
of America

Zhongke Feng
Institute of GIS, RS&GPS, College of Forestry,
Beijing Forestry University, Beijing, China

Joshua R. King
Biology Department, University of Central Florida,
Orlando, Florida, United States of America

Robert J. Warren
Biology Department, SUNY Buffalo State, Buffalo,
New York, United States of America

Mark A. Bradford
Yale School of Forestry and Environmental Studies,
Yale University, New Haven, Connecticut, United
States of America

Index